FISSION-TRACK DATING

Solid Earth Sciences Library

Volume 6

SPRINGER-SCIENCE+BUSINESS MEDIA, B.V.

Library of Congress-in-Publication Data

Wagner, Günther A.
   Fission track dating / by Günther Wagner and Peter Van den Haute.
      p.   cm. -- (Solid earth sciences library ; v. 6)
   Includes bibliographical references and index.
   ISBN 978-94-010-5093-7
   1. Fission track dating.   I. Haute, Peter Van den.   II. Title.
   III. Series: Solid earth sciences library ; SESL 6.
   QE508.W34   1992
   551.7'01--dc20   91-47967

ISBN 978-94-010-5093-7   ISBN 978-94-011-2478-2 (eBook)
DOI 10.1007/978-94-011-2478-2

© 1992 Ferdinand Enke Verlag. P.O. Box 10 12 54,
D-7000 Stuttgart 10
Originally published by Kluwer Academic Publishers in 1992
Softcover reprint of the hardcover 1st edition 1992

Contents

Preface ix

Introduction xi

1. Particle Tracks and Fission Tracks 1
  1.1. Particle Tracks in Solids: Generalities 1
  1.2. Structure of the Latent Track 1
    1.2.1. Track width 1
    1.2.2. Atomic structure 2
  1.3. Track Formation 5
    1.3.1. Particle-solid interactions 5
    1.3.2. Track-formation processes and theories 7
  1.4. Nuclear Fission and Formation of Fission Tracks 9
    1.4.1. Process of nuclear fission 10
    1.4.2. Latent fission tracks 12
    1.4.3. Spatial and areal density of fission tracks 13

2. Track Revelation and Observation 16
  2.1. Techniques of Track Revelation: Chemical Etching 16
  2.2. Etched Tracks in Glass 18
    2.2.1. Fundamentals 18
    2.2.2. Track density 18
    2.2.3. Track shape and size 25
    2.2.4. Effects of varying V_t 31
    2.2.5. Factors affecting track etching 35
  2.3. Etched Tracks in Crystals 36
    2.3.1. Fundamentals 36
    2.3.2. Track shape 39
    2.3.3. Track density 42
    2.3.4. Size distributions of track populations 47
    2.3.5. Factors affecting track etching 52
  2.4. Microscopic Observation 53
    2.4.1. Equipment and techniques 53
    2.4.2. Automated observation 57

3. Fission-Track Dating Method   59
   3.1. Natural Tracks and their Origin   59
   3.2. Principles of the Dating Method: Fundamental Age Equation   60
   3.3. Practical Age Equation   61
   3.4. Relevant Nuclear Parameters   63
         3.4.1. Decay-constant of ^{238}U spontaneous fission (λ_f)   63
         3.4.2. Cross-section of ^{235}U neutron-induced fission   65
   3.5. Dating Systems and their Calibration   66
         3.5.1. Absolute approach   66
         3.5.2. Age standard approach: the ζ-method   71
   3.6. Dating Procedures and Techniques   73
         3.6.1. Grain-population methods   74
         3.6.2. Grain-by-grain methods   75
   3.7. Practical Considerations   77
         3.7.1. Sample preparation and irradiation   77
         3.7.2. Track counting and measuring   80
   3.8. Data Analysis and Error Calculation   81
         3.8.1. Grain-population methods   81
         3.8.2. Grain-by-grain methods   85
         3.8.3. Error in the age determination   86
         3.8.4. Analysis of age groups   88
   3.9. Age Standards and Accuracy of Age Determination   90
   3.10. Data Presentation   94

4. Fading of Fission Tracks   95
   4.1. Causes of Track Fading   96
   4.2. Track Annealing under Experimental Conditions   98
         4.2.1. Annealing experiments   98
         4.2.2. Arrhenius diagram   100
         4.2.3. Length of annealed fission tracks   102
         4.2.4. Factors of influence on annealing   107
         4.2.5. Annealing kinetics   108
   4.3. Track Stability under Natural Conditions   111
         4.3.1. Bore-hole studies   112
         4.3.2. Outcrop studies   115
         4.3.3. Temperatures of effective track retention   117

5. Geological Interpretation   120
   5.1. Intersecting Probability of Faded Tracks   121
   5.2. Partial Annealing and Effective Retention of Tracks   123
   5.3. T–t-Path and Fission-Track Accumulation   126
   5.4. T–t-Path and Fission-Track-Length Distribution   129
   5.5. Types of Fission-Track Age   134
         5.5.1. A-type ages (formation and early cooling)   134
         5.5.2. B-type ages (cooling and uplift)   137
         5.5.3. C-type ages (complex and overprint ages)   142

5.6.	Age-Depth Profiles	145
	5.6.1. Stratigraphic profile	145
	5.6.2. Uplift profile	146
	5.6.3. Complex profile	148
	5.6.4. Modelled uplift age-depth profiles	151
5.7.	Correction of Track Fading	156
	5.7.1. Track-size-correction technique	157
	5.7.2. Plateau-correction technique	158

6. Applicability		161
6.1.	Time Span	161
6.2.	Geological Materials	161
	6.2.1. Allanite	162
	6.2.2. Amazonite	163
	6.2.3. Apatite	163
	6.2.4. Barysilite	165
	6.2.5. Bastnäsite	165
	6.2.6. Beryl	166
	6.2.7. Calcite	166
	6.2.8. Chlorite	167
	6.2.9. Epidote	167
	6.2.10. Garnet	169
	6.2.11. Glass	170
	6.2.12. Glauconite	175
	6.2.13. Hornblende	175
	6.2.14. Hübnerite	175
	6.2.15. Kyanite	176
	6.2.16. Mica	176
	6.2.17. Microlite	178
	6.2.18. Monazite	178
	6.2.19. Orpiment	179
	6.2.20. Quartz	179
	6.2.21. Scheelite	180
	6.2.22. Secondary lead minerals	180
	6.2.23. Sphene	180
	6.2.24. Stibiotantalite	182
	6.2.25. Tanzanite	182
	6.2.26. Vermiculite	183
	6.2.27. Vesuvianite	183
	6.2.28. Zeolite	184
	6.2.29. Zircon	184

7. Application		187
7.1.	Tephrochronology	187
7.2.	Post-Orogenic Uplift of Mountain Belts	191

	7.2.1.	General	191
	7.2.2.	Central Alps	192
7.3.	Epeirogenic Uplift of Basements		197
	7.3.1.	Transantarctic Mountains	197
	7.3.2.	Central European Basement	199
7.4.	Age and Amount of Displacement along Faults		203
7.5.	Thermal Evolution of Sedimentary Basins		206
	7.5.1.	Tejon Oil Field	208
	7.5.2.	Otway Basin	210
7.6.	Age and Thermal History of Ore Deposits		213
	7.6.1.	Mineralization ages	213
	7.6.2.	Thermal history of MVT deposits	214
7.7.	Meteoroid Impacts		217
	7.7.1.	Age of impact structures	218
	7.7.2.	Thermal evolution of the Ries impact crater	221
	7.7.3.	Tektites	223
7.8.	Sea-Floor Spreading		225
7.9.	Archaeological Application		228
	7.9.1.	Man-made glass	229
	7.9.2.	Obsidian	231
	7.9.3.	Fired stones and baked soils	235
	7.9.4.	Palaeolithic marker horizons	237

Appendix A: Etching Conditions for the Revelation of Fission
            Tracks   239

Appendix B: Annealing Properties of Fission Tracks in Minerals
            and Glasses   243

Appendix C: Fundamentals of Error Calculation in Fission-Track
            Dating   248

References   250

Index   275

Preface

It was at Besançon, during the *6th International Workshop on Fission-Track Dating* in September 1988, that we felt it opportune – after a quarter of a century since the first fission-track ages were reported – to collect the ideas and practices of this dating method and to condense them into one volume. Since 1975, when the classical monograph *Nuclear Tracks in Solids* by Fleischer, Price, and Walker, the three pioneers of solid-state particle-track detection and, in particular, of fission-track dating, was published, many new groups have entered the rapidly expanding field of fission-track dating. With the realization that fission tracks are a unique tool for deciphering the low temperature history of terrestrial materials, the demand for fission-track analyses rose greatly and is still growing. Not surprisingly, during the last two decades a vast body of data has accumulated and innovative concepts have been developed which are difficult to follow up, even for the specialist. Therefore, a comprehensive monograph on this subject seems to be timely.

The present book, however, is not aimed at the specialist; above all, it is written to provide an introduction to the newcomer and to geologists as well as archaeologists who want to learn about the potential of fission-track analysis. With this need in mind, we attempted to fill an information gap for students and nonspecialist scientists. But we hope that the specialist reader, too, will find some of the chapters rewarding.

As stated in the title, the book is essentially focused on uranium fission tracks and their use as a tool for dating terrestrial materials. Tracks created by other particles than the heavy fission fragments of uranium, do not receive much attention here. Underlying physical and mathematical principles are treated only as far as is necessary for understanding the fission-track dating method. An effort has been made to present a clear treatise on the subjects of track formation and etching but emphasis is placed on the fission-track dating method itself and its applications which are discussed in detail with the aid of several case studies. Only applications which aim at dating geological or archaeological material and at reconstructing the thermal and tectonic history of rocks are considered. Purely geochemical studies using tracks in order to map or analyze elements are not included.

Although we have attempted to present a balanced and comprehensive overview of the fission-track dating method and, thus, have written to some extent on behalf of the whole fission-track community, it seems inevitable that for one reason or

x *Preface*

other we are biased towards our own work at Heidelberg and Gent. With this respect, we have deliberately included previously unpublished results and ideas developed at our laboratories which we thought would enhance the text. At the same time, we apologize to all those colleagues who may rightly feel that their work is not adequately treated here and hope that this is interpreted as being due to our ignorance.

We are greatly indebted to our colleagues in nuclear science, Frans De Corte and Etienne Jacobs, for critically reading some of the crucial parts of the manuscript. Anthony J. Hurford is gratefully acknowledged for his highly appreciated review of the third chapter. Also, our younger colleagues, fission-track scientists Ewald Hejl, Raymond Jonckheere, Caroline Vercoutere and Irmtrud Lorenz, have contributed to the final shape of this book by reading and discussing sections of the text. Wolfgang Brandl, William Kelly, and John Westgate kindly placed unpublished photographs at our disposal, while several of the drawings were prepared by Martine Bogaert and Joachim Lutz. To all of these persons, we owe a debt of gratitude. Undoubtedly, the book benefited much from the work of other scientists involved in fission-track dating and nuclear-track studies. We gratefully acknowledge the publication of their figures and photographs through the courtesy of the various publishing houses who kindly permitted the reproductions.

GÜNTHER A. WAGNER
PETER VAN DEN HAUTE

Introduction

In 1963, two physicists, P. Buford Price and Robert M. Walker, who were at the time working at the General Electric Research Laboratory in Schenectady, developed a new method for dating geological materials based on the natural decay by spontaneous fission of the ^{238}U isotope. Compared to already existing dating methods based on natural radioisotopes, the new method was different in so far that it did not involve the measurement of isotopic abundances by mass spectrometry but relied on a simple counting of the individual damage tracks created by the spontaneous fission of ^{238}U atoms. Counting of the tracks could be done with a normal microscope because chemical etching of the track-bearing solid with a suitable reagent causes the tracks to be enlarged so that they become optically visible (Price and Walker, 1962a,b). Later, Price and Walker were joined by Robert L. Fleischer and by 1964 the common effort of these three scientists had led to several publications on the new dating method, including a list of about 30 mineral phases in which fission tracks had been successfully revealed (Fleischer and Price, 1964d).

The technique of optical revelation by chemical etching proved to be useful, not only for studying uranium fission tracks but also for investigating damage trails created by the passage of other kinds of charged particles through insulating solids. A new scientific method was born called *solid-state nuclear track detection* which rapidly found applications in numerous fields of research such as nuclear science and technology, lunar and cosmic ray studies, geology and biology (Fleischer, 1990).

Although it was only because of the work of Fleischer, Price, and Walker that the potential of the study of etched nuclear tracks became fully exploited, these authors were not really the first to have observed such tracks. Scientists at Harwell, England, had already published photomicrographs of uranium fission tracks in mica observed with transmission electron microscopy and had experimented with track revelation by chemical etching (Young, 1958; Silk and Barnes, 1959). Revelation by etching was a commonly used technique for studying defects in crystals during the 1950s and within the framework of these studies, uranium fission tracks had undoubtedly been observed by several scientists who, at the time, did not realize what they were looking at (e.g. Lovell, 1958). Surface etching has also been applied in the past for studying crystal symmetry and perhaps the first

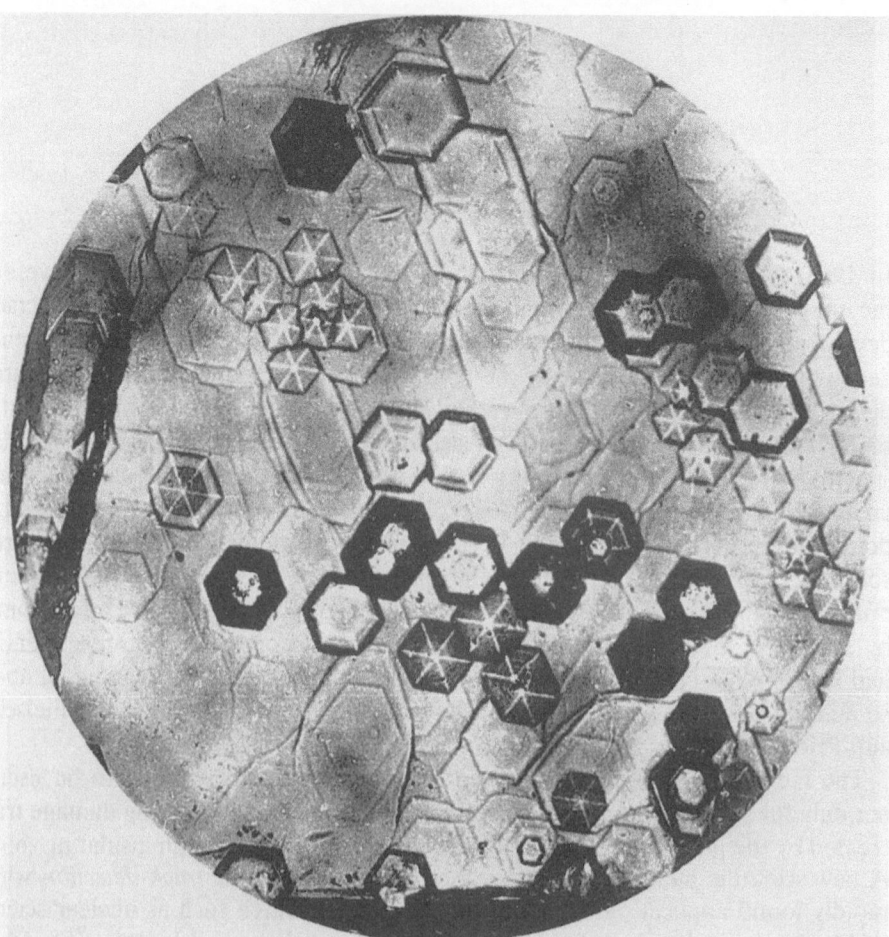

Figure 1. Apatite crystal from St. Gotthard, Switzerland. This photomicrograph was taken from Baumhauer's (1894) treatise on the etch method and is probably one of the earliest pictures of etched fission tracks in apatite. The base face of this crystal had been etched for 45 min in diluted sulphuric acid. Baumhauer distinguishes generally between *normal* and *anomalous* etch figures. Normal etch figures are those whose symmetry corresponds to that of the etched crystal face, i.e., in the case of the apatite basal face the normal etch figures have the shape of hexagonal pyramids. Anomalous etch figures have a lower symmetry than the etched face. According to Baumhauer, they may develop along channels that extend from the bottom of the etch pits in various directions into the crystal. From this description, it becomes clear that some of the anomalous etch figures are etched fission tracks. On the above photomicrograph, the oblique pyramids in the middle and to the right are such anomalous etch figures which probably developed along fossil fission tracks.

description of natural fission tracks was given by Baumhauer when, in 1894, he reported on the presence of anomalous etch figures in apatite (Figure I).

In 1975, Fleischer, Price, and Walker published a book entitled *Nuclear Tracks in Solids* in which they gave a comprehensive overview of the study of nuclear tracks and their applications, including a, for that time, relatively complete chapter on the fission-track dating method. In 1980, a first International Workshop entirely focused on fission-track dating was held in Pisa where much light was shed on the possibilities and problems of the method. As a response to this meeting, new ideas and techniques have been developed dealing with almost all problems, from calibration to geological interpretation. In the meantime, altogether six International Fission-Track Dating Workshops have been held. As a result, fission-track dating has taken a definite place among the other dating methods and because of this, there has been a dramatic increase in its use in geological applications.

In the recently published book *Solid State Nuclear Track Detection*, by Durrani and Bull (1987), which still covers the various applications of nuclear tracks, the chapter on fission-track dating merely attempted to expose the methodology and did not leave much room for discussing geological application. In the present work, we have tried to fill this gap and, at the same time, we have placed emphasis on the potential of the dating method for revealing the thermal history of rocks in relation to tectonic movements or other important geological events. This application probably represents the most significant contribution of the fission-track method to geology.

CHAPTER 1

Particle Tracks and Fission Tracks

1.1. Particle Tracks in Solids: Generalities

A fast charged particle, when passing through an insulating solid, creates a narrow trail of damage along its trajectory. This damage trail persists in the solid after the particle has come to rest and is called a *charged particle track* or *nuclear track*. The solid in which the tracks are registered is commonly called the *detector*. The length of a nuclear track is variable, ranging from less than 1 μm to several mm, depending mainly upon the charge and energy of the particle and upon the composition of the detector (glass, minerals, plastic). The width of the track is largely submicroscopic, in the order of a few nm. The trail of damage forms a site of preferential attack for chemical reagents. Chemical etching enlarges the tracks to such an extent that they become visible under an optical microscope (Price and Walker, 1962e,d). The unetched track which cannot be observed with an optical microscope, is commonly called the *latent track*. Optical revelation of latent tracks by chemical etching forms the subject of the next chapter.

Although the study of nuclear tracks has found a wide range of applications in different sciences, including the dating method treated in this book, there is still an incomplete knowledge of the structure of latent tracks and no general agreement has yet been reached on the processes which are responsible for their formation. Both problems are discussed in the following sections. The discussion essentially refers to tracks formed in inorganic detectors (such as minerals and glasses) by heavy charged particles like those which are released by the process of nuclear fission explained in Section 1.4. To some extent, the statements made also apply to tracks in plastics but it should be realized that the mechanisms of track formation and the structure of the latent track are different in these materials.

1.2. Structure of the Latent Track

1.2.1. *Track width*

Because of its extremely small size, the diameter of a nuclear track is difficult to determine exactly. In the early years, photomicrographs of latent uranium fission tracks in natural and synthetic mica were made with a transmission electron

1

microscope (TEM) using the electron beam in a diffraction contrast mode (Price and Walker, 1962a,b). The tracks showed up on the TEM pictures as dark lines having a width of ≈10 nm. This figure was interpreted as an upper limit of track width as the TEM image essentially results from diffraction of the beam at lattice distortions surrounding the central damage trail. From pictures taken after a short etch in hydrofluoric acid, a much smaller diameter of 2.5–4 nm was deduced for the central track channels (Price and Walker, 1962c). More recently, high resolution electron microscopy (Yada *et al.*, 1981, 1987) has enabled one to obtain lattice images of zircon crystals containing uranium fission tracks. On these images, the tracks have an appearance of distinct linear defects with a width of 1.5 to 4 nm depending upon their angle with the lattice planes (Figure 1.1).

More indirect methods have also been applied to measure track width. Electrical measurements yielded a value of 6–7 nm for the latent track diameter in muscovite mica (Bean *et al.*, 1970). In this method, the following technique was used. A thin track-bearing detector was placed as a dividing wall in an electrolyte which, at the same time, served as an etchant. The growth rate and the initial value of the diameter of the track holes were derived from the variation in conductance. In another approach, small-angle x-ray or neutron scattering (Albrecht *et al.*, 1982, 1984) has been applied to detectors (mica, plastic) previously bombarded with heavy ions of varying energies. The radii of damage calculated from these experiments amount to 2–6 nm, depending upon the mass and energy of the ions.

The general picture that is apparently revealed by the above experiments can be summarized as follows: Tracks of heavy charged particles in minerals consist of a highly damaged core zone, the width of which is 5 nm or less, surrounded by a zone of lesser damaged strained lattice material which may extend up to about 10 nm, some variation being possible depending upon the composition of the detector and the energy of the bombarding particle.

1.2.2. *Atomic structure*

The early TEM observations did not produce much information on the detailed atomic structure of nuclear tracks. It was generally assumed that the damage trail corresponded to a narrow zone of displaced atoms. Among other arguments, this was deduced from the fact that the tracks represent sites of preferential attack by chemical reagents and from the results of heating experiments which demonstrated that the latent damage is restored at high temperatures and that the activation energies involved are characteristic for diffusion processes on an atomic scale. The subject of *thermal track fading* (also called *track annealing*) is discussed more extensively in Chapter 4.

The lattice images of Yada *et al.* (1981, 1987), which attain a resolution of 0.15–0.2 nm, contain valuable information on the atomic structure of latent tracks (Figure 1.1). The crystal lattice appears to be completely destroyed in the track core and altered into an almost amorphous state of low density. In the vicinity of the track channels and around the track ends, point defects are observed, while the lattice planes appear to be strained at their intersection with the tracks. Lattice

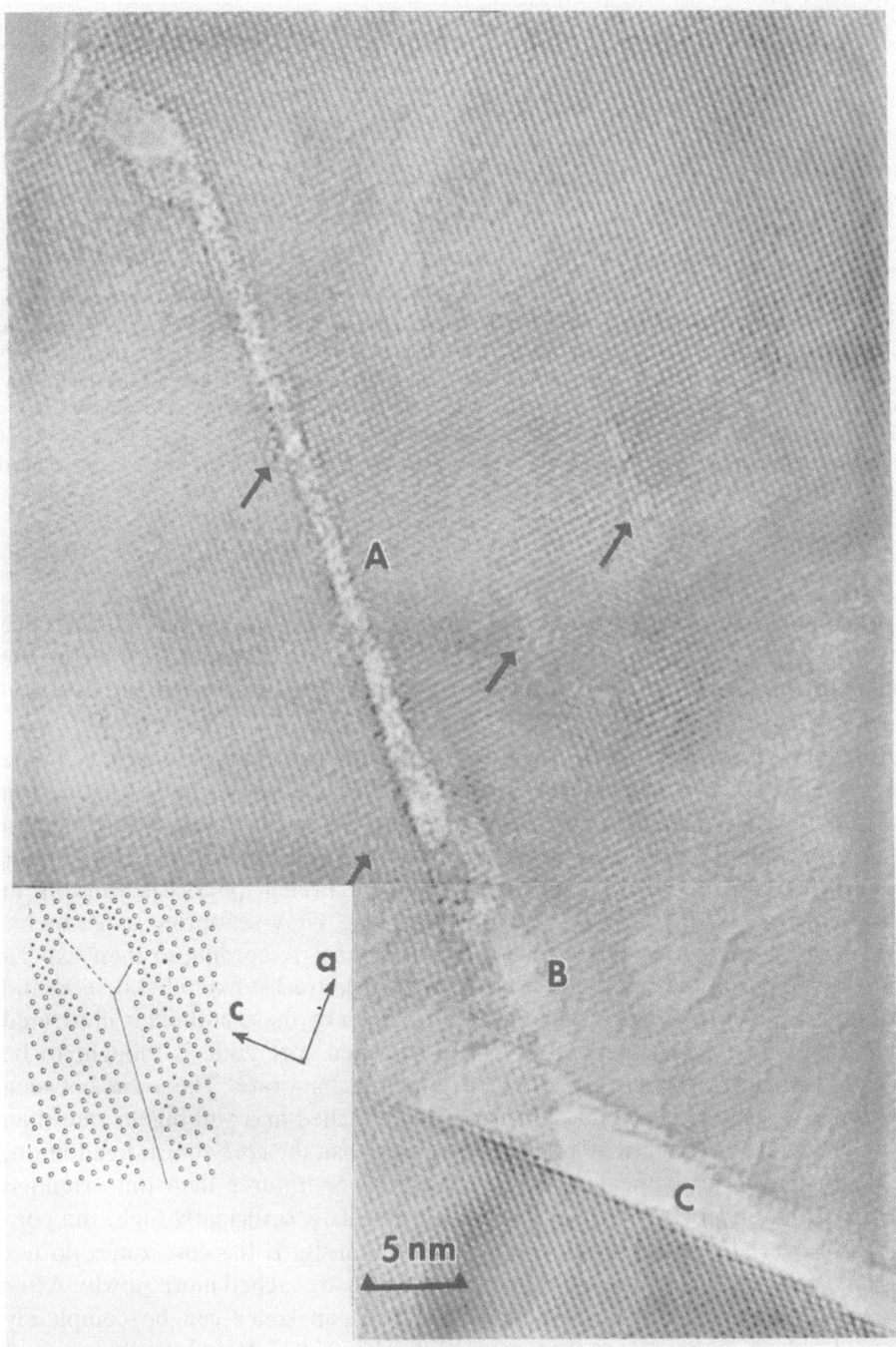

Figure 1.1. Three uranium fission-tracks (A, B and C) making different angles with the (100) plane of zircon (=plane of the picture) observed with high resolution electron microscopy. Arrows indicate sites of point defects. The inset visualizes the formation of the damage track according to Yada *et al.* (1987) together with two major crystallographic directions. (Electron micrograph obtained by courtesy of Dr K. Yada.)

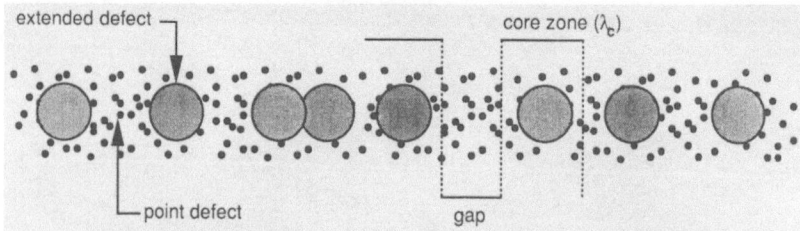

Figure 1.2. Schematic structure of a charged particle track according to the Orsay group (Dartyge *et al.*, 1981). The track consists of an array of extended defects surrounded by point defects. The linear density of the extended defects depends upon the charge (Z) and rate of energy loss ($-dE/dx$) of the particle and overlapping of the defects may occur when this rate is high. The size of the defects amounts to a few nm and mainly depends upon the charge of the particle. For each extended defect, there is an associated core zone with a size (λ_c) which can reach several tens of nm, characterized by a highly increased etch rate. The zones outside the core zones are called gap zones. They consist of point defects only, are etched at much slower rates, and anneal at lower temperatures than extended defects. (After Duraud, 1978.)

deformation around tracks had also been observed earlier from Moiré interference fringes (Morgan and Van Vliet, 1970). Broader zones of damage occur at the track ends. This is thought to be caused by the charged particle itself and by recoil nuclei.

Commonly, nuclear tracks are regarded as essentially continuous features. Some electron microscope studies, however (Morgan and Chadderton, 1968; Chadderton *et al.*, 1988), seem to indicate that in layered materials such as MoS_2 (Molybdenum sulfide), the latent tracks may have an intermittent character. A discontinuous structure has also been postulated by a group of French physicists working at Orsay (Dartyge *et al.*, 1978, 1981) from low-angle x-ray scattering experiments carried out on mica, plagioclase feldspar, and olivine. According to their experiments, two types of defects occur at charged particle tracks: extended defects and point defects. The structure of a latent track as seen by these authors is illustrated in Figure 1.2. To each extended defect, a so-called core zone is thought to be associated which is characterized by a very high etching rate. The areas between core zones which contain point defects only, are etched at a still higher rate than the undamaged material but at a much slower rate than the core zones. In addition, the point defects are annealed at much lower temperatures than the extended defects. If the linear density of the extended defects is sufficiently high, the core zones will overlap and the track will etch continuously. If the core zones do not overlap, the track will contain so-called 'gaps' which are etched more slowly. After a thermal treatment, the point defects in these gap zones can be completely removed, which makes them even more difficult to etch. Based on this type of structure, the double-humped length distributions of partially annealed Fe ion tracks observed in muscovite and the complex evolution of the track length distribution of unannealed Fe ion tracks with etching time observed in plagioclase, could be adequately explained.

The 'gap' model has not found general acceptance and has been criticized by a group of mainly German scientists (Albrecht *et al.*, 1985, 1986) who, also using small-angle x-ray in addition to neutron scattering, arrive at an essentially continuous structure of tracks which they described as long uninterrupted cylinders with a Gaussian radial density variation. They claim that, due to the high particle fluences used by Dartyge and coworkers, the tracks were strongly overlapping and the observation of two types of defects could be merely an effect of this overlapping. Recent work (Thiel *et al.*, 1988) also showed that particle tracks in olivine, which were apparently intermittent when observed with a normal transmission electron microscope, turned out to be continuous if high resolution TEM was used, the intermittency being merely an effect of contrast modulation.

In conclusion, it appears that an essentially discontinuous nature of tracks has not yet been truly demonstrated, even in minerals with a layered structure such as mica. On the other hand, discontinuities in the track etching behaviour have been observed especially when a thermal treatment was applied before etching. This observation not only holds for mica but also for other minerals such as apatite (Green *et al.*, 1986) and requires an explanation. The whole problem thus merits further study because it has to be taken into account that, in recent years, track-size measurements have become commonly used in fission-track dating in order to retrieve information on the cooling history of rocks (Chapter 5). A detailed knowledge of track structure in minerals will undoubtedly provide valuable information on the possibilities and limitations of this kind of application.

1.3. Track Formation

As fascinating as the study of particle track formation may be, it mainly concerns particle and solid-state physics which is beyond the scope of this book. The short discussion given below is therefore merely meant as an introduction and has been deliberately focused on qualitative description. The reader who wants a more quantitative insight into this subject is referred to the works of Fleischer *et al.* (1975) and of Durrani and Bull (1987) in which the physical processes capable of controlling track formation are discussed in more detail, and to volumes such as Lehmann (1977) which offer a more general theoretical treatise on the interaction of energetic charged paricles with solids.

1.3.1. *Particle-solid interactions*

When a heavy charged particle travels at a high velocity through a solid medium, it will undergo a series of interactions with the constituents of the solid (atoms, electrons) which will cause it to gradually lose its kinetic energy until it eventually comes to rest. The distance travelled by the particle is called the range. The range can be defined in various ways, e.g., as the vector connecting the starting and stopping point = vector range R_v) or as the total path length traversed by the particle between these both points (= total or linear range R). When the path followed by the particle is straight, which is generally the case for heavy particles such as fission fragments (Section 1.4.2), it can be stated that $|R_v| = R$.

The amount of energy lost by the particle per unit of distance $-dE/dx$ along its trajectory is called the stopping power. Two types of interactions are responsible for the deceleration of a track-forming charged particle: collisions with the lattice atoms of the solid (nuclear collisions, nuclear stopping power), and interactions with the electrons leading to their excitation or ejection out of their orbits which causes ionizaton of the lattice atoms (electronic collisions, electronic stopping power). One can write

$$-dE/dx = (-dE/dx)_n + (-dE/dx)_e.$$

Both processes thus contribute additively to the stopping power. They are, however, not simultaneously effective: at high energies, particles are predominantly slowed down by electronic interactions, while at lower energies collisions with the lattice atoms as a whole become dominant.

If a particle starts at very high energy, its velocity will exceed the orbital velocity of its electrons so that they will be stripped off. The particle will travel as a bare nucleus through the solid with a positive charge $Z_1 e$ and will gradually lose its energy (and speed) by interactions with the electrons of the lattice atoms. At a give moment, it will capture an electron and its charge will be reduced to $(Z_1 - 1)e$. The deceleration of the particle due to this type of interaction continues until most of the missing electrons are captured. Then, the elastic collisions with the lattice atoms as a whole take over as the major type of interaction.

During the stage of electronic stopping, the energy loss per interaction essentially depends upon the particle's energy per atomic mass unit. In other words, it is a function of the particle's speed rather than of its total energy or $(-dE/dx)_e = f(v)$. At very high energies $(-dE/dx)_e$ increases with decreasing velocity roughly as $1/v^2$, while at lower energies this behaviour becomes opposite and $(-dE/dx)_e$ is directly proportional to v. Hence, at a given velocity, the curve (Bragg curve) representing the stopping power has to reach a maximum value (Bragg peak) (Figure 1.3a). The energy loss per electronic interaction is always small when compared to the total energy of the fast particle so that its slowing down can be described as a friction type of stopping which causes virtually no deflection in the particle's trajectory.

In contrast to the electronic energy loss, the energy loss by nuclear collisions is proportional to the particle's energy rather than to its velocity or $(-dE/dx)_e = f(E)$. The energy loss per collision is also relatively important compared to the total energy of the particle and each collision may cause a substantial angular change of its course.

There is quite general agreement concerning the major importance of electronic interaction in track formation. When it comes to describing the specific mechanisms which actually cause the damage trails, rather divergent opinions have been formulated.

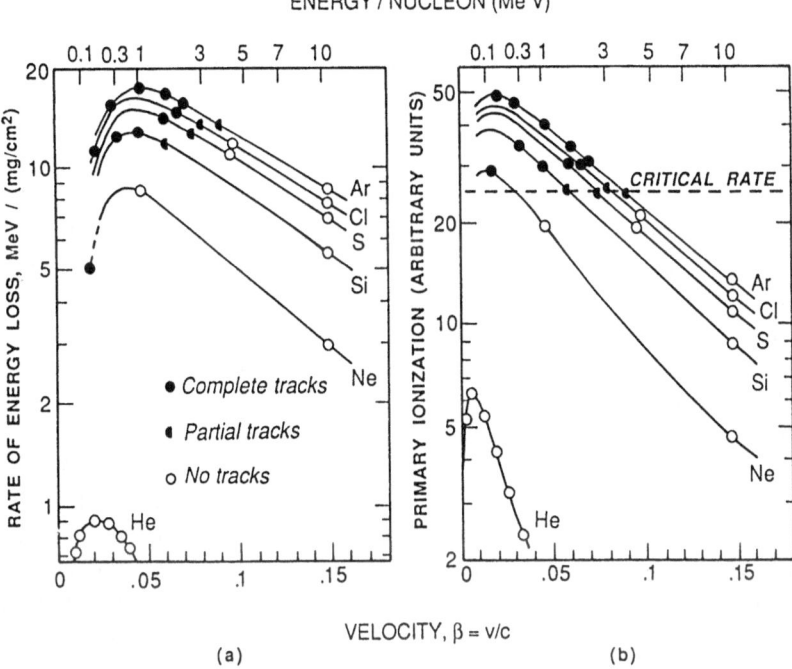

Figure 1.3. Results of a track registration experiment in muscovite mica. After bombardment with particles of varying charge and energy, the presence of etchable tracks is investigated. When the rate of the particle's total energy loss ($-dE/dx$) is plotted as a function of β (= particle velocity expressed as a ratio to light velocity), no distinct threshold or critical rate appears below which no tracks are registered (Figure 1.3a). Such a threshold does appear if the results are plotted on a dJ/dx (= rate of primary ionization) versus β diagram (Figure 1.3b) supporting the primary ionization as a major process of track formation. (From Fleischer *et al.*, 1967.)

1.3.2. *Track-formation processes and theories*

In the 1960s, several theories were formulated which tried to explain the formation of nuclear tracks in solids but progress in track-formation studies has been rather limited ever since, which is partially due to a lack of knowledge of the fine structure of nuclear tracks (Chadderton, 1988). At present, the question of the mechanism which couples the electronic interactions to the atomic motion, which results in the damage trail, still remains unsolved. Further, the equations which are to be used for an accurate mathematical treatment of the track-forming processes are still a matter of some dispute.

Commonly, the term *spike* has been used for describing the intense and very short event which takes place when an energetic charged particle traverses a solid. A number of mechanisms have been proposed as the major cause of the nuclear spike. In the earliest theoretical considerations by Fleischer, Price and coworkers (Price and Walker, 1962b; Fleischer *et al.*, 1964e) the rate of total energy loss ($-dE/dx$) was considered to be crucial in track formation and a minimum rate ($-dE/dx$)$_{crit}$ below which no tracks could be registered in the solid, was thought to exist. From later experiments (Fleischer *et al.*, 1965h, 1967, 1975) (Figure 1.3), the same authors abandoned the dE/dx criterion in favour of the criterion of primary specific ionization dJ/dx, i.e., the number of ions formed per unit distance along the pathway of the moving particle. Their theoretical model, which, invokes this process in track formation, is called the 'ion explosion spike' theory. The theory, although it does not explain all characteristics of nuclear tracks with equal convictive power, still acknowledges a relatively wide acceptance. According to the 'ion explosion spike' theory, track formation can be divided into three steps (Figure 1.4): (1) the charged particle induces a burst of ionization through electronic interactions and creates an array of positive ions in the lattice of the solid; (2) the adjacent ions repulse each other into interstitial positions and, at the same time, a series of vacancies are formed, finally; (3) the local lattice stress is spread more widely by elastic relaxation. It is the last step which enables the latent tracks to be directly observed with an electron microscope.

Among other things, the theory explains well why tracks are only formed in insulators and not in conductors and the idea of an ion explosion also found some support from the experiments of Seiberling *et al.* (1980) who registered abnormally high sputtering yields when bombarding uraniumtetrafluoride crystals with heavy ions. The theory, however, does not account for possible discontinuities in the track structure as were observed by the Orsay group. Hence, the model requires at least some modifications in order to explain these features. If two different types of defect really exist along a particle track, then two different ionization processes have to be involved in track formation or some threshold for energy loss should exist below which extended defects cannot be formed (Langevin and Duraud, 1982). An answer to this problem has been provided by Tombrello (1984a,b) who assumes that at separate sites along the particle's path ionization of tightly bound (K-shell) electrons occurs. The atoms which have been excited in this way, decay by Auger processes which are an intense source of electrons causing localized regions of high ionization. These processes are thought to trigger the track formation. Repulsive motion of the excited atoms still occurs but in such a way that, after a few mutual collisions, a local thermal equilibrium is reached (thermalized ion explosion spike) (Seiberling *et al.*, 1980). It can be noted here that fluctuations in energy loss and the possible role of Auger processes were already referred to much earlier (Lambert *et al.*, 1970; Maurette, 1970).

A quite different point of view is defended by Chadderton *et al.* (1988) who argue that discontinuities in the crystal lattice are the main cause for the observed intermittent character of tracks, especially in minerals with a layered structure.

Other theories have been developed for explaining track formation. Theories and ideas developed up to the beginning of the 1980s (e.g., Chadderton and Torrens, 1969; Benton, 1970; Katz and Kobetich, 1968, 1970; Paretzke, 1977; Dartyge *et al.*, 1981) are discussed in Fleischer *et al.* (1975) and Durrani and Bull (1987). The last decade has acknowledged a revived interest in the track-formation problem leading to several new publications. We mention some of these for the interested reader: Ritchie and Claussen (1982), Tombrello (1984a,b), Albrecht *et al.* (1985), Chadderton *et al.* (1988), and Groeneveld (1988).

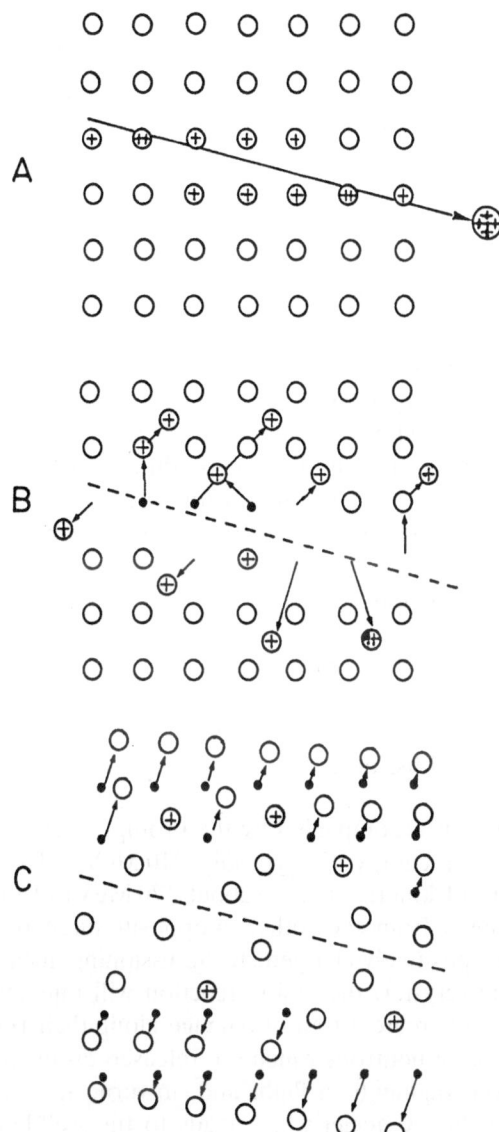

Figure 1.4. The three stages of track formation according to the 'ion explosion spike' theory of Fleischer, Price, and Walker. (A) Ionization of the lattice atoms by the moving charged particle. (B) Due to Coulomb repulsion, the ions are displaced from their lattice sites into interstitial positions; at the same time, a number of vacant lattice sites are created. (C) The stressed region of the track relaxes, elastically straining the surrounding undamaged lattice. (After Fleischer *et al.*, 1975.)

1.4. Nuclear Fission and Formation of Fission-Tracks

This section, which offers a description of the process of nuclear fission and the damage tracks resulting from this process, is, like the preceding section, also meant to provide the reader unfamiliar with this matter with some minimum information thought to be necessary for a basic understanding of the formation of fission tracks. The interested student is referred to classic handbooks on nuclear physics and chemistry such as Vandenbosch and Huizenga (1973) and Friedlander *et al.* (1981) which offer a more elaborate introduction to this subject and from which most of the information given below has been retrieved.

1.4.1. *Process of nuclear fission*

Nuclear fission is one of the several modes of disintegration which occur among heavy, unstable (radioactive) nuclides. In the fission reaction, the unstable nucleus splits into two daughter nuclides of roughly the same size. Fission reactions occur both spontaneously and artificially by bombardment with neutrons, protons or other particles as well as by irradiation with γ-rays. The fission reaction was discovered by Hahn and Strassmann in 1939 when they bombarded ^{235}U atoms with low-energy neutrons.

Examples of fission reactions are

$$^{252}Cf \rightarrow {}^{108}Ru + {}^{140}Xe + 4n + Q$$
(spontaneous fission),
$$^{235}U + n \rightarrow {}^{236}U \rightarrow {}^{90}Kr + {}^{143}Ba + 3n + Q$$
(neutron induced fission).

Each fission reaction is accompanied by the prompt release of several neutrons and a large amount of energy Q (typically 210 MeV). This energy is mainly liberated in the form of kinetic energy (about 170 MeV) of the two fission fragments which travel away from each other in opposite directions at high speed, as they are both heavily positively charged. If the fissioning nuclide is a lattice atom of an insulating solid (crystal) the fission reaction will thus take place internally and both fragments will create a trail of damage along their trajectory, called the *fission track*. The prompt neutrons which are released during fission, mainly originate from the fission fragments in flight and, in turn, are capable of producing new fissions of other heavy nuclei which leads to the well-known nuclear chain reaction on which applications such as nuclear power plants (controlled reaction) or the atomic bomb (uncontrolled reaction) are based. Occasionally, a fission reaction is accompanied by the emission of an α-particle (^{4}He), a proton, or other light particles. The fission fragments have a relatively high neutron/proton ratio, and are therefore radioactively unstable and decay by a series of β^--emissions to stable isobars.

Fission reactions are predominantly binary, meaning that the nucleus is split into two daughter fragments. These fragments are not always the same, even if they result from fission of the same nuclide. The spontaneous ^{252}Cf and the induced ^{235}U fission reaction quoted above therefore represent only one of the many possible splitting schemes of these nuclei. As shown by the examples and as is most often the case, the daughter fragments are of unequal mass. The fission

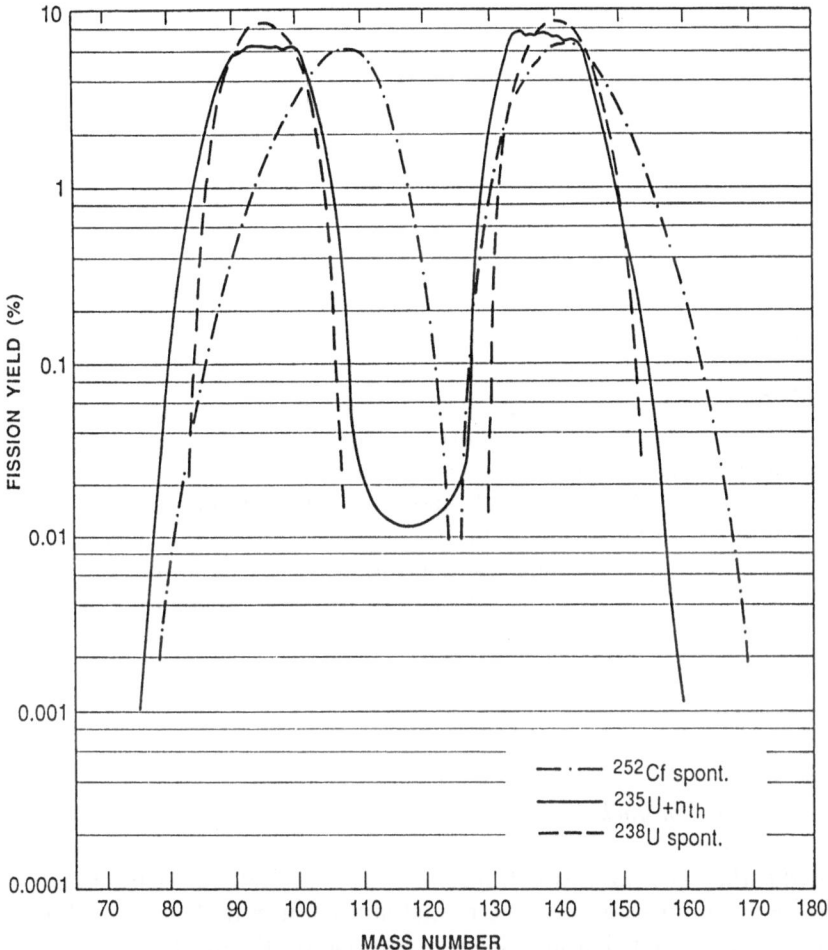

Figure 1.5. Mass distribution curves of fission fragments for the thermal neutron induced fission of ^{235}U and for the spontaneous fission of ^{238}U and ^{252}Cf. (After von Gunten, 1969.)

reaction is therefore said to be asymmetric. Figure 1.5 shows the mass distribution curve for the thermal neutron-induced fission of ^{235}U together with those for the spontaneous fission of ^{238}U and ^{252}Cf. It can be seen from this figure that the mass ratio M_H/M_L of the heavy to the light fragment in extreme cases amounts to a value $\geqslant 2$, the most frequent split being situated at $M_H/M_L \approx 1.4$.

Spontaneous fission only occurs with very heavy nuclides of atomic number $Z \geqslant 90$ and atomic mass $A \geqslant 230$. These nuclides are isotopes of elements belonging to the actinide series (Th, Pa, U, Np, Pu, etc.), most of which also disintegrate by other processes such as α-decay. Only ^{232}Th and two isotopes of U (^{235}U and ^{238}U) commonly occur in measurable concentrations as primary constituents in natural substances. The transuranium actinides merely result from artificial production, except maybe for ^{244}Pu (total half-life = $8.1\ 10^7$ a) which seems to have been measured in very minute concentrations. Two isotopes of protactinium (^{231}Pa and ^{234}Pa) as well as a number of Th isotopes and ^{234}U are produced in nature as

Table 1.1. Abundances and half-lives of the four major naturally occurring nuclides suffering spontaneous fission

	Relative abundance (compared to ^{238}U)	Total half-life (years)	Spontaneous fission half-life (years)
^{232}Th	4[d]	1.40×10^{10}[a]	1.0×10^{21}[c]
^{234}U	5.44×10^{-5}	2.46×10^{5}[b]	1.5×10^{16}[b]
^{235}U	7.25×10^{-3}	7.04×10^{8}[a]	1.0×10^{19}[b]
^{238}U	1	4.47×10^{9}[a]	8.2×10^{15}[b]

[a]Steiger and Jäger (1977),   [b]Holden (1989),   [c]Baard *et al.* (1989),   [d]Geochemical average.

intermediate members of the ^{235}U, ^{238}U, or ^{232}Th α-decay series. Some of these nuclides, besides other modes of decay, also suffer spontaneous fission but, except for ^{234}U, they do not have a significant abundance compared to those of their primary parents, because of their short total half-lives ($<10^5$ a).

In Table 1.1, the abundances, total half-lives and partial half-lives for spontaneous fission of ^{232}Th, ^{232}U, ^{235}U, and ^{238}U are displayed. From this table, it can be readily deduced that virtually all fission reactions and, hence, all fission tracks observed in natural solids, are due solely to the fission of ^{238}U atoms. The two other uranium isotopes and ^{232}Th have too low abundances and/or much longer spontaneous half-lives to produce any significant amount of natural spontaneous tracks compared to ^{238}U tracks.

1.4.2. *Latent fission tracks*

As stated earlier, the two fission fragments expel each other in opposite directions and create a trail of damage in the crystal lattice. Both trails meet each other at the locus of the fissioned atom, giving rise to one continuous fission track. When the parent nuclide splits, both fragments receive an equal momentum or $M_H v_H = M_L v_L$. Hence, not only the mass but also the kinetic energy $E_K (= \frac{1}{2} M v^2)$ is unequally divided over the two fragments in such a way that $E_{KH}/E_{KL} = v_H/v_L = M_L/M_H$. The light fragment thus receives the highest velocity, will be capable of causing ionisation over a longer distance, and has a longer range than the heavy fragment ($R_L > R_H$). This is illustrated in Figure 1.6 which schematically shows the variation of the total stopping power with distance along the trajectory of the two fission fragments.

The initial total kinetic energy of both fragments amounts to 170 MeV on average, but actually varies from about 160 to 190 MeV, depending upon the mass split. For ^{235}U fission induced with thermal neutrons, the most energetic reactions occur for $M_H/M_L \approx 1.25$. The average energy/amu amounts to ≈ 0.7 MeV, which places the fission fragments in the upper-left part of both diagrams in Figure 1.3. In this region, the rate of energy loss along the trajectory of the fission fragment is decreasing, as is illustrated in Figure 1.6.

From the above considerations, a fission-track can thus be regarded as a more or less continuous trail of damage, the intensity of which can be expected to

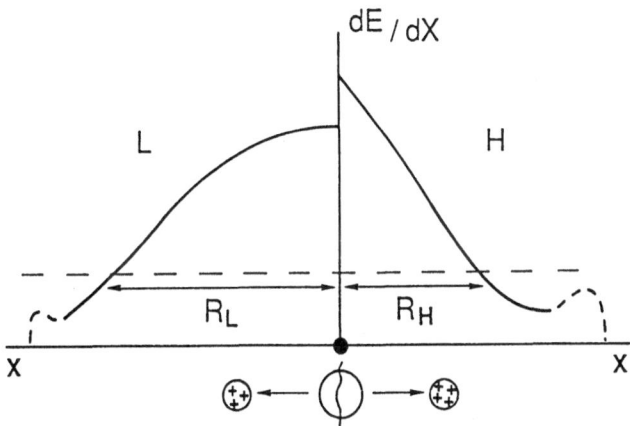

Figure 1.6. Schematic illustration of the variation of the rate of energy loss along the trajectory of the light (L) and heavy (H) fragment released during fission. Energy loss is due to electronic interactions along the major part of the trajectory; nuclear collisions only become important at the end (dashed sections). The horizontal dashed line corresponds to the registration limit of an unspecified detector. R_L and R_H correspond to the etchable ranges of the damage tracks created by the light and heavy fragments, respectively.

decrease when moving away from the locus of the uranium atom which does not usually coincide with the track centre.

1.4.3. *Spatial and areal density of fission tracks*

In applications such as fission-track dating or quantitative uranium analysis, the kernel of experimentation is an accurate determination of the number of fissioned ^{235}U or ^{238}U nuclei per unit of volume in the detector. This number is obviously equal to the number of fission tracks counted per unit of volume, due to the one-to-one relationship track – fissioned atom. If a technique of optical revelation is applied such as etching, only the tracks which cross the etched surface of the detector will be observed and counted under the microscope, as will be discussed in more detail in the next chapter. The number of tracks per unit of volume (i.e., the spatial track density) has thus to be derived from the number of tracks counted per unit of area in the observed surface (i.e, the areal or planar track density).

The theoretical relation between the areal and spatial track density can be easily derived taking into account the following basic assumptions:

- the fissioned atoms are homogeneously distributed throughout the volume of the solid,
- all tracks in the solid are of equal length l and the locus of the fissioned atom is at the centre of each track (or $l = 2R$),
- the tracks have isotropically distributed orientations, meaning that they exhibit no preferential direction, the probability of track formation in the detector being the same in all directions.

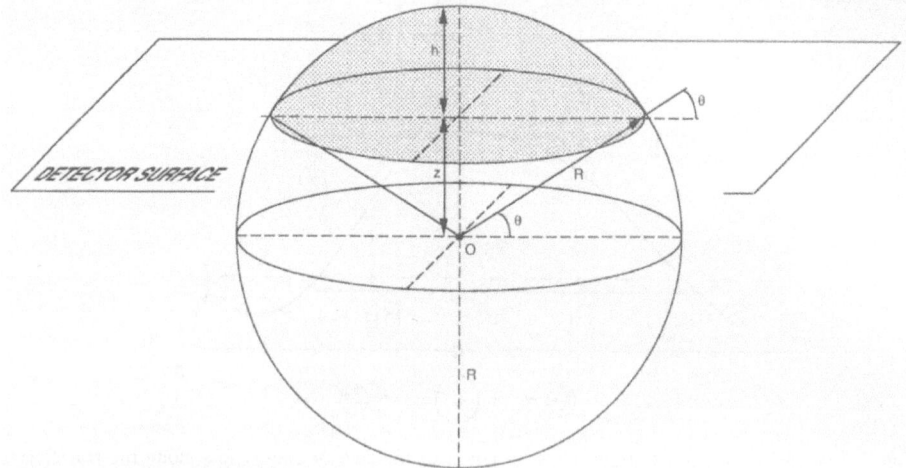

Figure 1.7. When an internal uranium atom at point 0, situated at depth z beneath the detector surface fissions, only those tracks whose upper ends fall in the hatched zone of the sphere with radius R will cut the detector surface (R = length of the track of one fission fragment). The probability of a track for cutting the detector surface is given by $2\pi Rh/2\pi R^2$, i.e, the area of the sphere zone/the area of the semi-sphere representing all possible track orientations.

To some extent, these assumptions represent a simplification of reality. The major simplification comes from the second assumption which is obviously incorrect when the asymmetry and variability of the fission process are considered together with the anisotropic properties of crystalline detectors. The variation of the total track length is, however, relatively small for a given type of detector and a given nuclide.

If N_f = the number of fissioned nuclei per unit volume, the number of nuclei situated within a layer of unit area and of elementary thickness dz is given by $N_f\,dz$. When this layer is situated at a distance z ($\leqslant R$) from the detector surface, only tracks which make a dip angle $\geqslant \arcsin z/R$ will cut this surface (Figure 1.7). Because the probability of track formation is equal in all direcions, the probability $P(z)$ of a track for making an angle $\geqslant \theta = \arcsin z/R$ is given by the ratio of the surface area of the hatched sphere zone in Figure 1.7 to the surface area of the entire semi-sphere with radius R (representing all possible directions) or

$$P(z) = 2\pi Rh/2\pi R^2 = R(R - R\sin\theta)/R^2 = (1 - \sin\theta). \tag{1.1}$$

The number of tracks which cut the detector surface is thus given by $(1 - \sin\theta)$ $N_f\,dz$ and the areal track density of latent tracks ρ_l is then obtained from

$$\rho_l = \int_0^R (1 - \sin\theta)N_f\,dz = \int_0^R N_f(1 - z/R)\,dz$$

which, after calculation, yields the simple relation

$$\rho_l = N_f R/2 \quad (\text{or} = N_f l/4). \tag{1.2a}$$

In case the investigated surface is an internal surface situated within the volume

of the detector, tracks can originate from nuclei at both sides of the surface an
the density is given by

$$\rho_l = N_f R \quad (\text{or } = N_f l/2). \tag{1.2b}$$

Both of the above equations can be combined into

$$\rho_l = g N_f R. \tag{1.3}$$

g represents the geometry factor and, depending upon whether the surface is
internal or external, $g = 1$ (so-called 4π geometry) or $g = 1/2$ (2π geometry).

The above relationship between spatial and areal track density represents the
simplest possible theoretical approach of the track density problem in track detectors. Nevertheless, it is of fundamental importance for a good understanding of
the fission-track dating method and the idea of measuring track lengths for the
interpretation and/or correction of fission-track ages (as will be discussed in Chapter 5) is, in principle, based on this relationship. The densities and lengths of
fission tracks which are actually observed under the optical microscope are, however, strongly dependent of the process of chemical etching which is used for
revelation of the latent tracks, a subject which is discussed in the next chapter.

CHAPTER 2

Track Revelation and Observation

2.1. Techniques of Track Revelation: Chemical Etching

Due to their limited width of a few nm, a direct observation of latent fission tracks is only possible with transmission electron microscopy (TEM) and the early observations of tracks in mica were made with this method (Silk and Barnes, 1959; Price and Walker, 1962a,b) using the electron beam in a diffraction contrast mode. In order to render the damage trails visible under a normal optical microscope, several techniques of track enlargement, also called *visualization* or *revelation techniques*, have been developed. An overview of these techniques and their application in the various fields of nuclear track detection has been given by Fleischer *et al.* (1975) and Monnin (1980). A first group of revelation techniques are the so-called *decoration techniques*. Examples are segregation and precipitation of metals along the damage trails (Fleischer and Price, 1963b) and polymerization followed by subsequent dyeing (Monnin and Blanford, 1973; Somogyi *et al.*, 1979). Decoration techniques are essentially used on plastics and have not found application in fission-track dating.

The second group of revelation techniques are the *etching techniques*. These are destructive methods as they use the damage along the tracks as a preferential site of attack and removal of detector material. While decoration techniques can be capable of revealing tracks within the volume of a detector, etching always starts at a detector surface and essentially reveals the tracks that intersect that surface. At the time of introduction of the fission-track dating method, revelation of defects through etching was already a major subject of study, especially in semiconductor physics. Several methods were investigated such as chemical etching and electrolytic etching. Of these methods, only chemical etching has become currently applied in fission-track dating. On polymers, the electrochemical etching technique (ECE) developed by Tommasino in 1970 (see Tommasino *et al.*, 1980a,b; Tommasino, 1980; Durrani and Bull, 1987 for an overview) is also used in some specific fields of research (e.g., particle dosimetry). In ECE, an alternating electric field of high frequency is applied across the detector while it resides in an appropriate chemical etchant. In this way, the tracks are greatly enlarged as tree-like figures with a size of a few hundred μm, allowing an easy assessment of low-track densities at low optical magnification.

This chapter is essentially confined to track revelation by chemical etching.

However, some of the findings and statements formulated here will be valid for other etching techniques as well. Chemical etching principally results from a simple immersion of the mineral or glass in a suitable chemical reagent, i.e., a reagent that causes a preferential attack of the tracks so that they become enlarged. Here, it is necessary to have good control on the concentration and temperature of the reagent and on the etching time. Appropriate etching conditions are mainly selected on an empirical basis. Appendix A contains a list of geologically important glasses and minerals together with correspondent etchants, successfully used to reveal fission-tracks.

Although in practice chemical etching is quite simple, the process of track revelation through this technique is complicated and – especially in minerals – not completely understood. Etching affects the tracks profoundly, not only in a qualitative way, i.e., their shape and size, but also in a quantitative way, i.e., their areal density, meaning that the density of etched tracks ρ will deviate significantly from the density of latent tracks ρ_l ($= gN_fR$, Section 1.4.3).

In the first place, the etchant is generally not capable of revealing the entire range of a fission fragment, which means that the etchable length of a fission-track is shorter than the combined physical range of the light and heavy fragment (Figure 1.6). Because it is only the etchable part of a latent track which matters if revelation through etching is considered, the terms range (R) and length (l) will, unless specified, refer to the etchable range and the etchable length of the latent damage trails during further discussion. Secondly, as early as 1964, Fleischer and Price noted that only a part of the latent tracks that cross a glass surface are revealed by etching, depending upon the ratio of the general or bulk etching rate V_g of the glass to the etching rate V_t of the tracks themselves. Later (Fleischer and Hart, 1972; Fleischer et al., 1975), these findings were integrated into an etching efficiency factor η. In addition to the above effect, bulk etching will cause a layer with a certain thickness to be removed from the glass or crystal after a certain etching time. As a consequence, tracks situated well beneath the original surface but with their upper ends falling within this layer, will also be etched and observed. Khan and Durrani (1972b) were the first to acknowledge this phenomenon in glass and defined a prolonged etching factor $f(t)$ which they also quantified theoretically and experimentally. Both the etching efficiency factor η and the prolonged etching factor $f(t)$ are adopted in this book and considered in some detail as they are necessary for a precise understanding of the process of track revelation by chemical etching and its consequences on the fission-track dating method.

A final important point is that with respect to track etching, glasses behave very differently from crystals. Glasses are amorphous substances that etch with the same rate (same V_g) in all directions and, therefore, they are called isotropic detectors. The track etching rate V_t is also not much higher than the bulk etching rate V_g; in silicate glasses the ratio V_t/V_g typically ranges from 1.5 to 3. The etching behaviour of crystals, on the contrary, is like their other physico-chemical properties governed by their highly ordered internal atomic structure. Depending upon the crystal symmetry, both the mineral surface and the tracks that cut across it, will be etched in a characteristic way. Furthermore, V_t is generally much higher than V_g, commonly by a factor of 10 or more, V_g itself being rather variable due to the above-mentioned anisotropy. Because of this contrasting behaviour, track etching in crystals and glasses is discussed separately.

2.2. Etched Tracks in Glass

2.2.1. *Fundamentals*

In glass, fission tracks are typically revealed by the chemical etchant as conical pits whose intersections with the etched surface are circular or oval, depending upon the dip angle of the track (Figures 2.6 and 2.7). With prolonged etching, these cones gradually become larger in size and they are pointed (sharp-bottomed) as long as the etchant does not reach the track end. When it does so, deepening of the cones stops and the bottom becomes rounded (flat-bottomed). Tracks with largely rounded bottoms can be difficult to identify in volcanic glass and vitreous tephra because they resemble pores and bubbles. In fission-track dating, therefore, it is common practice to consider the pointed tracks only. Due to important bulk etching, new tracks that were previously confined within the glass are also systematically brought to the surface and added to the total track population. These new tracks appear normally as cones of smaller size than the original surface tracks.

The track etching characteristics of glass can be theoretically described to a good approximation with the use of the assumptions formulated in Section 1.4.3 completed with following

- the track etching rate V_t is constant along the entire length l of each track,
- the bulk etching rate V_g is identical everywhere in the glass (in practice, this implies a constant chemical composition on a microscopic scale).

Basing ourselves on these assumptions, we will now discuss how the etching process affects the areal density and the shape and size of fission tracks.

2.2.2. *Track density*

Etching efficiency factor. In order to be effectively revealed by the etchant, a track has to intersect the glass surface at an angle that exceeds a certain minimum value. This minimum angle is called the critical angle of incidence θ_c and is equal to arcsin V_g/V_t. This can be derived from Figure 2.1 which shows that if V_t is split into its horizontal and normal (or vertical) components V_{th} and V_{tn}, a track will only be revealed if $V_{tn} > V_g$ or, in other words, if the dip angle θ of the track $>$ arcsin $V_g/V_t = \theta_c$ (Fleischer and Price, 1964c). The critical angle can be experimentally determined, e.g., by bombarding glass detectors with collimated beams of fission fragments from an artificial ^{252}Cf source, under decreasing angles till no more tracks are revealed by the etchant (Khan and Durrani, 1972a). These

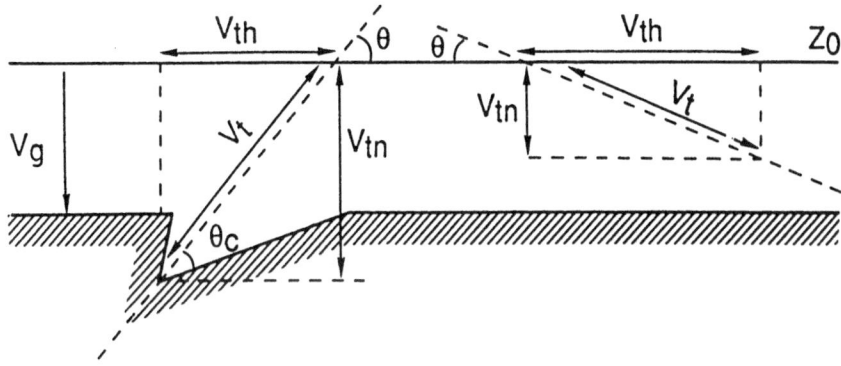

Figure 2.1.   In order to be revealed by the etching process, the normal component (V_{tn}) of the track etching rate V_t must exceed the bulk etching rate of the glass V_g. Taking into account that $V_{tn} = V_t \sin \theta$, this implies that the angle of incidence θ of the track must exceed arcsin V_g/V_t. This minimum angle of incidence is called the critical angle θ_c. The track at the right does not meet this criterion and is not etched.

determinations typically yield values of 25–35° for natural glasses such as tektites and obsidians.

Taking into account the critical angle and using the same reasoning as in Section 1.4.3, a track situated within the glass at a distance z ($\leqslant R$) from the surface, will have a probabilty $P(z)$ for crossing the surface *and* being revealed by the etchant which is given by

$$P(z) = 1 - \sin \theta = 1 - z/R$$

for tracks originating at a distance z between $R \sin \theta_c$ and R from the surface,

$$P(z) = 1 - \sin \theta_c$$

for tracks originating at a distance z between 0 and $R \sin \theta_c$.
Hence, the density of etched tracks is given by

$$\rho_0 = g2N_f \left[\int_0^{R \sin\theta_c} (1 - \sin \theta_c) \, dz + \int_{R \sin \theta_c}^{R} (1 - z/R) \, dz \right],$$

which, after further calculation, yields

$$\rho_0 = gN_f R \cos^2 \theta_c \quad \text{or} \quad \rho_0 = \rho_l \cos^2 \theta_c, \tag{2.1}$$

where, as in Section 1.4.3, $\rho_l = gN_f R$ and $g = 1$ or 0.5 depending upon whether the glass surface is external or internal. The factor $\cos^2 \theta_c$ is called the etching efficiency factor η of the tracks that cross the original glass surface.

Effect of prolonged etching. Because, after an etching time t, a layer with thickness $h = V_g t$ will be removed from the glass, all confined tracks whose upper ends are situated within this layer will also have been reached by the etchant. They will be effectively revealed if their dip angle exceeds θ_c. The density of the added tracks steadily increases with h and is obviously independent from the registration geome-

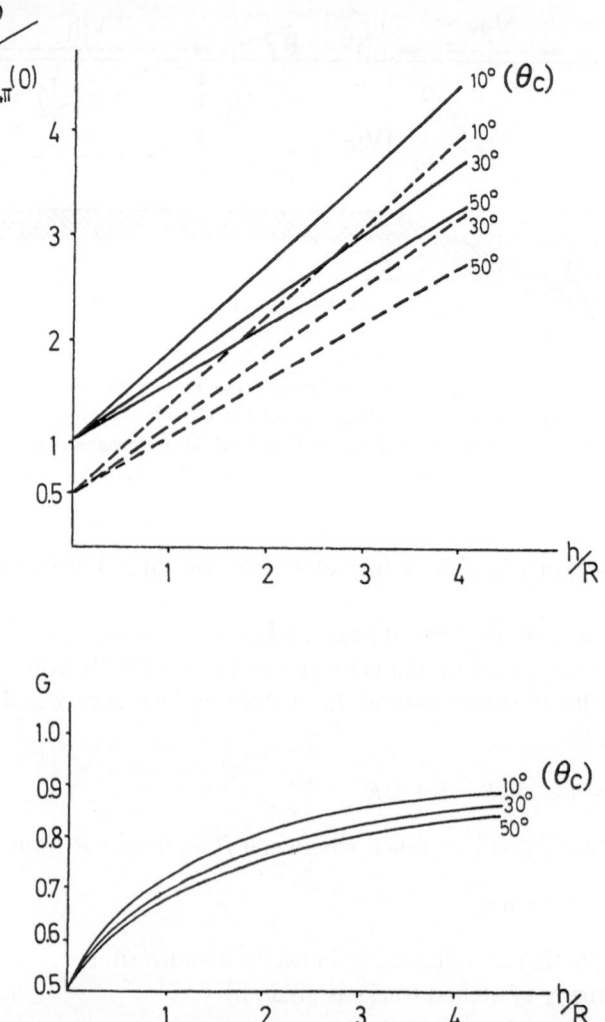

Figure 2.2. (*top*) Theoretical increase of the total track density with prolonged etching for various values of θ_c if the track etching rate V_t is supposed to be constant along the track. The solid lines correspond to an internal glass surface (4π registration geometry), the dashed lines to an external surface (2π registration geometry). Track density is expressed as a ratio to the initial density in an internal surface, prolonged etching by the thickness of the removed glass layer h relative to the etchable range R. (*bottom*) Corresponding variation of the geometry ratio (i.e., the ratio of the track density in an external surface to that in an internal surface).

try (external or internal) of the original glass surface. It is given by (Somogyi and Nagy, 1972; Green and Durrani, 1978)

$$\Delta\rho = N_f h(1 - \sin\theta_c).\tag{2.2}$$

The factor $1 - \sin\theta_c$ represents the etching efficiency factor η_a of the added tracks.

The total track density as a function of prolonged etching is given by (2.1) + (2.2) and is illustrated in Figure 2.2. If the total track density is expressed

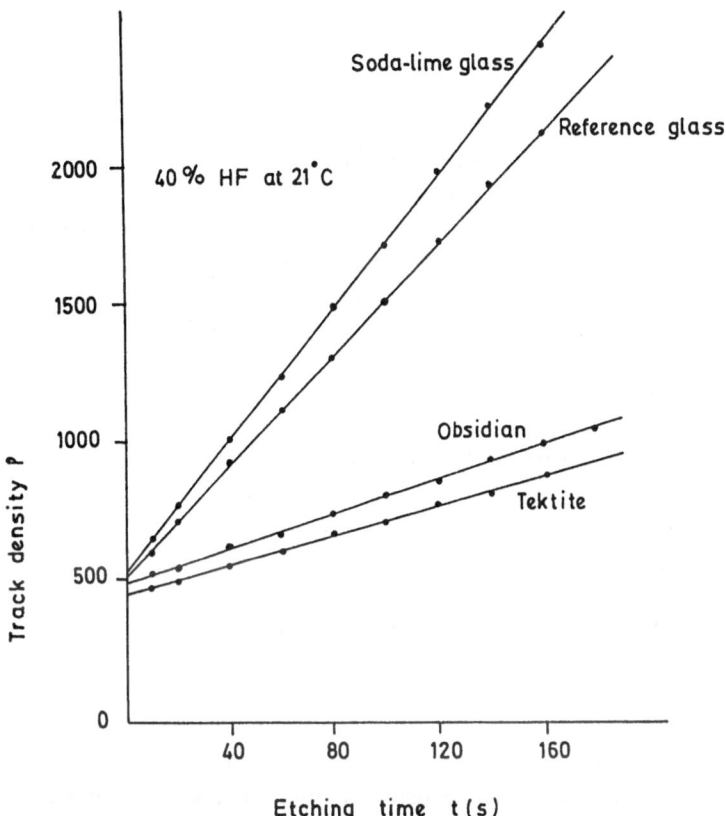

Figure 2.3. Experimentally registered linear increase in total fission-track density with prolonged etching for various types of glass (from Durrani and Khan, 1972b). The reference glass refers to U2 glass manufactured at Corning, New York (Schreurs *et al.*, 1971) which contains 43 ppm of U (^{235}U/^{238}U ratio = 3.6 × 10^{-3}).

as a product of the latent track density ρ_l ($= gN_fR$), an initial efficiency factor η ($= \cos^2 \theta_c$) and an etch time factor (Khan and Durrani, 1972b), then one obtains the following linear expression in h ($= V_g t$)

$$\rho = \rho_l \eta \left[1 + \frac{h}{gR(1 + \sin \theta_c)} \right]. \tag{2.3}$$

In this expression the factor between square brackets represents the etch time factor $f(t)$. A linear increase in total track density has been found in several experimental studies (e.g., Khan and Durrani, 1972b; Van den haute, 1985) clearly demonstrating that bulk etching continuously adds new tracks to the track population (Figure 2.3.). The evolution of the ratio G of the track density in an external surface to that in an internal surface, can be calculated starting from Equation (2.3) after substitution of the appropriate values for g and ρ_l. As can be seen from Figure 2.2, this ratio steadily increases from its initial value of 0.5 to 1 at infinite etching times.

If only the pointed tracks are considered, the total track density must be reduced with the tracks that became rounded (flat-bottomed) during etching. In contrast to the density of the added tracks, the variation of the density of the flat-bottomed tracks does depend upon the registration geometry of the original glass surface. Two etching depths are crucial here, namely h_R ($= R \sin \theta_c$) and $2h_R$. These depths correspond to a distance y_t travelled by the etchant along the tracks themselves of, respectively, R and $2R$ ($=$ total track length).

If $0 \leqslant h \leqslant h_R$, no flat-bottomed tracks are formed in an originally external glass surface, because all tracks originate from uranium atoms localized beneath the surface and, hence, have an etchable length $\geqslant R$. In an originally internal surface, the density of flat-bottomed tracks is given by (Van den haute, 1985)

$$\rho_{fb} = N_f h \cos^2 \theta_c / 2 \sin \theta_c. \tag{2.4}$$

When $h_R \leqslant h \leqslant 2h_R$, a formerly external surface also starts to exhibit flat-bottomed tracks, their density being given by

$$\rho_{fb} = N_f (h - h_R) \cos^2 \theta_c / 2 \sin \theta_c. \tag{2.5}$$

Equation (2.4) remains valid for an internal surface.

The different stages in the variation of the pointed-track density with prolonged etching are illustrated in Figure 2.4. In an internal surface, ρ_{sb} is obtained by subtracting Equation (2.4) from the total track density given by Equation (2.3). If Equation (2.4) is compared with Equation (2.2.), it will be noticed that $\rho_{fb} > \Delta\rho$ for all values of θ_c. Hence, the density of pointed tracks, in principle, always decreases with etching times. In an external surface, ρ_{sb} equals ρ, as long as $h \leqslant h_R$, while it is given by (2.3)–(2.5) when $h_R \leqslant h \leqslant 2h_R$, which yields the same value as for an internal surface. As a consequence, once $h \geqslant h_R$, the geometry ratio equals 1 and remains so throughout further etching (Figure 2.4). Several experiments (Edwards, 1967; Reimer *et al.*, 1970; Van den haute, 1985), show that this equality is actually reached after varying etching times depending upon the type of glass and the etching conditions (Figure 2.5).

When h equals $2h_R$, all tracks that crossed the original pre-etched glass surface have become rounded, meaning that the only tracks that exhibit pointed bottoms are added tracks. Their density is given by

$$\rho_{sb} = 2N_f R (1 - \sin \theta_c) \sin \theta_c. \tag{2.6}$$

Equation (2.6) is independent of etching time because once all primary tracks have become rounded and etching is continued, the etchant now starts to round off the added tracks at the same rate as it reveals new ones.

Relatively few experiments on the variation of pointed-track density with etching time have been published (Edwards, 1967; Van den haute, 1985). In an internal surface, instead of a decreasing trend as theoretically expected (Figure 2.4), often the density is observed to remain relatively constant during most of the etching time (Figure 2.11). This is probably due to the fact that V_t actually varies along a track instead of being constant as will be discussed in Section 2.2.4.

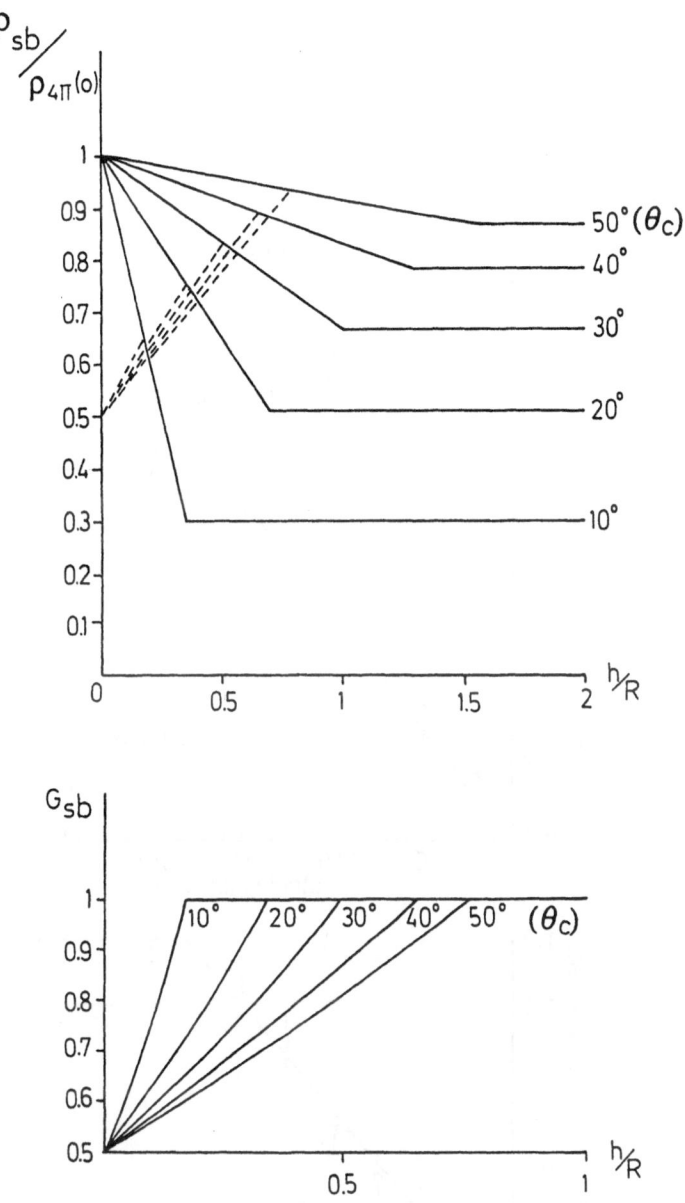

Figure 2.4. (*top*) Theoretical variation of the density of pointed (or sharp-bottomed) tracks with prolonged etching for various values of θ_c (V_t is supposed to be constant along the track). The solid lines correspond to an internal glass surface, the dashed lines to an external surface. The variation in an external surface is only indicated for the first etching phase ($h \leqslant h_R = R \sin\theta_c$); when h equals h_R it becomes identical to that in an internal surface. In an internal surface, the track density decreases linearly until $h = 2h_R$ and then stabilizes. (*bottom*) Corresponding variation of the geometry ratio. The ratio increases to reach a plateau value 1 when $h = h_R$ (from Van den haute, 1985).

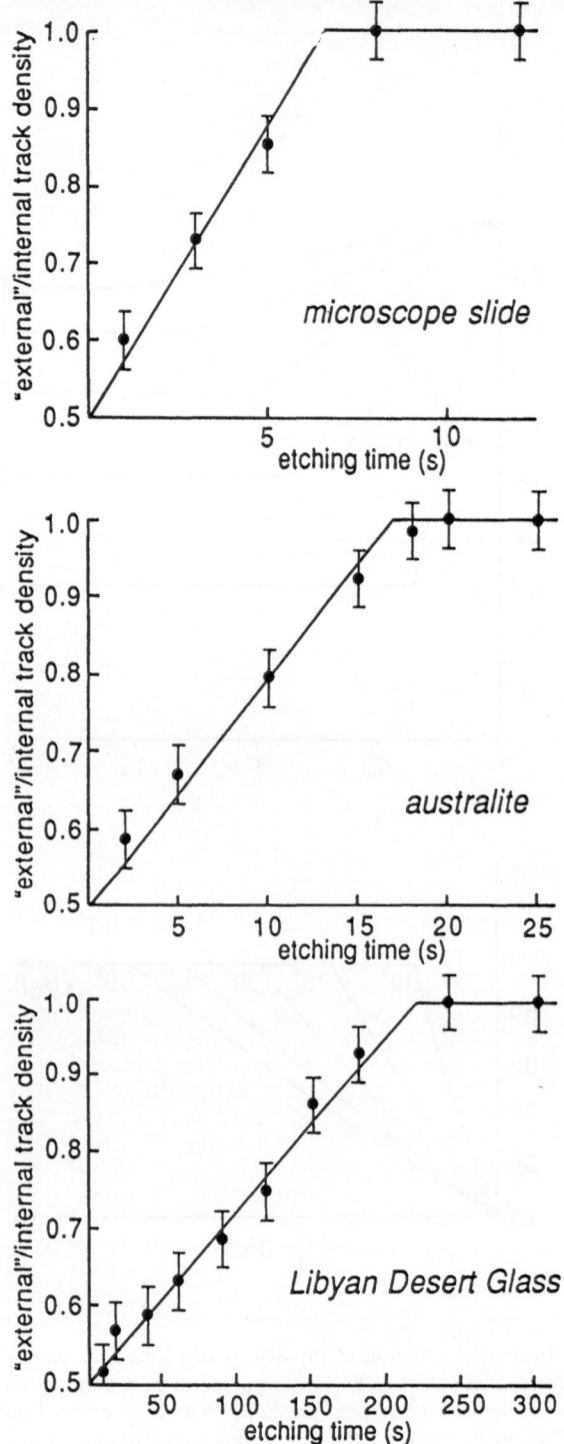

Figure 2.5. Experimentally registered variation of the geometry ratio with etching time for various types of glass (from Reimer *et al.*, 1970). Etching was carried out in HF (48%) at 25°C.

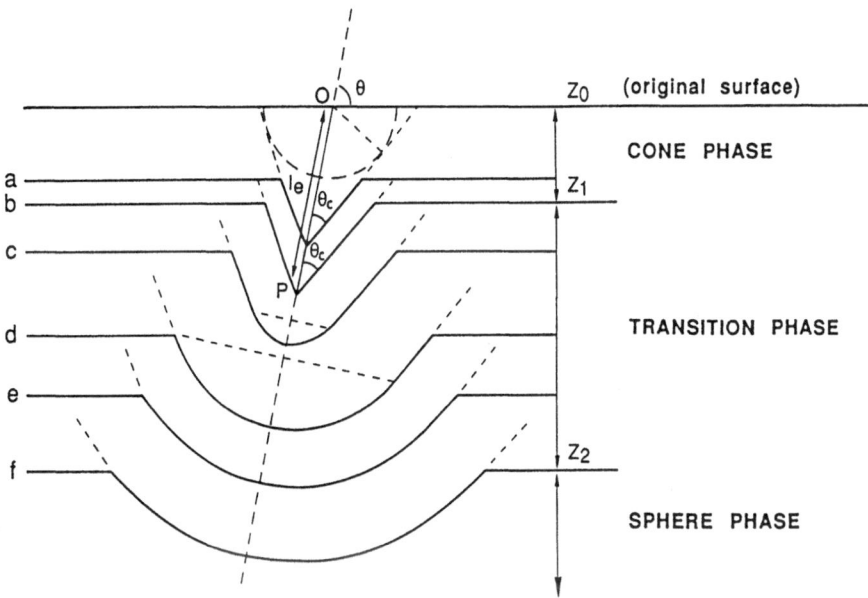

Figure 2.6. The three phases in the evolution of the shape of an etched track in glass with prolonged etching (simplified after Ali and Durrani, 1977). During the initial cone phase, the damage trail is attacked at a rate V_t and the bulk of the material at a rate V_g, revealing the track as a circular cone with angle θ_c. When the etchant reaches the end of the track at P (corresponding to level z_1) preferential etching along the direction of the track stops and the transition phase sets in during which further widening of the track is accompanied by a progressive rounding of its bottom. At level z_2 the track shape has become entirely spherical, a shape which persists throughout further etching.

2.2.3. *Track shape and size*

Single track geometry. Under the assumption of a constant V_t and a constant and isotropic V_g, each track is etched as a circular cone which has the latent trail of damage as its axis and a semi-cone angle equal to θ_c ($= \arcsin V_g/V_t$). The figures of intersection of these cones with the etched glass surface are ellipses and their excentricity depends upon the dip angle of the tracks. For normally incident tracks, the intersection is circular. With prolonged etching, a track becomes continuously larger and its geometry changes, the conical shape only being preserved during the initial stage of etching.

Following Ali and Durrani (1977), three major phases can be distinguished in the evolution of the etched-track profile. These phases are, respectively, the conical phase, the transition phase, and the spherical phase and are illustrated in Figure 2.6. Photomicrographs of etched fission-tracks at various stages, of etching are shown in Figure 2.7. The evolution of track geometry is outlined here briefly, the discussion being mainly limited to those parameters which are relevant to fission-track dating. There are several publications which treat nuclear track geometry more thoroughly (Price and Fleischer, 1971; Henke and Benton, 1971; Somogyi and Szalay, 1973; Fleischer *et al.*, 1975; Ali and Durrani, 1977; Somogyi, 1980; Durrani and Bull, 1987) and the interested reader is referred to these works for further information.

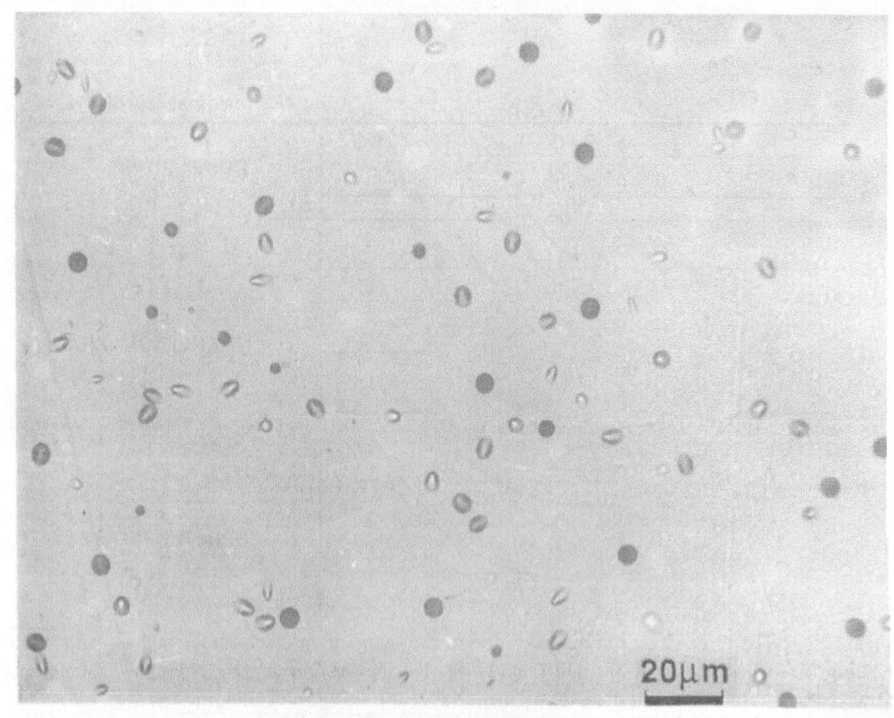

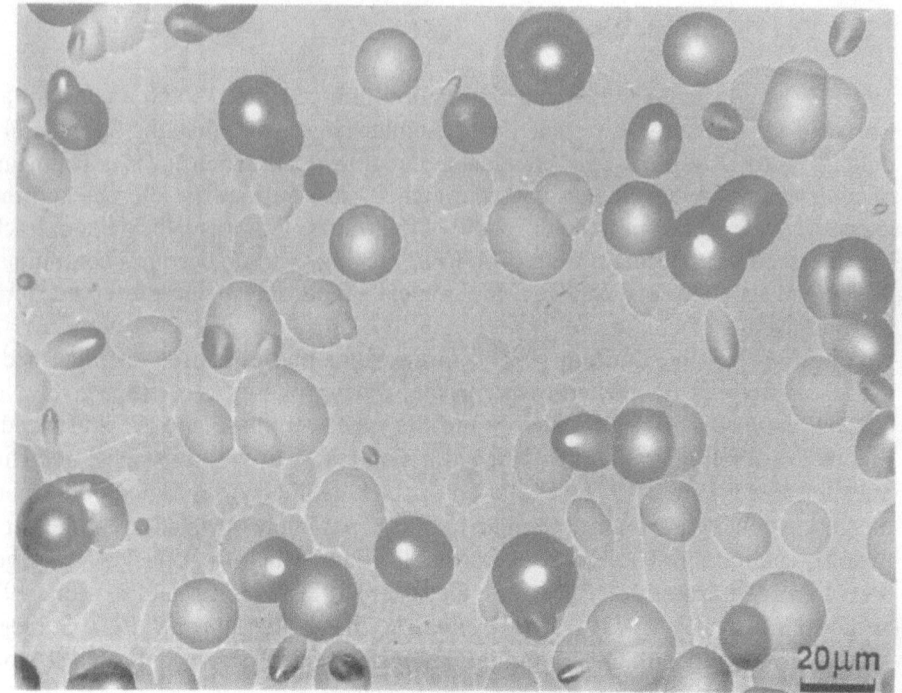

If l_e (= OP in Figure 2.6) is the etchable length of a track and θ its dip angle with respect to the glass surface, the first phase corresponds to a thickness of the removed glass layer $h \leqslant l_e \sin \theta_c$ (i.e., between levels z_0 and z_1 in Figure 2.6). During the first phase, the etch pit has the typical pointed-cone shape with an elliptical surface opening and the following dimensions:

length of the axis of the etch cone

$$l = h \left(1/\sin \theta_c - 1/\sin \theta\right), \tag{2.7}$$

depth of the cone

$$d = h \left(\sin \theta / \sin \theta_c - 1\right), \tag{2.8}$$

major axis of the elliptical opening

$$a = 2h \cos \theta_c / (\sin \theta + \sin \theta_c), \tag{2.9}$$

minor axis of the elliptical opening

$$b = 2h \sqrt{(\sin \theta - \sin \theta_c)/(\sin \theta + \sin \theta_c)}. \tag{2.10}$$

Tracks standing perpendicular to the surface ($\theta = 90°$) have a circular opening with $a = b = 2h \cos \theta_c / (1 + \sin \theta_c)$, while for such tracks it will also be noticed that $d = 1 = h \left(1/\sin \theta_c - 1\right)$. When θ equals θ_c, then $a = h \cot \theta_c$ while $b = 0$, meaning that the track loses its visibility, which demonstrates in another way the significance of the critical angle of etching θ_c.

The second phase corresponds to $l_e \sin \theta_c \leqslant h \leqslant l_e [\sin \theta_c + (\sin \theta - \sin \theta_c)/1 - \cos(\theta - \theta_c)]$. It starts when the etchant has reached the end of the etchable track length l_e. At this moment, the track depth reaches a value $d = l_e(\sin \theta - \sin \theta_e)$. From now on, no more preferential etching occurs along the direction of the track and, hence, the further evolution of the etch pit is governed by V_g only. This causes the shape of the track to alter progressively from a cone into the segment of a sphere, while the track depth remains constant (Figure 2.6). At first, the bottom of the track becomes rounded (=spherical) while the surface opening remains an ellipse whose dimensions are given by Equations (2.9) and (2.10). As etching proceeds, the spherical bottom gradually becomes more important in the total track profile and at a certain stage (level d in Figure 2.6) depending upon the dip angle θ, it will reach the etched glass surface. From now on, the contour of the track opening becomes partly elliptical and partly circular and, with further

---

Figure 2.7. ^{235}U fission tracks in SRM 612 glass (National Bureau of Standards, Washington) etched with HF (16%) for 75 s (top picture) and 250 s (bottom picture). In the top picture the tracks are in the cone phase with circular or elliptical openings depending upon their angle of incidence with the etched surface. In the bottom picture, the tracks are of much larger size with shapes typical for the transition phase or even the sphere phase; the smaller tracks which are still in the cone phase are added tracks which were revealed at some later stage of etching. Both pictures were taken at the same magnification. In fission track dating, etching conditions yielding track shapes like those in the top figure are to be preferred.

etching, the circular part gradually grows at the expense of the elliptical part, until the third and final stage is reached, when

$$h = l_e[\sin \theta_c + (\sin \theta - \sin \theta_c)/1 - \cos(\theta - \theta_c)]$$

and the track is completely altered to a segment of a sphere with an entirely circular surface opening (level f or z_2 in Figure 2.6). The equations describing the dimensions of these more evolved etch pits can be found in Ali and Durrani (1977) and Somogyi (1980).

Size distributions of track populations. Detailed studies on single-track geometry are of major importance in the identification and physical characterization of nuclear particles (Fleischer *et al.*, 1975). In fission-track dating, all tracks which are observed originate from the same physical process: the fission of uranium isotopes (Chapter 3) and one could expect that variations in track shape or size, apart from those related to etching conditions, are relatively small. Some variation related to the energy of the fission fragments does, however, exist and etch pit size studies in glass were successfully used to determine the energy spectrum of ^{252}Cf fission fragments, the best results being obtained with artificial phosphate glass (Khan and Durrani, 1973a; Aschenbach *et al.*, 1974).

Environmental effects which act upon the stability of the latent fission tracks such as annealing, have, however, a much greater influence, as will be pointed out in Chapter 4. Thermally affected tracks clearly display reduced sizes compared to intact tracks, when etched under the same conditions. Fission-track size studies have therefore the potential to yield information on thermal effects suffered by a natural glass in its geological past. In order to retrieve this information, the fission-track geochronologist focuses on the size characteristics of the entire track population in a sample, rather than on single track geometry.

In glasses, the major axis of the elliptical surface openings of the etched tracks (often simply called the track diameter) is easy to measure and, therefore, commonly used as a representative parameter of track size. The shape of the diameter distribution of the conical tracks with pointed bottoms and its evolution with prolonged etching has been studied by Van den haute (1985) using Equation (2.9) as a basis for calculation (Figure 2.8). If $0 \leqslant h \leqslant 2h_R$, the diameter distribution consists both of original and added tracks and is composed of two parts: a flat, horizontal part ranging from diameter values 0 to $2h \cos \theta_c/(1 + \sin \theta_c)$ and a peak situated between $2h \cos \theta_c/(1 + \sin \theta_c)$ and $2h \cot \theta_c$. The peak is mainly fashioned by original tracks cutting the pre-etched glass surface, while the flat part is built up by added tracks only. With prolonged etching the flat part gradually lengthens while the peak shrinks to disappear completely when $h = 2h_R$, i.e., the moment that all original tracks have become rounded. The track diameter distribution is then composed of added tracks only and has the approximate shape of a flat trapezoid situated between the diameter 0 and $2R \cos \theta_c$. It does not change anymore with further etching.

During the first stage of etching ($h \leqslant h_R$) the diameter distributions in external surfaces exhibit a more shallower peak than those in internal surfaces, the height of which, however, remains constant because no tracks are rounded in an external

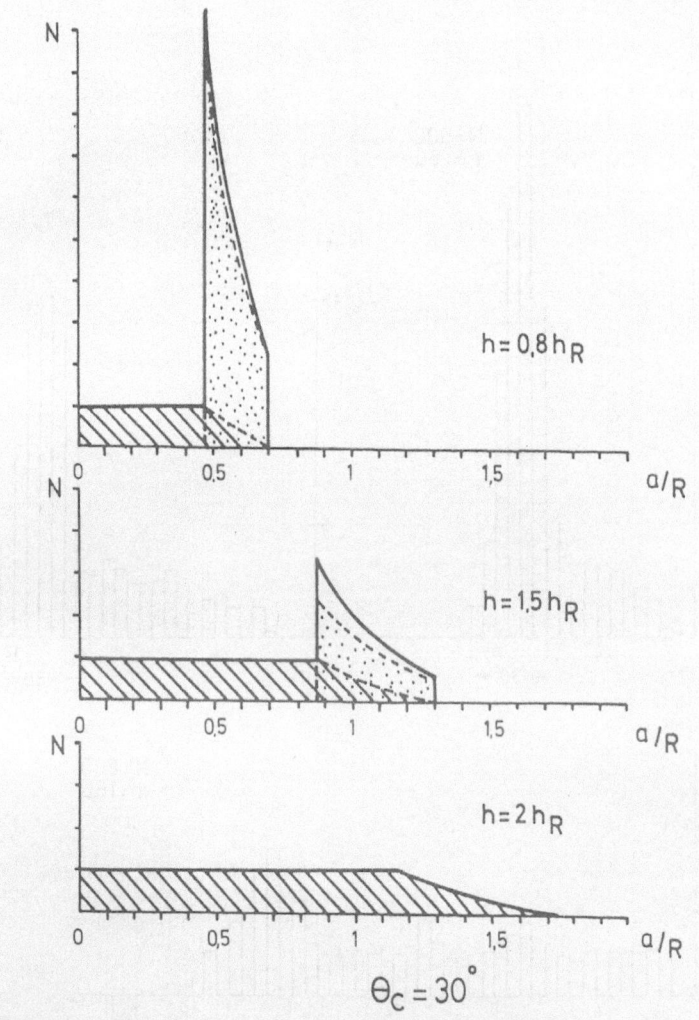

Figure 2.8. Theoretical evolution of the diameter (= major axis of the elliptical opening) distribution of pointed tracks with prolonged etching in an internal glass surface (the V_g/V_t ratio of the glass was taken to be 0.5 corresponding to $\theta_c = 30°$ and V_t is assumed to be constant). The solid line encompasses the complete distribution, the dotted area represents the distribution of the original surface tracks, the hatched area the distribution of the added tracks. When h reaches $2h_R$, the distribution turns flat and becomes composed of added tracks only. This shape remains unchanged by further etching. (From Van den haute, 1985.)

surface as long as $h \leqslant h_R$. When h equals h_R, the diameter distributions become identical in both types of surface and remain so during further etching. Experimentally registered evolutions of diameter distributions with etching time appear to be in good agreement with this model (Figure 2.9).

The variation of the average diameter of a track population with prolonged etching is illustrated in Figure 2.10. Calculations of this parameter using somewhat different approaches can be found in Somogyi and Nagy (1972), Hashemi-Nezhad and Durrani (1981a), and Van den haute (1985). The reader is referred to these

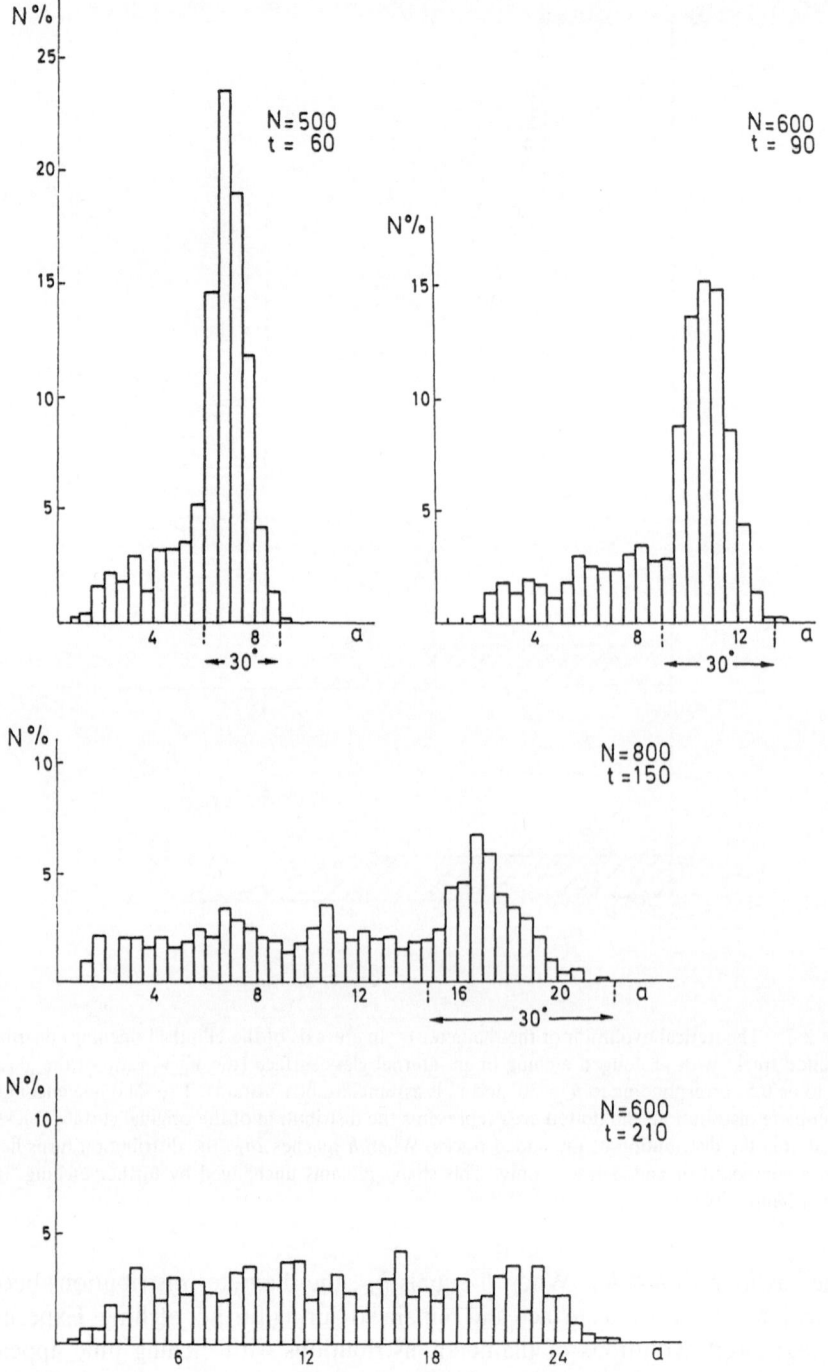

Figure 2.9. Experimentally registered evolution of the diameter distribution of pointed ^{235}U fission tracks with prolonged etching (16% HF, 25°C) in NBS-SRM 612 glass (internal surface). N = number of tracks measured, t = etching time in seconds. One diameter unit corresponds to $0.63\,\mu$m. The theoretically expected diameter limits of the peak in the distribution are indicated for $\theta_c = 30°$. (From Van den haute, 1985.)

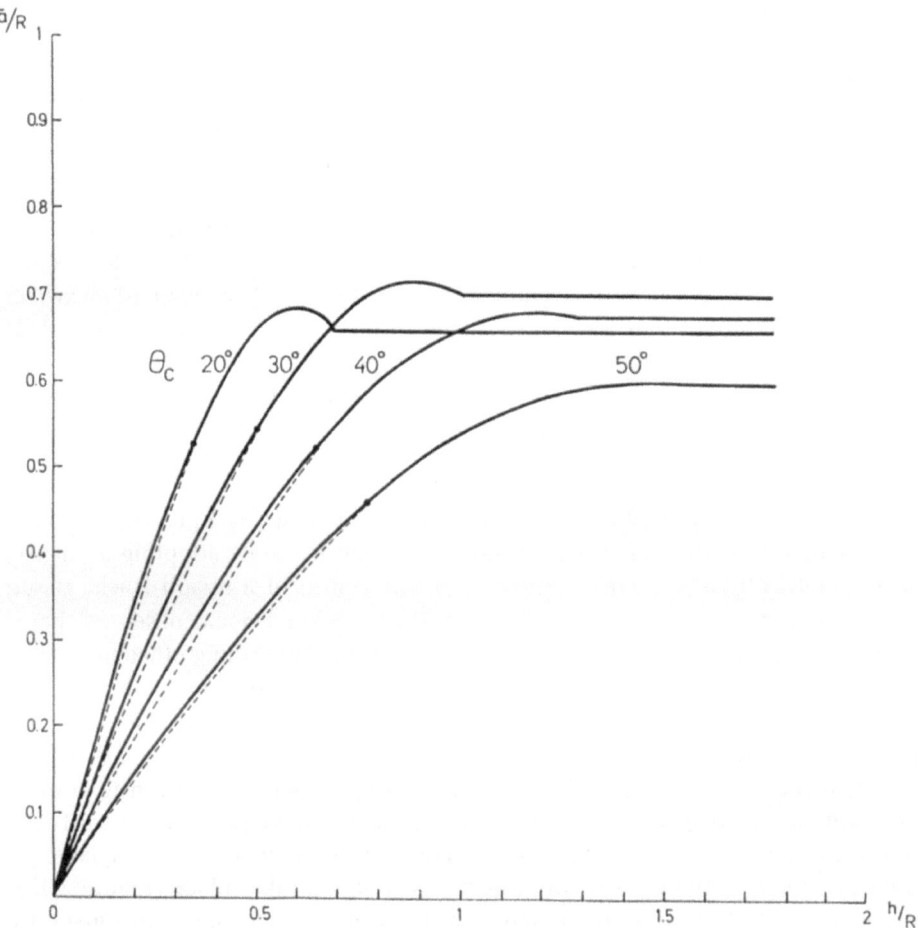

Figure 2.10. Calculated variation of the mean-track diameter (as a ratio to the etchable range R) of pointed tracks with prolonged etching in an internal (solid line) and an external (dashed line) glass surface. When h equals h_R the mean diameter becomes identical in both surfaces, a situation which is maintained during further etching. The mean diameter stabilizes when $h = 2h_R$, which corresponds to the moment that all original tracks have become rounded.

papers for the relevant equations. The average diameter varies according to curves which are nearly linear during the first stage of etching, but which later on display a maximum that is most clearly pronounced for glasses with high V_t/V_g ratios (small θ_c). As the diameter distribution stabilizes when h equals $2h_R$, the same applies to the average diameter.

2.2.4. *Effects of varying V_t*

The main constraint on the above approach comes from the assumption of a constant V_t. While the assumption of a constant and isotropic V_g holds for most glasses unless they exhibit important variations in composition on a microscopic scale, the assumption of a constant V_t along the entire track length represents an oversimplification of reality. Indeed, the variation of the track-etching rate along the damage trail constitutes one of the keystones in particle identification with the track-detection technique. The fragments released during U fission whose energies fall in the range 0.5–1 MeV/nucleon create tracks with a generally decreasing rate of ionization along their trajectory and, hence a similar behaviour is expected for V_t. A varying V_t, although it does not drastically alter the principles outlined above, may have a considerable effect on some etched track parameters and usage of the equations based on a constant V_t for calculating quantities such as the etching efficiency factor, the etchable track length, etc., can lead to incorrect results. It also implies that the critical angle of incidence has to be redefined. If V_t decreases along the range of a fission fragment then this angle can be considered as $\theta_0 = \arcsin V_{tmax}/V_g$, V_{tmax} being found approximately in the centre of a fission track at the locus of the fissioned nuclide.

Track density. Some insight into the consequences of a varying V_t can be obtained by recalculating the parameters governing track density while adopting a linearly and symmetrically decreasing V_t away from the centre of a fission track. Based on this assumption, Fleischer and Hart (1972) recalculated η and observed that it is reduced from $\cos^2 \theta_c$ (Equation (2.1)) to $1 - \sin \theta_0$. The etching efficiency factor η_a of the added tracks on the other side is independent of the variation of V_t and is just like η, given by $1 - \sin \theta_0$ (Van den haute, 1983). It can be noticed here that part of the tracks whose centre is situated beneath the pre-etched glass surface but which cut through it, now belong to the group of added tracks instead of to the population of original tracks. Indeed, due to the increase of V_t towards the track centre, it now occurs that the etchant has to remove a certain layer h_c (Somogyi, 1980) of the glass before effective revelation takes place (Figure 2.13).

The variation of the total track density with prolonged etching is now given by

$$\rho = \rho_0 + N_f h (1 - \sin \theta_0), \tag{2.11}$$

which, as for a constant V_t, represents a linear increase with etching time. The linearity of ρ therefore cannot be regarded as an indication of a constant V_t. If the variation is expressed as a product $\rho_l \eta f(t)$, then we obtain

$$\rho = \rho_l \eta (1 + h/gR) \tag{2.12}$$

and the prolonged etching factor only depends upon the etchable range R and not upon V_t.

The variation of the density of sharp-bottomed tracks with prolonged etching is illustrated in Figure 2.11 for some values of θ_0. The curves are clearly different from the simple linear decrease found for a constant V_t (Figure 2.4). When all tracks that cut through the pre-etched glass surface have become rounded, the pointed-track density again stabilizes. This moment arrives when a thickness $h_R = 2R \sin \theta_0 \ln \sin \theta_0/(\sin \theta_0 - 1)$ is removed from the glass surface (i.e. when $y_t = 2R$).

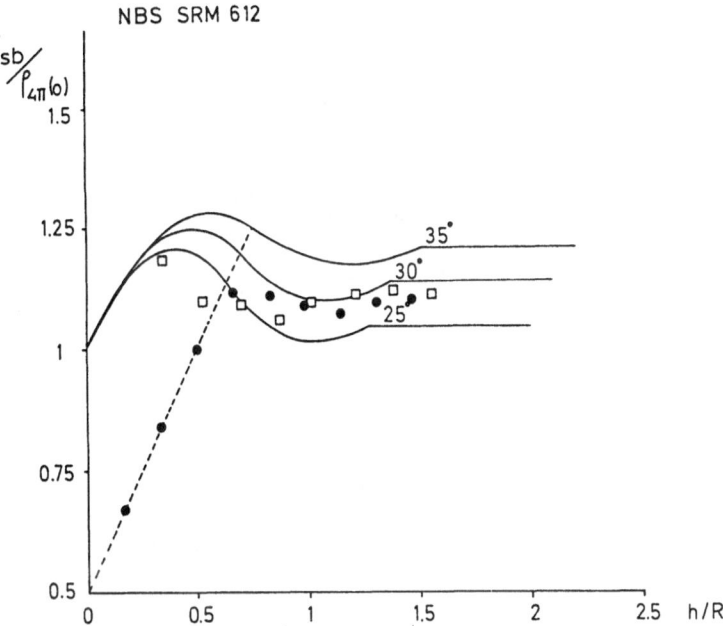

Figure 2.11. Experimental variation of the pointed-track density (^{235}U fission tracks) with prolonged etching in an internal (open squares) and an external surface (solid circles) of SRM 612 glass. The solid curves refer to the calculated variation for an internal surface (for some specific values of θ_0) assuming that the track etching rate V_t decreases linearly and symmetrically away from the track centre. The dashed line corresponds to the calculated variation for an external surface (this variation is identical for all values of θ_0). When $h = h_R$, the density in both surfaces becomes equal and further variation in an external surface is identical to that in an internal surface. The above curves provide a better fit to the experimental data than the curves of Figure 2.4 based on a constant V_t. (From Van den haute, 1983.)

Experimental studies which often reveal that the density of pointed tracks remains relatively constant with etching time (Edwards, 1967; Fleischer *et al.*, 1975; Van den haute, 1985) are more adequately explained by a linear than by a constant V_t (Figure 2.11). Determinations of V_t carried out for induced ^{235}U fission tracks in plastic (Endo and Doke, 1973) and for implanted ^{252}Cf fission fragment tracks in glass detectors (Fleischer and Hart, 1972; Fleischer *et al.*, 1975) give us more direct information on the real variation of V_t along a fission fragment track and reveal a clear decrease with decreasing residual range (Figure 2.12). Detailed experimental studies of the variation of V_t/V_g along internal uranium fission-tracks in glass and models using this variation for describing the variation of track density during etching have, however, not been published.

Track geometry. A varying V_t causes the shape of an etched track to deviate from a simple cone to a more complex figure generated by revolution of a higher order curve around the track axis, the track profile being convex inwards if V_t increases or concave if V_t decreases along the damage trail.

Internal uranium fission-tracks, which are combined tracks created by two fission

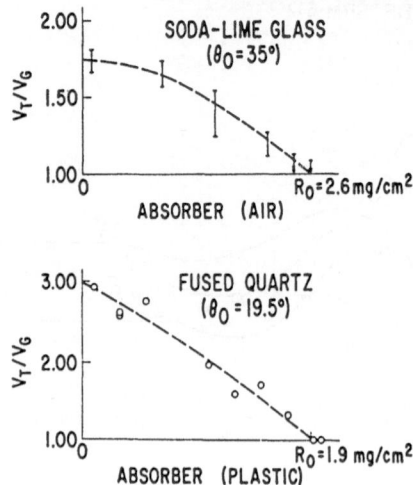

Figure 2.12. Experimentally registered variation of the V_t/V_g ratio along the tracks of externally implanted ^{252}Cf fission fragments in two glasses. The data illustrate the decrease in track etching rate V_t with decreasing residual range. (From Fleischer and Hart, 1972.)

fragments, will thus show varying track profiles, depending upon where they intersect the glass surface (Figure 2.13). As for a constant V_t, the high order cone will gradually evolve to a sphere segment with prolonged etching. The effect of a varying V_t upon the equations describing single nuclear track geometry have been treated in several papers (Somogyi and Szalay, 1973; Fleischer *et al.*, 1975; Somogyi, 1980; Durrani and Bull, 1987). In practice, numerical computation using iteration procedures is required for calculating track-size parameters such as the major diameter of the surface openings.

The effect of a varying V_t upon the diameter distribution of a population of etched fission-tracks and its evolution with prolonged etching has not been studied

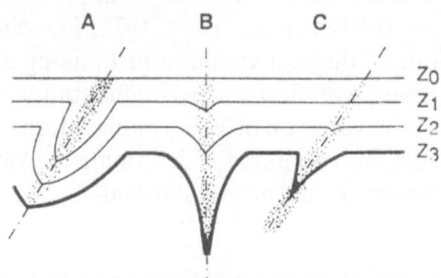

Figure 2.13. When the damage varies along the particle trajectory, the track will be etched as a cone that is concave (A) or convex (B, C) inward depending upon whether V_t decreases or increases with depth (modified after Fleischer *et al.*, 1975). The stippling is meant to indicate the variation in V_t and does not refer to the actual diameter of the damage. Note that track C, because of a too low V_t at its upper end, is not etched immediately but only when the surface is at level z_2.

in detail. Because of fission-track symmetry it could be expected that a varying V_t does not greatly change the shape of such a distribution. Experimental distributions at least exhibit shapes that seem to be well-approximated by the constant V_t theory, as can be deduced from a comparison of Figures 2.8 and 2.9.

2.2.5. *Factors affecting track etching*

Glasses are commonly etched with acid (typically HF) solutions of various concentrations at temperatures between 20 and 30°C (Appendix A). Because the density and shape of the etched tracks are strongly dependent upon etching conditions, a precise control on these conditions is necessary and a well-stabilized thermostatic bath together with precise equipment for measuring the temperature of the etchant and the etching time (Monnin, 1980) is not a superfluous luxury if well reproducible results are expected. In this respect, it is often advisable to employ diluted HF solutions (20% or lower) in order to achieve longer etching times which can be more accurately reproduced than shorter ones. Etching should also be carried out in fresh solutions and the use of the same solution for etching a large number of samples should be avoided because an increasing concentration of etch products in the reagent can affect the etching rate of the glass.

With prolonged etching, an etch product layer may form on the glass surface and protect it from the reagent, causing a progressive lowering of V_g. As a consequence, etching in three steps of 20 s interrupted by cleaning of the glass surface may therefore not give the same result as etching in one step of 60 s (Khan, 1973a; Yadav and Sharma, 1979). Stirring of the solution and more effectively brushing of the glass surface during etching inhibits the formation of this layer and in this way keeps V_g more constant. Although the formation of an etch product layer does not generally represent a serious problem in fission-track dating, it is clear that a careful standardization of etching conditions should be pursued in order to improve the reproducibility of a fission-track age determination.

The role of environmental effects on track revelation has been reviewed by Fleischer *et al.* (1975) and Khan (1980b). Of these effects, the thermal effect is by far the most important. The process of thermal annealing can be understood as a gradual reduction of both the etchable length of the tracks and their etching rate V_t (Chapter 4). Through sufficient heating, fission-tracks can, in this way, be completely erased, i.e., unrevealable by chemical etching. Apart from this effect which is of major importance for the fission-track dating method, heating of a glass can also affect its bulk etching rate V_g meaning that tracks freshly created in a preheated glass will etch differently from tracks created in the same but thermally untreated glass. This effect is ascribed to the devitrification phenomena and becomes progressively important with heating time (Khan and Ahmad, 1975). Low temperature annealing ($< 100°C$) carried out in diluted acid or basic aqueous solutions instead of air, also appears to affect V_g to some extent (Yadav and Sharma, 1979). Finally, high fluences of particles such as energetic protons may affect both the bulk and track etching rate in glass (Durrani *et al.*, 1973). This last effect however, is only of some importance if fission-track dating of extra-terrestrial glasses is considered.

2.3. Etched Tracks in Crystals

2.3.1. *Fundamentals*

In crystals, etched fission tracks have complex shapes which, together with other parameters such as etching efficiency, can hardly be described in terms of a simple V_g/V_t ratio as is done for glass. The needle-like aspect of the tracks which is commonly observed during the early stages of optical revelation, indicates that the track etching rate V_t is generally much higher than the bulk etching rate V_g. In minerals, an accurate determination of V_t is, however, not simple, because due to the narrowness of the etched track channels, factors other than radiation damage, such as the rate of diffusion of the etchant and/or of the etch products, can play an important role in their rate of deepening and widening (Perron and Maury, 1986). Determinations of V_t based on the visible increase in track length during etching (Thiel and Külzer, 1978; Watt and Durrani, 1985) are likely to be influenced by these effects and, therefore, can yield underestimated values of V_t. The most accurate V_t determinations are probably obtained with the etch-anneal-etch method as introduced by Green *et al.* (1978). In this method, first a short etch is applied which leaves the tracks below the limits of optical visibility, then the crystals are heated in order to erase the remaining unetched portions and, finally, full revelation of the formerly slightly etched parts is carried out. This method yields relatively high values of V_t. For instance, a V_t/V_g ratio $>10^2$ can be deduced for ions approaching fission fragments (1 MeV/nucleon Xe) in orthopyroxene (Green *et al.*, 1978) or even $>10^4$ for U fission fragments in muscovite mica (Belyaev *et al.*, 1988).

Similar to isotropic detectors a variation V_t may also occur along the trajectory of a fission fragment, related to variation in particle energy loss (Price *et al.*, 1973; Green *et al.*, 1978; Belyaev *et al.*, 1988) which is now likely to be affected by the internal arrangement of the lattice atoms in the crystal.

In crystalline solids, the bulk etching rate is also highly variable. Each crystallo-graphic plane exhibits its own rate of etching with respect to a given reagent so that the simple overall V_g can be better replaced by $V_{[h,k,l]}$ which represents the etching rate of the specific plane that is considered, i.e. the rate normal to $[h, k, l]$. As far as the etching efficiency and etch time factor are considered, it is merely the etching rate of the crystallographic plane that corresponds to the etched surface (here simply called $V_{[s]}$) that is relevant but, in the evolution of the track shape, the etching rate of other planes revealed along the track channel also plays an important role.

If we turn to the process of optical revelation itself, another substantial differ-ence between minerals and glasses exists. Whereas in glass there is an important first phase where the visible evolution of the etched track results from an interac-

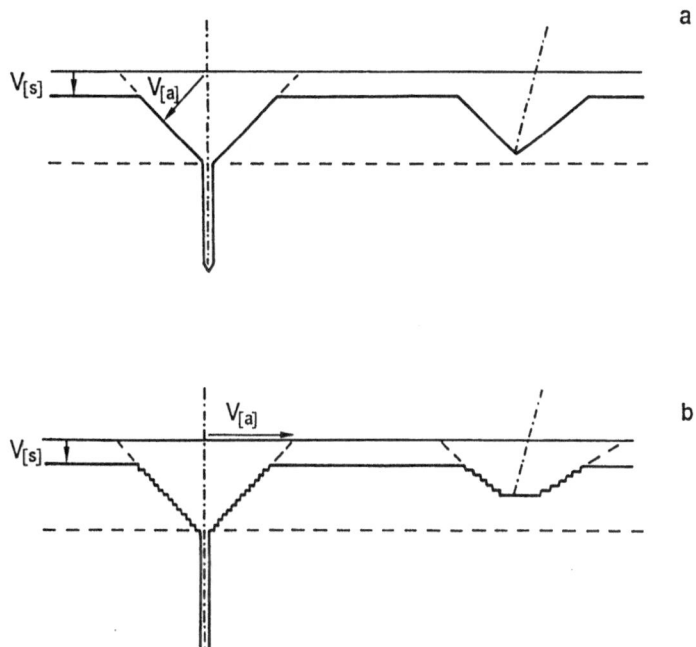

Figure 2.14. Two alternative processes of track etching which result in the formation of a funnel shaped track with a surface etch pit: (a) the walls of the pit are formed by preferential etching at a rate $V_{[a]}$ of crystal planes standing at an angle to the surface; when the end of the track is reached, a sharp-bottomed polygonal pit is left. (b) The pit walls result from a layer-by-layer removal parallel to the etched surface at a rate $V_{[a]}$ and are pseudoplanes, as is indicated by their terraced profile; when the track end is reached and etching is continued, a flat-bottomed polygonal pit develops.

tion of V_g with V_t, the *visible* evolution of fission tracks in minerals is almost entirely governed by bulk etching only. Indeed, due to a very high V_t, the etchant will almost have reached the end of the tracks before they become optically discernible. Optical revelation can therefore be regarded to result from a progressive widening of the track channel, i.e., a process of radial enlargement. This enlargement is governed by the $V_{[h,k,l]}$ components of the crystal planes that make up the track walls and should attain minimally 100× (i.e., from a diameter of $\approx 2 \times 10^{-3}$ μm to $\approx 2 \times 10^{-1}$ μm) in order to make the tracks visible under the microscope. It cannot be rejected that the initial phase of etching during which the track length is seen to increase rapidly, to a certain extent depends upon V_t but, as stated above, this phase can also be influenced by diffusion processes rather than by the true track-etching rate.

Simultaneous to the process of radial enlargement of the track channel, another feature is often observed which is specific to crystals: the tracks are attacked at their intersection with the etched surface by an etching process which has its own specific rates and which causes the formation of the often-observed surface etch pits or openings. It is thus responsible for an additional variation in track shape, typically yielding more or less funnel-shaped tracks (Figures 2.14 and 2.15).

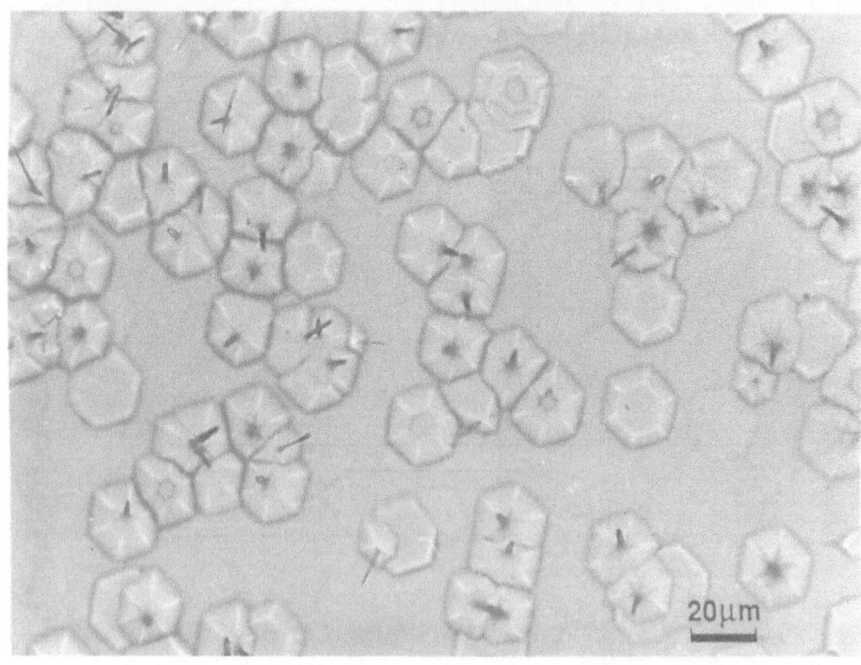

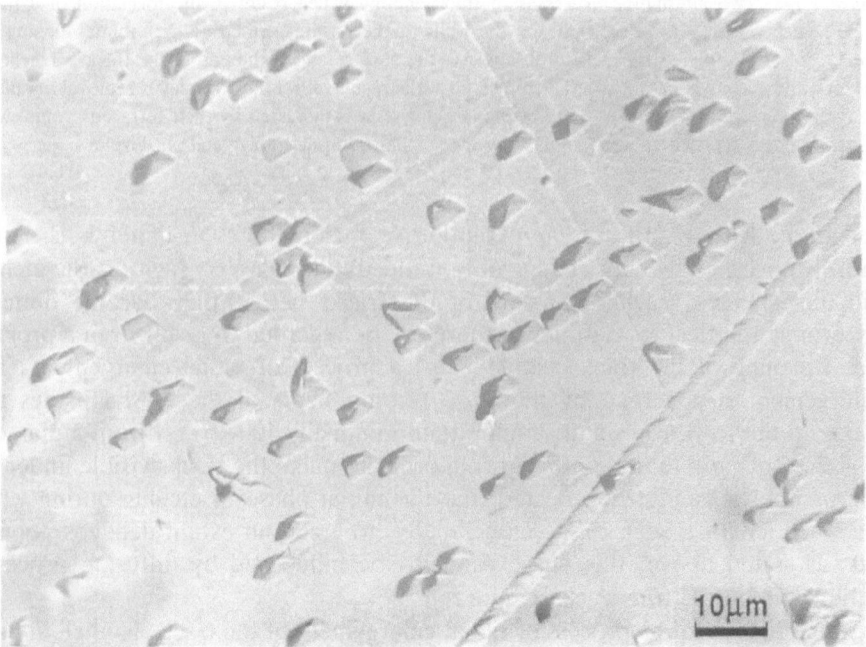

Figure 2.15 (*top*) Funnel shaped tracks with narrow channels and distinct hexagonal openings (type B pits) on the basal plane of apatite; the pits which exhibit flat bottoms correspond to shorter tracks that are completely etched through (etching conditions: 4N HNO$_3$ at 25°C for 1 min). (*bottom*) Tracks with type A pits etched in a pyramid plane of apatite; the picture was taken with reflected light to give a better view of the surface pits while the channels remain practically invisible (etching conditions: 0.4N HNO$_3$ at 25°C for 5 min).

2.3.2. *Track shape*

As stated above, track revelation results from a dual etching process acting simultaneously along the track channel and at the etched surface. The surface attack of
fission tracks is responsible for the formation of surface openings and occurs in a
way which is similar to the attack of other crystal defects. Part of the work done
in the 1950s and early 1960s on dislocations and other defects (see Gatos (ed.)
(1959) and Holmes (1962) for an overview) is therefore still relevant to fission-
track studies. For instance, it is well known that for a given mineral, some crystal
faces when etched with the same reagent, exhibit surface etch pits at defects while
others do not. The shape of the pits varies with crystallographic orientation of the
etched surface. As an example, on the basal face of apatite, etched defects,
including fission tracks, exhibit clear surface openings with the shape of inverted
hexagonal pyramids (Figure 2.15) while no clear openings can be observed on
prism faces (Figure 2.17) (Lovell, 1958; Wagner, 1968, 1969a).

Surface openings result from preferential etching at the locus of the emergence
of the defect (or track) of crystal planes which may or may not form a closed
figure (Figure 2.14; Faust, 1959). In the first case (type A pits), the pit walls are
low indices planes having a smooth appearance (e.g., pits on the pyramidal planes
of hexagonal crystals), while in the second case (type B pits), pit walls develop
which correspond to higher indices planes or pseudoplanes and which are sometimes seen to be terraced (e.g., pits on basal planes of hexagonal crystals) (Figure
2.15). The shape of the pits is further influenced by the orientation of the track,
e.g., by being pointed in the track direction (especially type B pits) and as long
as the track end is not reached), their depth often appears to be practically
constant.

Crystal surfaces carrying tracks with surface openings, generally correspond to
slow and not to fast etching planes as is sometimes thought. The formation of a
surface pit depends upon the relative increase in surface energy created by the
defect. This increase will be more marked on planes of low surface energy which
typically exhibit slow etching rates. The opposite does not seem to be always true:
in the [001] plane of mica, tracks do not exhibit openings and yet this plane has
a very low etching rate. In apatite, however, if etched with HNO_3, the absence
of openings clearly corresponds to the faster etching planes, as can be derived
from Table 2.1.

Some crystal faces show surface openings for tracks with specific orientations
only. Possible explanations of this observation are (Figure 2.16):

(a) if deepening of the surface opening in the direction of the track occurs at
a rate V_d, then pits can only form when the dip angle of the track exceeds
arcsin $V_{[s]}/V_d$ (this explanation is similar to the one quoted for the track-
etching efficiency in the glass, Figure 2.1);

(b) no visible opening can develop if its growth rate is inferior to the rate of
radial enlargement of the track channel; due to anisotropy in the rate of
radial enlargement, it can happen that for some orientations the track
channels widen relatively faster and, thus, surface pits will not show up,
while in other directions, widening is slower and, hence, pits will appear.

Table 2.1. Track etching characteristics of apatite[a]

Plane[b]	$V_{[s]}$[c] (μm/min)	Polishing scratches (width in μm)	Etched surface	Track openings
[0001] basal	0.08	Broad (14.5–16.5)	Broad flat texture	Regular hexagonal pyramids (diam. = 15.5 μm)
[10$\bar{1}$0] prism	0.68	Narrow (1–2) almost disappeared	Dull, appearance of dense hillocky texture	None
[11$\bar{2}$0] prism	0.38	Narrow (1.5–3.5)	Smooth	Elongated hexagonal pyramids (max. diam. = 7 μm) distinct at tracks perpendicular to c-axis only
[10$\bar{1}$1] pyramid	0.62	Narrow (1–2) tending to disappear	Dull, appearance of dense hillocky texture	None
[11$\bar{2}$1] pyramid	0.32	Variable (4–12) depending upon orientation	Smooth to flatly textured	Irregular tetragonal pyramids (max. diam. = 10 μm)

[a] Etching experiment performed on Durango apatite using 2.5% HNO_3 at 25°C for 5 min.
[b] Internal planes obtained by cutting large crystals with a 0.1 mm diamond saw; a small deviation from the quoted orientation is possible.
[c] Surface etching rate evaluated after an etching time of 30 min.

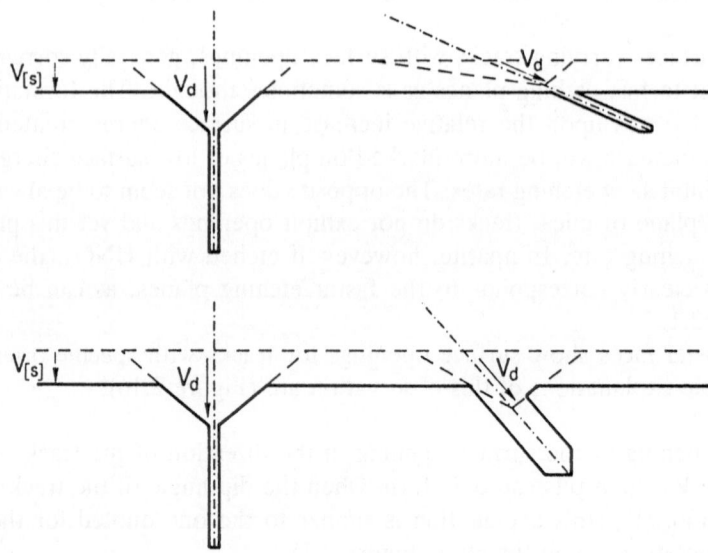

Figure 2.16. Two reasons why tracks may fail to develop surface openings in certain directions: (a) if the rate of pit deepening V_d is constant in the direction of the tracks, only tracks where the vertical component of V_d (i.e. normal to the etched surface) exceeds $V_{[s]}$ will develop openings; hence openings will not show up for low dip angle tracks. (b) Widening of the channel can be highly anisotropic and may occur faster than pit development in some directions; only those orientations where the track channels are etched as narrow needles develop a distinct surface opening.

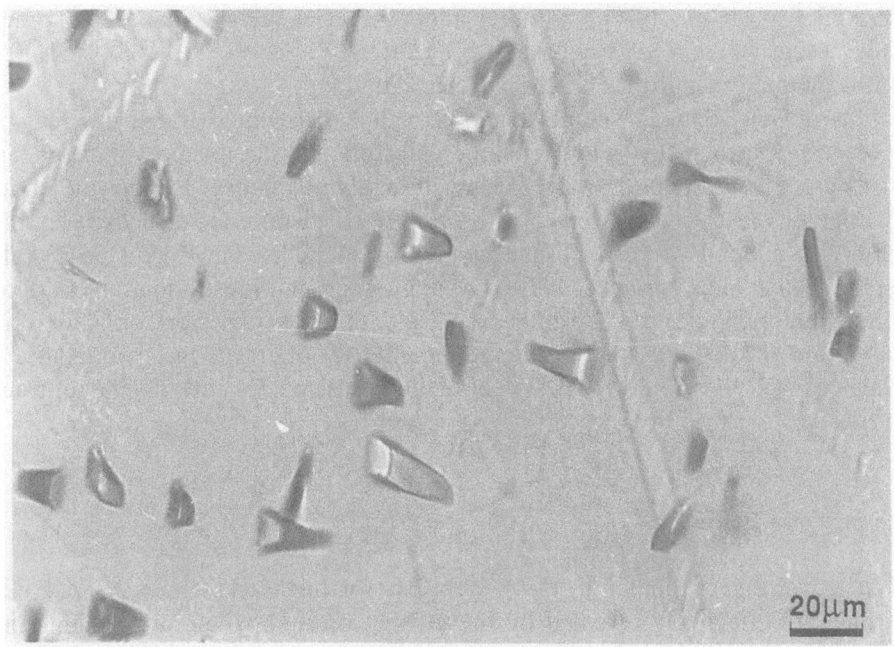

Figure 2.17. Typical knife-blade shaped tracks resulting from anisotropic widening of the track channels in an apatite prism face; except for a kink in some of the track profiles, no distinct surface pits are developed here (etching conditions: 4N HNO$_3$ at 25°C for 1 min).

Surface etch pits inform us about the presence of defects but do not say much about their nature. In other words, the surface pit is a rather undiagnostic part of a fission track. The true diagnostic part is the channel which is revealed by the process of radial enlargement and which may represent the only visible part of a track if no surface pit develops. Initially, the channels have the appearance of fine needle-like figures. With prolonged etching, the influence of crystal anisotropy becomes more evident and when large differences exist in the etching rate of the various crystal planes that form the channel walls, the often described knife-blade or dagger shapes will gradually appear (Figure 2.17) (Gleadow, 1981). The length of the channels rapidly grows during the initial stages of etching, i.e., until the track end is reached. Then, further deepening of the track bottom may be *very slow or even absent*. As a consequence, due to a high $V_{[s]}$ of the etched surface (see the next section) or because of fast deepening of their surface openings, the length of the channel-like part of the tracks can effectively decrease with prolonged etching (Thiel and Külzer, 1978). Gradually, the surface openings may completely encompass the tracks and turn flat-bottomed when their end is reached, thus demonstrating in this way the absence of terminal etching at the track's end (Figures 2.14 and 2.15). Tracks which have completely lost their channel-like parts can hardly be recognized as such.

For a given mineral, the length of the etched channels may further depend upon track orientation. Such anisotropy in track length has clearly been registered in quartz (Khan and Ahmad, 1976), feldspar (Thiel and Külzer, 1978) and even in mica (Belyaev *et al.*, 1980a). In other mineral species such as apatite, it is practically absent (Green, 1981b; Laslett *et al.*, 1984; Green *et al.*, 1986), at least if fresh, thermally unaffected tracks are considered.

It is obvious that no simple equations can be presented which describe the geometry of etched tracks in crystalline detectors and the reader will understand why the above descriptions have remained highly qualitative. Besides the photomicrographs in this book (see also Section 6.2), pictures of etched tracks in a variety of minerals were published in 1964 by Fleischer and Price (1964d) (also reproduced in Fleischer *et al.*, 1975). Other photographic illustrations are scattered throughout the literature. Worth mentioning within the framework of fission-track dating are the pictures of Fleischer *et al.* (1964c) of tracks in zircon, those of Wagner (1968, 1969a) of tracks in apatite, and those of Gleadow (1978a) of tracks in sphene.

2.3.3. Track density

Etching efficiency. Similar to glass, only part of the tracks that cross a crystal surface are effectively revealed by the etching process. Therefore, etching efficiency in minerals has also been approached in terms of a $V_{[s]}/V_t$ ratio and a corresponding critical angle of incidence θ_c. Taking into account their relatively high track etching rate, crystals generally exhibit higher etching efficiencies than glasses. Because both $V_{[s]}$ and V_t are variable in crystals, the value of θ_c will, however, depend upon the crystallographic orientation of the etched surface and, for a given surface, even upon the orientation of the track. Hence, simple equations describing the etching efficiency η cannot be given and the available studies on this matter are essentially experimental.

Direct determinations of θ_c, e.g., by bombardment with collimated beams of fission fragments under varying angles (Khan and Durrani, 1972a; Belyaev *et al.*, 1980b), yield values lower than 10°, which are in accordance with the narrow track channel profile seen in most crystals. Calculations of θ_c (or η) based on $V_{[s]}/V_t$ measurements and the equations used for isotropic detectors, may yield values differing from those obtained by direct measurements, which indicates that these equations are probably invalid (Belyaev *et al.*, 1988). Determinations of the $V_{[s]}/V_t$ ratio based on track-size parameters such as diameter and length (Singh *et al.*, 1986), are to be regarded with caution and may lead to erroneous results, especially if the tracks exhibit distinct surface openings.

A practical way of determining the track etching efficiency of a mineral surface is obtained by comparing its track density (e.g., created by irradiation with a fission fragment source) with that registered on the [001] cleavage plane of muscovite mica. Muscovite is a very suitable reference material because of the ease with which the tracks can be etched and recognized in the [001] plane and because this plane exhibits an etching efficiency close to unity due to its extremely low $V_{[s]}$. Its value of η is, however, known with a somewhat limited accuracy and values ranging from 0.94 to 0.99 have been reported (Gold *et al.*, 1968; Wall, 1986; Roberts *et al.*, 1984).

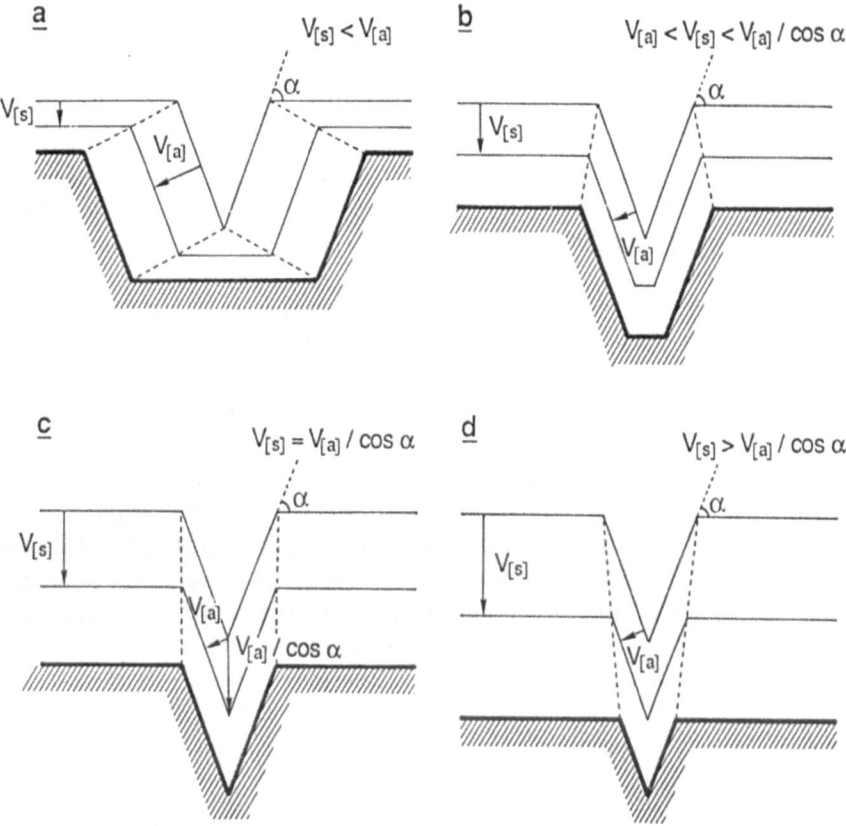

Figure 2.18. Schematic evolution of the profile of a polishing scratch with continued etching for varying etching rates $V_{[s]}$ of the etched crystal surface; if $V_{[s]} < V_{[a]} \cos \alpha$, the scratch profile turns broad and flat bottomed (a), (b); if $V_{[s]} = V_{[a]} \cos \alpha$ it remains unchanged (c) while it becomes sharper and shallower if $V_{[s]} > V_{[a]} \cos \alpha$, i.e., for the most rapidly etching faces (d).

Besides quantitative determinations, the bulk etching rates of the different crystal planes can be estimated in a relative way from the appearance of polishing scratches after etching. After initial removal of the locally strained material by the etchant, scratches rapidly obtain a shape dictated by competitive etching of the surface plane at a rate $V_{[s]}$ and the planes excavated by the scratching particle, which are generally at high angles to the surface plane. In contrast to what is often thought (Naeser *et al.*, 1980; Gleadow, 1978b, 1981), faces showing very broad scratches have a relatively low $V_{[s]}$ while faces with narrow or sharp polishing scratches correspond to relatively faster etching planes. This is illustrated in Figure 2.18. If, for simplicity, the walls of a scratch are thought to correspond to crystal planes with an etching rate $V_{[a]}$, the speed of motion of these planes in a vertical direction, i.e., perpendicular to the surface plane, is of prime importance (Irving, 1962). This speed is given by $V_{[a]}/\cos \alpha$. If it is equal to the etching rate of the surface plane, then the scratch width will remain unchanged with etching time. If $V_{[s]} > V_{[a]}/\cos \alpha$, then the scratch becomes narrower but, at the same time also shallower and tending to disappear, whereas the rapid broadening of a scratch only occurs if $V_{[s]} < V_{[a]}$. This evolution is an application of a general law of crystal etching which states that, in concave situations, slow etching planes grow at the expense of the rapid etching planes which tend to eliminate themselves.

Together with broad polishing scratches and the presence of tracks with large surface openings, a generally smooth or flat texture of the etched surface is also indicative of a low $V_{[s]}$ and all these features are often seen to occur together. To the contrary, crystal surfaces which develop a sharp and dense hillocky texture are high $V_{[s]}$ planes and exhibit tracks without openings together with narrow polishing scratches that rapidly tend to disappear with etching time (Table 2.1). Natural cleavage planes of minerals such as the [001] plane of muscovite are of low surface energy and, hence, are typically low $V_{[s]}$ planes.

As stated at the beginning of this section, etching efficiency may not only vary from one crystal plane to another, but also within the same etched surface from one direction to another. This has been observed from angular distributions registered on tracks etched in specific crystal surfaces which show that, in certain directions, less tracks are revealed than in others (Dorling *et al.*, 1974; Khan, 1977, 1980a; Gleadow, 1978a,b; Naeser *et al.*, 1980). The variation in track frequency from one orientation to another seems to be correlated with a corresponding variation in track length (Khan and Ahmad, 1976) which is not surprising if it is taken into account that track length is the major factor relating the planar and spatial density of fission tracks in a solid (Section 1.4.3). In extreme cases, the etchant may even completely fail to reveal tracks in a given orientation.

Effect of prolonged etching. The evolution of track density with etching time can principally be divided into three major phases (Figure 2.19): (1) an initial phase of zero density during which the tracks are still under the limit of visibility, (2) a phase of rapid increase during which tracks are essentially developed that cross the pre-etched mineral surface (phase of underetching), and (3) a final phase of much slower crease governed by the addition of tracks confined within the crystal and revealed by the gradual removal of material from the crystal surface, i.e., tracks whose upper ends fall within a layer $h = V_{[s]}t$ (phase of overetching).

The strong variability in the shape of the tracks and other factors related to crystal anisotropy tend to complicate this simple evolution. As stated earlier, in planes where the tracks exhibit surface openings, a deepening of these openings occurs at the expense of the channels and causes the tracks to progressively turn into broad pits (Figures 2.14 and 2.15). Because track recognition is based on the channel-like parts, these pits are not usually identified as tracks (Gleadow and Lovering, 1977). The alteration of the surface tracks into pits occurs more rapidly than the addition of new tracks due to the low $V_{[s]}$ of such planes and, as a consequence, the track density can be observed to decrease (Gleadow, 1978a) during the third etching phase.

In intermediate $V_{[s]}$ planes, surface pits are small or absent and the channel-like parts of the tracks remain recognizable for a much longer time (e.g., knife-blade shapes). Hence, while in this case no substantial loss of surface tracks occurs,

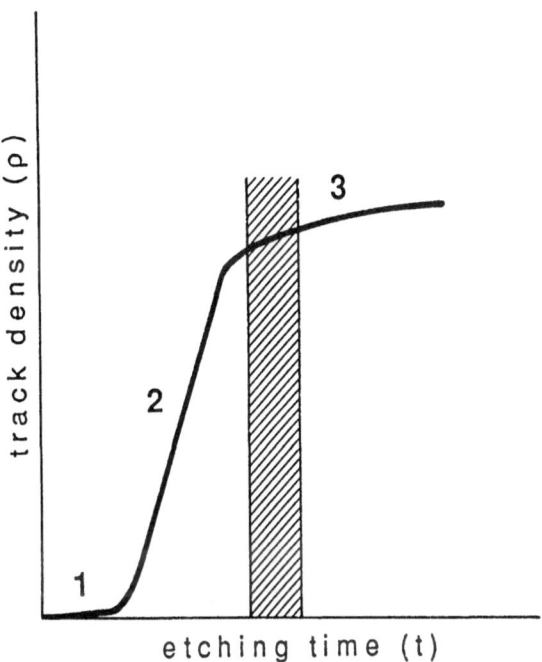

Figure 2.19. The three stages of track etching in crystals: (1) no visible tracks are developed; (2) revelation of surface tracks; (3) revelation of added confined tracks. The hatched area corresponds to optimum etching times for fission-track dating purposes.

the systematic addition of new tracks from below the surface will effectively cause the total track density to increase as expected.

In very high $V_{[s]}$ planes, the surface tracks also remain recognizable as channels for a long time and the addition of new tracks occurs rapidly. This, however, does not imply that these planes will exhibit the most marked increase in track density with prolonged etching. Once the surface tracks are completely etched, further deepening at their bottoms is often very slow and, hence, they will be effectively digested by the rapid removal of material at the surface, comparable to polishing scratches which also disappear with prolonged etching (Figure 2.18d). Addition of new tracks by bulk etching will often only balance this loss and, as a result, the total track density in high $V_{[s]}$ planes can remain relatively constant with etching time. Figure 2.20, which shows the evolution of etched track density on three different apatite planes, provides an experimental illustration of the above.

From this discussion it becomes clear that, in track counting operations, some tracks are not just missed because they are not visibly etched but because they are not recognized as such. In other words, we are dealing with a problem of observation efficiency rather than of etching efficiency. The role of the observation factor in fission-track counting is further discussed in Section 2.4.

Further complications have to be reckoned with. A first one is related to the development of a surface texture which gradually becomes coarser with etching time. Especially if the texture consists of a dense field of sharp hillocks, a dramatic fall of track density can be seen if these irregularities attain a size of several μm,

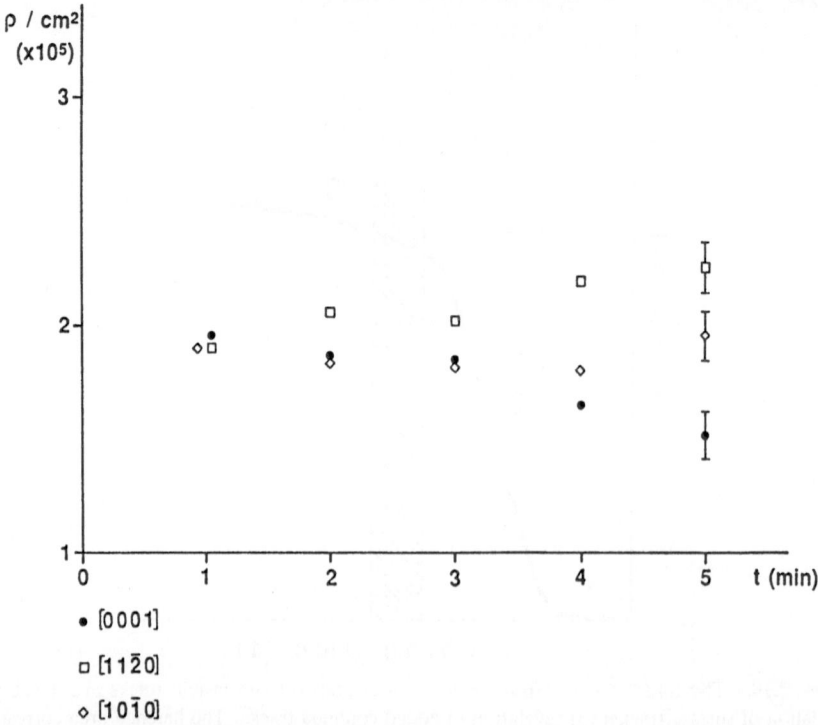

Figure 2.20 Evolution of (spontaneous ^{238}U) fission-track density with etching time during the third phase of etching in three different internal crystal planes of Durango apatite (etchant: 0.4N HNO$_3$ at 25°C); track determination was based upon the presence of a distinct channel-like part (compare this figure with the data of Table 2.1).

which causes a complete breaking up of the tracks (Gleadow, 1978a). Another complication occurs on surfaces with very high densities ($\approx 10^7$ tr/cm^2) of tracks that exhibit large surface openings. In such cases, the attack of the bulk may become the result of the growth of these openings rather than of surface removal by $V_{[s]}$.

Other complications are related to crystal anisotropy acting within the same surface: e.g., tracks with different orientations can be revealed at different rates. Fortunately, a marked decrease of anisotropy is often seen with prolonged etching and a complete isotropic track distribution may be revealed if etching is continued long enough (Gleadow, 1981; Sumii *et al.*, 1987).

It should be noted that the evolution of track density in an external mineral surface can be substantially different from that in an internal surface. Although this has been debated (Green and Durrani, 1978), the geometry ratio may significantly differ from the theoretical value 0.5 (Reimer *et al.*, 1970) and is dependent upon the etching conditions. This is due to the fact that in an external surface, the tracks have a length which is minimally equal to R ($= l/2$), while in an internal surface lengths down to 0 μm occur. As stated earlier, in low $V_{[s]}$ faces addition of confined tracks is slow and the tracks exhibit distinct openings. These openings rapidly encompass the short tracks in an internal surface rendering them unrecognizable and this causes the observed track density to decrease. Contrarily, in an external surface, the tracks are longer and will exhibit recognizable channels for longer periods of time. The density will thus remain relatively constant. On rapidly etching faces, another process takes place: the short tracks are consumed by the lowering surface itself because of the slow terminal etching at the track's end. In an internal surface, the addition of new tracks will more or less balance this loss which results in a rather constant track density, while in an external surface, no such loss occurs and the track density will effectively increase.

In conclusion, it can be stated that once the surface tracks are revealed, the variation in track density with prolonged etching occurs at a much slower rate than in glass, but it cannot be neglected. Within the scope of fission-track age determination, an optimal degree of track etching can therefore be defined. It corresponds to the development until the beginning of the third phase (Figure 2.19) is reached but not very much further (Gleadow and Lovering, 1977). If track development is complicated by anisotropy effects, the transition between the second and third phases may not be abrupt but gradual. Finally, it will be understood that because the track morphology in minerals is highly variable, subjective factors such as skill, experience, and care of the observer, can play a considerable role in track determination.

2.3.4. *Size distributions of track populations*

In minerals just as in glass, the influence of environmental effects on fission tracks is reflected in their size after etching. While in glass, the major diameter of the etched cones is used as a source of information on these effects, it is the length of the etched channels which is the more relevant parameter in minerals. Again, it is the length characteristics of entire track populations which are analyzed rather than individual tracks. Although the earliest investigations of environmental effects (mainly thermal effects, Chapter 4) on track length populations were carried out soon after the introduction of the fission-track dating method (Maurette *et al.*, 1964), it took 15 years (Dakowski, 1978) before the first systematic analysis was published dealing both theoretically and experimentally with the length distributions of fresh, unaltered tracks. A basic understanding of the length characteristics of intact track populations is, however, of prime importance before an attempt can be made to interpret thermally altered populations.

Two types of etched tracks can be considered for track-length investigations. The first type is represented by the normal surface tracks which cut the mineral surface and which are also counted towards an age determination; the second type are the confined tracks which are entirely localized in the interior of the crystal but which are revealed by the etchant because they intersect a surface track or a crack or cleavage plane that emerges at the crystal surface (Figure 2.21). Confined tracks are therefore also called TINTs (Track IN Tracks) or TINCLEs (Tracks IN CLEavage) (Bhandari *et al.*, 1971) (Figure 2.22).

The length distributions of confined tracks and surface tracks can theoretically be derived as follows. For simplicity, all tracks are assumed to be of equal length

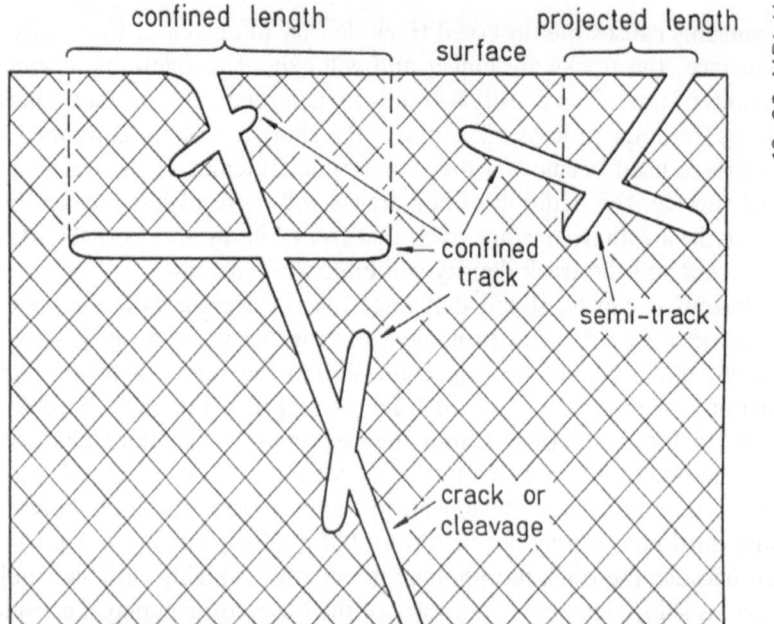

Figure 2.21. Two types of tracks are used for track-length studies: confined tracks and surface or semi-tracks; of the surface tracks, the projected length is measured, of the confined tracks, only those which are horizontal (i.e., parallel to the etched crystal surface) are considered. The three confined tracks on the left are TINCLES while the one on the right is a TINT.

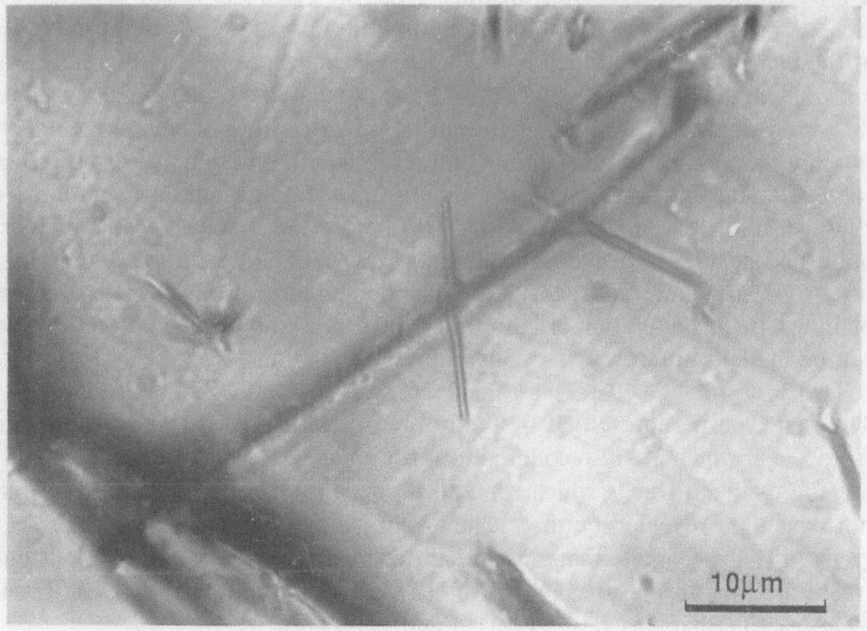

Figure 2.22. Horizontal confined fission-track revealed along a crack in apatite from Fish Canyon tuff (Colorado, U.S.A.).

l and the assumptions of spatial track homogeneity and isotropy (Section 1.4.3) are taken into account. The distribution of the dip angles θ, (with respect to a horizontal plane) of confined tracks with all possible orientations and situated within a layer with thickness dz will be given by (N = number of tracks per unit of volume):

$$n(\theta)\, d\theta\, dz = N \cos\theta\, d\theta\, dz. \qquad (2.13)$$

The length of a confined track projected onto the horizontal plane corresponds to $l_p = l \cos\theta$. Hence $dl_p = l \sin\theta\, d\theta = \sqrt{l^2 - l_p^2}\, d\theta$, and by substituting these two equations in Equation (2.13), the projected length distribution of the confined tracks is obtained

$$n(l_p)\, dl_p\, dz = \frac{N l_p}{l\sqrt{(l^2 - l_p^2)}}\, dl_p\, dz. \qquad (2.14)$$

Following Dakowski (1978), the dip-angle distribution of the surface tracks (internal surface) is obtained by integrating Equation (2.13) over dz between the limits 0 and $l \sin\theta$ or

$$n(\theta)\, d\theta = N \cos\theta\, d\theta \int_0^{l \sin\theta} dz = N(l/2) \sin 2\theta\, d\theta. \qquad (2.15)$$

Further calculation yields that the distribution of the true etchable lengths of the surface tracks is given by

$$n(l_e)\, dl_e = N/2\, dl_e \qquad (2.16)$$

and the distribution of the projected lengths of the surface tracks by

$$n(l_p)\, dl_p = N(1 - l_p/l)\, dl_p. \qquad (2.17)$$

Both distributions are defined between the lengths 0 and l. The equations show that the distribution of the true etchable lengths is horizontal with all lengths being represented in equal amounts, while the projected length distribution has a triangular shape with maximum frequencies at zero lengths (Dakowski, 1978; Green, 1981b). Note that these equations only take into account the tracks which cut the pre-etched surface and not those which may have been added as a result of bulk etching. The theoretical shapes of the length distributions for the different types tracks are illustrated in Figure 2.23. In this figure, the curve for each distribution has been constructed assuming that the true track length varies around a mean value l according to a normal Gauss curve with standard deviation $0.05l$ (Figure 2.23a). This more closely corresponds to reality than assuming that the tracks are all of equal length. In fission-track dating, both confined tracks and projected length measurements of surface tracks are used as a source of information. Generally, only confined tracks which are nearly horizontal are measured. This allows an accurate sampling of the original track length distribution (Figure 2.23b). A certain sampling bias, however, cannot be avoided because the confined tracks which are observed are generally TINCLEs crossing cleavage planes or cracks and the probability of a track for crossing such a plane is directly pro-

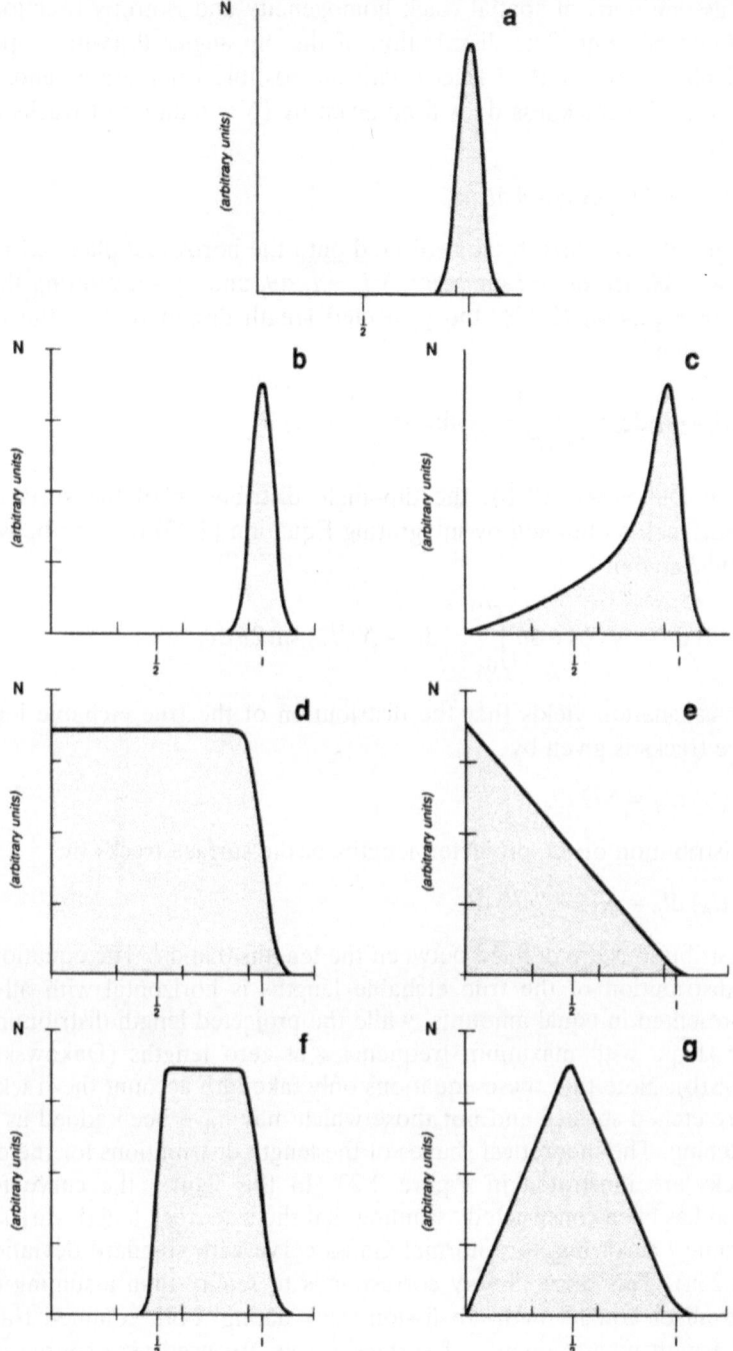

Figure 2.23. Starting from a true track-length distribution with a Gaussian shape (mean length = *l* and relative standard deviation = 5%) as shown under (a), the length distributions that will effectively be observed are illustrated in case: horizontal confined tracks are measured (b), projected lengths of confined tracks with all possible orientations are measured (c), true etchable lengths of surface tracks are measured in an internal (d), or in an external surface (f), projected length of surface tracks are measured in an internal (e) or in an external surface (g). Curves are drawn after model calculations based on an isotropic track revelation. (After Jonckheere, unpublished.)

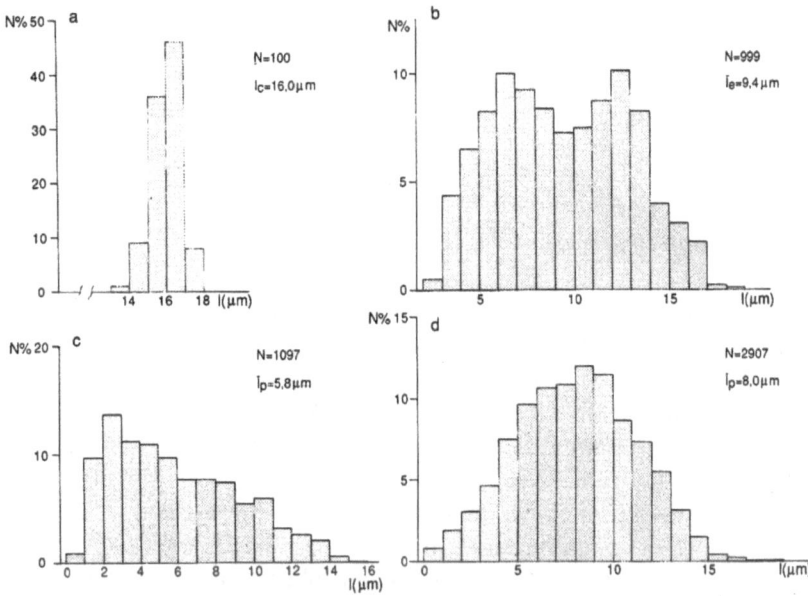

Figure 2.24. Examples of observed length distributions of induced [235]U tracks in apatite for: horizontal confined tracks (a) (Van den haute, unpublished), true etchable lengths of surface tracks in an internal surface (Dakowski, 1978), projected lengths of surface tracks in an internal (c) and an external surface (d) (Van den haute and Jonckheere, unpublished). For each distribution, the number of tracks measured and the average length is given. All distributions were registered on Durango apatite except (a) which was registered on Fish Canyon tuff apatite.

portional to its length (Laslett *et al.*, 1982). Hence, the amount of shorter tracks will always be underestimated compared to the amount of longer tracks. The projected length distributions (Figure 2.23e,g) exhibit a much stronger sampling bias and, as a consequence, it is practically impossible to retrieve the shape of the original length distribution from them (Laslett *et al.*, 1982). Relevant information is therefore mainly retrieved from the average length and the relative amount of tracks with long projected lengths (Section 5.5.2). In theory, the average projected length closely approaches $l/3$ in an internal and $l/2$ in an external surface. Principally, true etchable lengths of surface tracks are at least as informative as projected lengths, but have been rarely measured. The reasons are mainly practical: in addition to a measurement of the projected length, they require a precise determination of the vertical depth of each track which is not easily accomplished. The average length of the true etchable length distribution theoretically amounts to $l/2$ for an internal and to $3l/4$ for an external distribution, as can be simply derived from Figure 2.23d,f.

Most of the experimental track-length studies have been carried out on apatite (Figure 2.24). The length distributions of intact confined tracks in this mineral, exhibit a shape which is close to a normal distribution with an average length of about 16 μm and a standard deviation of 0.8–1 μm (Gleadow *et al.*, 1986; Figure 2.24a). The early distributions that have been reported of projected lengths also show rather symmetrical shapes (Wagner and Storzer, 1972, 1975) but it is likely that in these distributions the amount of short tracks has been largely underestimated, probably because of insufficient etching conditions. Practically all recently reported distributions clearly exhibit the expected triangular shapes, albeit with a relative deficiency of very short tracks ($< 2 \, \mu m$) (Van den haute, 1984; Wagner, 1988; Figure 2.24c).

2.3.5. *Factors affecting track etching*

The process of track revelation is not only related to crystallographical properties, but also to the chemical properties of the etchant. For a given crystal surface, some reagents appear to etch the tracks more anisotropically than others. As an example, a change in pH of the WN (Appendix A) etchant which is used for olivine, causes significant differences in track revelation (Krishnaswami *et al.*, 1971; Davie and Durrani, 1978; Perron and Maury, 1986). Also, the shape and orientation of surface openings may depend upon the etchant. Caution should thus be taken when comparing track densities in minerals attacked with different etchants. This is especially true for zircon and sphene which are etched both with basic and acidic reagents. Even a simple change in the concentration of the etchant may have an influence on the characteristics of the etched tracks, e.g., for apatite, small but distinct differences can be seen in the shape of the etched tracks depending upon whether the attack was made with concentrated or diluted HNO_3. In fission-track dating, etchants which isotropically reveal the tracks and with the highest possible efficiency, are always to be preferred.

Another factor which may affect the track-etching characteristics of a mineral is the amount of radiation damage (mainly due to α-decay) accumulated in the crystal lattice. This kind of damage occurs in high U and Th minerals and causes the tracks to be revealed in a more isotropic way and, hence, with increased efficiency. If the sample is heated at a certain temperature, the damage disappears and track etching again becomes anisotropic. The effect has been extensively described in sphene but does not seem to occur in apatite (Gleadow, 1978, 1981). The decrease of the optimum etch time for track revelation with an increasing density of spontaneous ^{238}U tracks in zircon (Krishnaswami *et al.*, 1974; Gleadow *et al.*, 1976; Amin, 1988) is also interpreted to be due to accumulated radiation damage. In sphene, a certain level of radiation damage corresponding to a spontaneous ^{238}U track density $\approx 5 \times 10^6$ tr/cm^2 appears to be required to achieve an isotropic track revelation (Gleadow, 1978a). In zircon, on the other hand, isotropic etching is already observed at spontaneous ^{238}U track densities as low as 10^3–10^4 tr/cm^2 (Watanabe, 1988).

Besides the Th and U content which indirectly influence track revelation through radiation damage, other factors relating more directly to the microchemical composition of the etched crystals (e.g., the kind and concentration of substituents) may possibly bear an influence on their track etching behaviour but have not yet been thoroughly studied.

In minerals, just as in glasses, the tracks are susceptible to thermal fading which causes a progressive shortening of fission tracks at their ends and, eventually, a

Figure 2.25. Scanning electron micrograph of a replica of etched fission tracks in a prism plane of apatite (apatite from Durango, Mexico; etching conditions 4N HNO_3 at 25°C for 1 min).

fragmentation at stronger annealing conditions (Dartyge *et al.*, 1981; Green *et al.*, 1986). It has great consequences for the fission-track dating method for and the interpretation of fission-track ages, as will be discussed in Chapter 5. Recent investigations (Durrani and James, 1988; James and Durrani, 1988) indicate that the temperature of the mineral at the moment of track registration may also bear some influence on the etchable track length and on V_t. The results of this study are mainly relevant to tracks in meteorites.

2.4. Microscopic Observation

2.4.1. *Equipment and techniques*

In fission-track studies, scanning electron microscopy (SEM) is sometimes used for a detailed investigation of track size and shape. SEM is typically carried out on replicas which were taken from the sample after etching and which were coated with a conductive film (Figure 2.25). Occasionally, SEM is used for determining track densities in a previously etched and coated mineral surface itself (Seitz *et al.*, 1973). Because the track channels cannot be seen, it is the surface outcrops of the tracks which are counted, a method which yields reliable results provided there are no other spurious defects which intersect the sample surface. Track counting by SEM can be useful if the track densities are high ($>5 \times 10^6 \, \text{tr/cm}^2$). In such a case, the tracks would strongly overlap by the time they become visible under the optical microscope. For observation by SEM, a short etch is sufficient, yielding tracks that are still in an early stage of development with tiny surface openings (Monnin, 1980). The technique has been frequently applied in track studies of extra-terrestrial materials (Fleischer *et al.*, 1975).

In fission-track dating, optical microscopy is the method of observation which is routinely used. Often, the dating technique is adopted by geologists or mineral-ogists who are accustomed to polarizing microscopes equipped with circular rotat-ing stages. These microscopes are, however, not ideal for fission-track analysis. A microscope which is equipped with both reflected and transmitted light and which carries a rectangular scanning stage with coaxial x, y controls operated with one hand, is a much more suitable instrument. A one-handed control of the stage leaves the other hand free to adjust the focusing knob, an operation which is almost continuously done during track identification. High-power (preferably plan-apochromatic) objectives are needed for track counting and track-size measure-ments so that a total magnification up to 1500× or higher can be achieved. It is worthwhile possessing several types of high power (100×) objectives such as the dry and oil immersion types for use with and without cover glass. Dry objectives are especially suitable for measuring confined tracks in minerals but they can also be used for counting (Gleadow *et al.*, 1986). In combination with reflected light, they enable a close inspection of polished and etched surfaces (polishing scratches, track openings, evaluation of crystallographic orientation) and they can be used for searching horizontal confined tracks which show a bright internal reflection. Oil immersion, on the contrary, obliterates all surface irregularities, including the track openings, so that the channels of the surface tracks are more easily distin-guished (Wagner, 1968) (Figure 2.26). For observation with oil immersion, a more gentle etching is often to be preferred, while for observation with dry objectives, etching can be stronger.

Identification of fission-tracks is based on the following relatively simple criteria (Fleischer and Price, 1964d; Fleischer *et al.*, 1975):

– they form line defects of a limited length ($<20 \, \mu\text{m}$),
– they are straight,
– they exhibit no preferred orientation,
– they disappear after suitable heating.

These criteria basically refer to the characteristics of latent tracks. They can be applied to etched tracks in minerals on the obvious conditions that they exhibit clear channels and that revelation was not strongly anisotropic. Spurious defects are generally recognized by their curvature or by the fact that they occur in parallel groups or arrays (Figure 2.27).

In glass, fission tracks show up as randomly oriented conical etch pits with

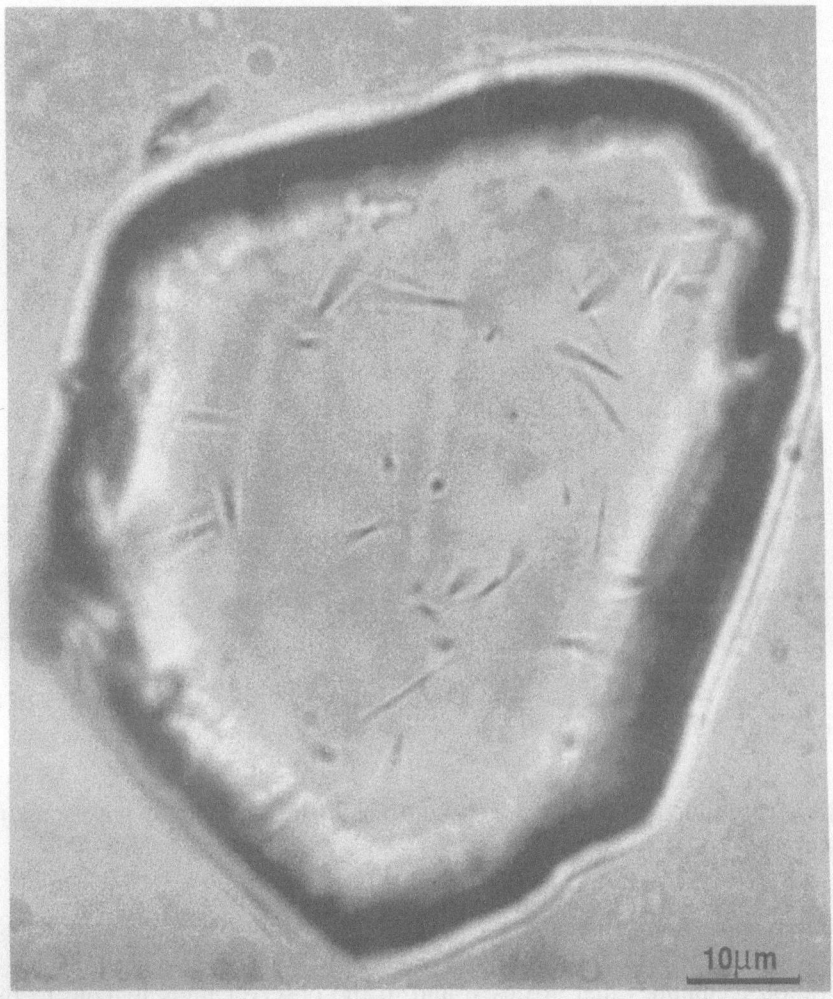

Figure 2.26. Etched fission tracks in a basal plane of apatite from the Odenwald (Germany) observed with oil immersion (etchant: 65% HNO₃ at 20°C for 15 s).

elliptical cross-sections which can be easily identified if they are sharp-bottomed (Fleischer and Price, 1964c). For track observation in glass, the use of reflected light has the advantage that it does not show interfering subsurface irregularities such as inclusions and crystallites (Durrani *et al.*, 1971). Tracks with circular openings can be sometimes confused with etched pores. If incident light is used, they can be recognized by the fact that they are dark, due to the internal reflection in the track cone, while the more rounded pores exhibit bright centres.

At first sight, fission-track counting often appears to be a trivial job but, as his work proceeds, every fission-track geochronologist realizes that skill and experience do play an important role in track identification and that it takes some time before precise and reproducible counts and size measurements are obtained (Green, 1985; Bigazzi *et al.*, 1988). In every new kind of material that is investi-

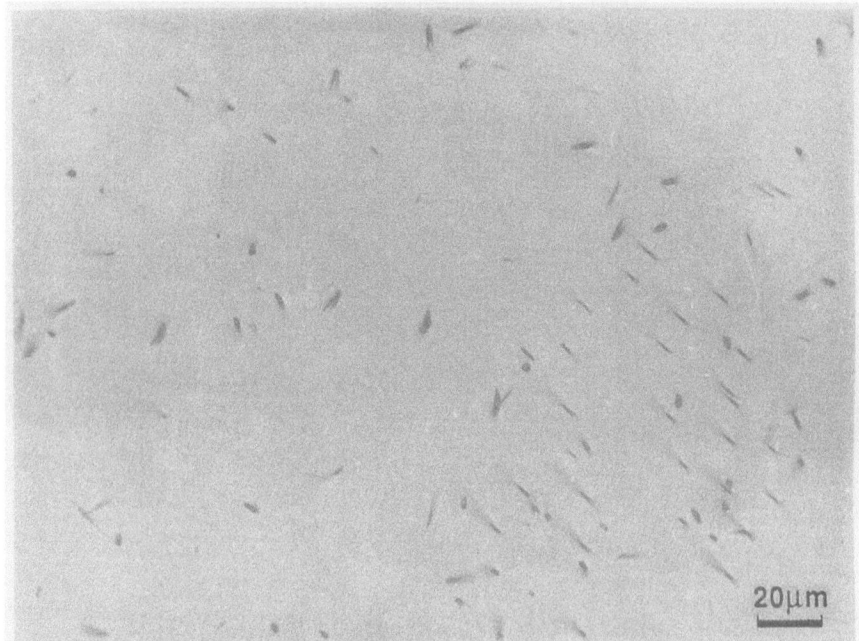

Figure 2.27.   Etched fission tracks and spurious track-like defects in Durango apatite; the defects form a cluster at the right of the picture and exhibit a parallel orientation.

gated or every time a new etchant is used, track identification may pose some new problems.

The observation factor. From the above considerations, it will be understood that track identification and counting is subject to some variability linked to the observation techniques that are used and to the observer himself. Due to these effects, precise and accurate determinations of absolute track densities are difficult to establish and only in plastic detectors and high-quality micas are track counts reproducible within a few percent. The results of interlaboratory comparisons on commonly used reference minerals in fission-track dating (Naeser *et al.*, 1981) clearly demonstrate that large differences can exist between absolute track densities reported by different laboratories. Even when exactly the same samples are scanned, a difference in track density of 20% between observers is not uncommon depending upon the quality of the sample with respect to track identification. In high quality minerals, the variability can be reduced to 10% or less (Van den haute and Chambaudet, 1990), but it rapidly increases in minerals or glass rich in track-like spurious defects. Other factors which impede a precise track-density determination are the degree of heterogeneity of the track-density distribution in the sample and the shape of the etched tracks in the investigated surface. Taking into account that the channel is the diagnostic part of the track, crystal faces in which the tracks exhibit large openings are more problematic with respect to a quantitative track identification than faces in which the tracks are free of such openings. Also tracks whose channels have strongly anisotropical shapes changing from one direction to another, are more difficult to count accurately.

In practice, the reaction of the observer is to count only the tracks which he is sure of and to omit the uncertain defects. Hence, not all tracks that have been revealed by etching will be recognized and counted. It is therefore logical to acknowledge a systematic observation factor $q \leqslant 1$ in any track-counting effort. The exact value of the q factor is difficult to determine. Besides the microscopic conditions, and the characteristics of the investigated surface with respect to track identification, it depends upon the experience, skill, and care of the observer. The observation factor also applies to fission-track age determinations; its significance is discussed in more detail in the next chapter.

2.4.2. *Automated observation*

Counting and measuring fission tracks under the microscope is a labour-intensive and time-consuming task which, as has been outlined above, is susceptible to observer-bias. Automatic observation systems which improve the speed of track analysis and eliminate personal bias would undoubtedly contribute greatly to the dating method. Several types of automation systems have been developed, some of them being currently used in nuclear-track studies (Gold *et al.*, 1984). The most promising systems are the digital image analysis systems working with a TV camera attached on top of the microscope and connected to a microprocessor which translates the picture into a high resolution digital image. Until a decade ago, the capabilities of commercially available systems were rather limited but their improvement has been spectacular ever since. Ultra-sensitive, high-resolution (up to 2000 TV lines) cameras and modern image processors yielding 1024×1024 pixel digitized images and distinguishing 256 grey levels, together with 16 or 32 bit control computers and high disc storage capacities are now currently available in desk top versions and allow a rapid interactive processing of the microscope image. Identification and selection of tracks using such a system is typically based on their optical density and on their shape characteristics. Image analysis systems proved to be efficacious for counting and measuring tracks at relatively low magnifications in simple detectors such as plastics, artificial glasses, and mica which are virtually free from disturbing background irregularities. They have not found common application in fission-track dating. Specific features such as inclusions or spurious defects and overlapping tracks which are common in natural minerals, are easily recognized by the human eye but often cause problems for the automatic system, resulting in a significant reduction of its effective speed of operation. In addition, the high-power objectives that are used have a very shallow depth of field which makes it necessary to move the focusing knob up and down to obtain a sharp view of the entire track and this makes track length measurements by automatic systems almost impossible. Recently developed systems which are able to produce composite images of the tracks obtained by combining the images collected at consecutive focus settings, may provide an answer to this problem (Wadatsumi *et al.*, 1988; Wadatsumi and Masumoto, 1989). Nevertheless, a gen-

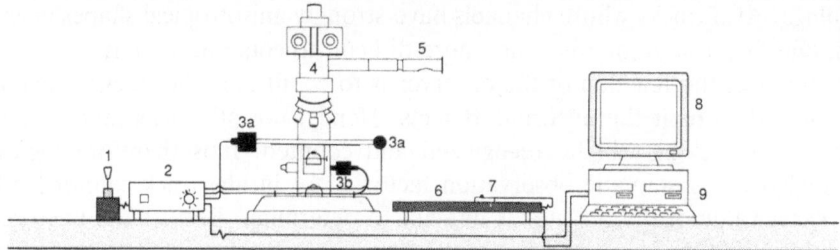

Figure 2.28. Schematic illustration of equipment used for track counting and track size measurements: (1) tri-axial joystick for manual control of motorized stage; (2) controller unit of motorized stage; (3a) step motors for movement in X, Y direction; (3b) step motor for movement in the Z direction (focus); (4) microscope; (5) drawing tube attachment; (6) high resolution digitizing tablet; (7) cursor with centred LED; (8) and (9) personal computer and monitor connected to tablet and stage controller.

eral use of image analysis systems in fission-track dating is not to be expected in the near future. This also has to do with the relatively high costs of the equipment, the gain of time and effort being relative if the even more time-consuming task of mineral separation is taken into account.

At present, in fission-track dating laboratories, data acquisition is commonly done in a semi-automatic way with the aid of a high-resolution digitizing tablet connected to a micro-computer. The apparatus set-up is schematically illustrated in Figure 2.28. Track identification remains entirely with the observer. Track size measurements are carried out by pointing the track ends on the tablet with a cursor equipped with a LED in the centre of the cross hair and observed in the viewfield of the microscope with the aid of a drawing tube attachment. The equipment can also be used for track counting. The measurements are transferred to the on-line computer and stored on disc which allows a rapid treatment with common statistical programs. A further kind of automation which found its application in fission-track dating concerns the use of computer-controlled motorized microscope scanning stages (Smith and Leigh-Jones, 1985) (Figure 2.28). The modern versions of these stages are driven by step motors which attain a resolution of 0.1 μm and allow a rapid and precise positioning at any specific spot in a detector or sample mount, which is particularly useful in the external detector method of dating (Section 3.6). If used in combination with image analysis systems, a fully automatic scanning of large plastic or mica track detectors becomes possible (Gold *et al.*, 1984).

CHAPTER 3

Fission-Track Dating Method

3.1. Natural Tracks and their Origin

In 1962, Price and Walker discovered that micas contain natural tracks which can be optically revealed by etching with hydrofluoric acid. This discovery started a search for natural tracks (also called *fossil tracks*) in different kinds of materials and only two years later, natural tracks had been observed in about 10 different types of minerals and in several glasses (Fleischer and Price, 1964d). The major question that was immediately raised related to the origin of these natural tracks. Due to their low sensitivity for track registration, it could be readily stated that in mineral and glassy substances, tracks could only be created by heavy charged particles (Price and Walker, 1963a). In terrestrial materials, such heavy particles are typically produced by the spontaneous or induced fission of heavy nuclides which are present in the minerals themselves (Section 1.4). Induced fission reactions occur only very rarely in the Earth's crust. In order to take place, they require a considerable concentration of heavy elements such as uranium residing for a certain time in a flux of energy (e.g., γ-rays) or particles (e.g., neutrons or α-particles). The Oklo uranium mine in Gabon (West-Central Africa) which was once the site of a natural nuclear reactor, represents one of the very few cases where such induced fissions effectively occurred in significant numbers (Bodu *et al.*, 1972; Petrov, 1977). It can thus be stated that practically all natural tracks which are observed in terrestrial materials originate from spontaneous fission reactions which we know almost only occur with ^{238}U nuclides, as has been outlined in Section 1.4.1. In other words, natural tracks can essentially be regarded as the products of the decay by spontaneous fission of ^{238}U atoms localized within the minerals themselves. This statement forms the basis of the fission-track dating method.

If extra-terrestrial rocks are considered, the situation becomes more complicated and a number of other track sources such as cosmic ray effects or spontaneous fission of extinct isotopes such as ^{244}Pu have to be considered. Extra-terrestrial tracks form a special subject of study which is beyond the scope of this book. A short remark can be made, however, with respect to two special types of natural tracks besides ^{238}U fission tracks which have also been observed in terrestrial materials. The first type are the α-recoil tracks. These are tracks of very short length (approx. 0.02 μm) which result from the recoil of heavy nuclei of the U–

Pb and Th–Pb series when they emit α-particles. α-recoil tracks were discovered by Huang and Walker in 1967. The second type are the α-interaction tracks which have a length between 0.1 and 1 μm and which were reported for the first time in terrestrial minerals by Price and Salamon in 1986. They are thought to be created by recoiling Al or Si atoms struck by high-energy α-particles emitted by ^{212}Po, a member of the ^{232}Th–^{208}Pb series. Both α-recoil and α-interaction tracks have so far only been observed in mica. Dating of mica with α-recoil tracks has been proposed but the reliability of the results has been questioned (Hashemi-Nezhad and Durrani, 1981b).

3.2. Principles of the Dating Method: Fundamental Age Equation

In principle, the fission-track method is no different from other isotopic dating methods based on the decay of a naturally radioactive parent to a stable daughter atom. This decay is a first-order reaction and therefore occurs at a rate which is proportional to the number of parent atoms (N_P) remaining at any time, or

$$dN_P/dt = -\lambda N_P, \tag{3.1}$$

where λ is called the decay-constant which is expressed in a^{-1}. Integration of this equation while taking into account that $(N_P)_0$ is the original number of parent atoms at the time $t = 0$ yields

$$N_P = (N_P)_0\, e^{-\lambda t}, \tag{3.2}$$

which is the classic equation describing the variation of the number of parent atoms with time.

In order to determine the age t of a sample, the quantities that are generally measured are the present amount (per unit of volume) of parent atoms N_P and daughter atoms N_D. Therefore, Equation (3.2) should be rewritten in the more convenient form

$$(N_P)_0 = N_P\, e^{\lambda t},$$

which taking into account that

$$N_D = (N_P)_0 - N_P$$

can be modified to

$$N_D = N_P(e^{\lambda t} - 1). \tag{3.3}$$

This is the basic equation of the majority of isotopic-dating methods including the fission-track method.

In the fission-track method, it is spontaneous fission tracks instead of daughter isotopes that are measured as a product of the decay of ^{238}U. This parent not only decays by spontaneous fission but also by α-emission (to ^{206}Pb). This second reaction represents one of the keystones of the U/Pb dating method. If λ_d, λ_α and λ_f are, respectively the total decay-constant, the decay constant for α-emission, and the decay constant for spontaneous fission, it can be stated that $\lambda_d = \lambda_\alpha + \lambda_f$.

According to Equation (3.3), the total number of decayed ^{238}U atoms after a time t is given by $^{238}N(e^{\lambda_d t} - 1)$, where ^{238}N represents the present number of ^{238}U atoms. The number of decays that are due to spontaneous fission stands in fixed proportion (λ_f/λ_d) to the total number of decays of ^{238}U. Hence, the number of spontaneous tracks N_s that will have accumulated (per unit of volume) is given by

$$N_s = \lambda_f/\lambda_d {}^{238}N(e^{\lambda_d t} - 1). \tag{3.4}$$

Because the decay constant for spontaneous fission is several orders of magnitude lower than the constant for α-decay ($\approx 8.5 \times 10^{-17}\,a^{-1}$ versus $1.5 \times 10^{-10}\,a^{-1}$), it can be stated that $\lambda_d = \lambda_\alpha$. If Equation (3.4) is solved explicitly for t, we then obtain

$$t = 1/\lambda_\alpha \ln[(\lambda_\alpha/\lambda_f)(N_s/{}^{238}N) + 1]. \tag{3.5}$$

In principle, the calculation of a fission-track age is thus based on the determination of the number of spontaneous fission tracks and the determination of the number of ^{238}U atoms per unit of volume in the sample. For determining the quantity ^{238}N, a procedure is used which is also based on fission-track counting. By irradiating the sample in a nuclear reactor with a fluence (ϕ) of thermal neutrons, fission will be induced in the ^{235}U atoms, the number N_i of such fissions being given by

$$N_i = {}^{235}N\sigma\phi, \tag{3.6}$$

where σ refers to the cross-section of ^{235}U for fission induced with thermal neutrons (Section 3.4.2).

Except for some rare situations such as the above-mentioned Oklo mine, the relative abundances of the uranium isotopes are practically constant in nature. The ^{235}U/^{238}U ratio can thus also be regarded as a constant and is called I. Hence, we can write:

$$N_i = {}^{238}N I \sigma\phi. \tag{3.7}$$

Combination of Equations (3.5) and (3.7) finally yields

$$t = 1/\lambda_\alpha \ln[(\lambda_\alpha/\lambda_f)(N_s/N_i) I \sigma\phi + 1], \tag{3.8}$$

which is the fundamental age equation of the fission-track method. The measurement of a fission-track age is now reduced to the determination of the ratio of a spontaneous to an induced track density and the determination of the thermal neutron fluence.

3.3. Practical Age Equation

In Equation (3.8), both N_s and N_i are expressed as numbers of tracks per unit of volume, i.e., as spatial track densities. In practice, the tracks which are counted under the microscope are those which cross the investigated sample surface. Using the theoretical relation between the planar and spatial track density derived in Section 1.4.3 and taking into account the effects of track etching and the observation factor, for the observed spontaneous and induced track densities, respectively, we can write

$$\rho_s = g_s N_s R_s \eta_s f(t)_s q_s, \tag{3.9a}$$

$$\rho_i = g_i N_i R_i \eta_i f(t)_i q_i, \tag{3.9b}$$

where $g_{s,i}$ = the geometry factor,
     $R_{s,i}$ = the average etchable range of a fission fragment track in the investigated material,
     $\eta_{s,i}$ = the etching efficiency factor,
     $f(t)_{s,i}$ = the etch time factors,
     $q_{s,i}$ = the observation factor.

The geometry factor g refers to the initial geometry of the pre-etched sample surface which, by definition, is constant and $= 1$ for an internal or $= 0.5$ for an external surface. In the same material, the etchable ranges of spontaneous and induced tracks are practically equal or $R_s = R_i$ (Togliatti, 1965; Bhandari *et al.*, 1971). The values of η, $f(t)$) and q depend upon the techniques that are used for revelation and observation of spontaneous and induced tracks. Substitution of Equations (3.9a) and (3.9b) in Equation (3.8) finally yields

$$t = 1/\lambda_\alpha \ln[(\lambda_\alpha/\lambda_f)(\rho_s/\rho_i) QGI\sigma\phi + 1]. \tag{3.10}$$

This is the practical age equation in which the spatial track densities have been simply replaced by the observed planar track densities. In this equation

$$G = g_i/g_s \quad \text{and} \quad Q = \eta_i f(t)_i q_i / \eta_s f(t)_s q_s.$$

The factor G refers to the initial geometry ratio of the surfaces investigated for counting the spontaneous and induced tracks and normally $= 0.5$ or $= 1$ depending upon the dating procedure that is used (Section 3.6). The factor Q can be considered as a procedure factor. If revelation of spontaneous and induced tracks is identical and both types of tracks are counted under identical conditions of observation then it can be stated that $Q = 1$.

For young samples the above equation can be simplified to

$$t = 1/\lambda_f(\rho_s/\rho_i) QGI\sigma\phi, \tag{3.11}$$

which yields the same result as Equation (3.10) within 1%, as long as $t < 100$ Ma.

If we adopt the following values for the nuclear parameters

$\lambda_\alpha = 1.55125 \times 10^{-10} \, a^{-1}$,
$\lambda_f = 8.46 \times 10^{-17} \, a^{-1}$,
$I = 7.2527 \times 10^{-3}$,
$\sigma = 570.8$ barn (1 barn $= 10^{-24} \, cm^2$; Section 3.4.2),

and introduce them, respectively, into Equations (3.10) and (3.11), we obtain the numerical age equations:

$$t = 6.446 \times 10^9 \ln[7.607 \times 10^{-18}(\rho_s/\rho_i)\, QG\phi + 1], \tag{3.12}$$

and

$$t = 4.908 \times 10^{-8}(\rho_s/\rho_i)\, QG\phi. \tag{3.13}$$

The values of λ_α (Jaffey *et al.*, 1971) and I (Cowan and Adler, 1976) are recommended by the IUGS Subcommission on Geochronology (Steiger and Jäger, 1977) and generally accepted. No general agreement has been reached yet about the values of the two other parameters, λ_f and σ, and different values are used by different investigators (Hurford, 1986a). Therefore, these parameters are briefly discussed in the following section.

3.4. Relevant Nuclear Parameters

3.4.1. *Decay-constant of* ^{238}U *spontaneous fission* (λ_f)

Since the discovery of the disintegration of ^{238}U by spontaneous fission (Flerov and Petrzhak, 1940), numerous determinations have been carried out of the decay constant λ_f using different methods: direct measurements with ionization chambers or rotating bubble chambers, radiochemical methods including measurements of fission products, accumulation of natural fission tracks in mica or other detectors (U-detector sandwich methods), and analyses of samples of a known age. One of the earliest, relatively precise measurements using the ionization chamber, yielded $(8.60 \pm 0.29) \times 10^{-17}\,a^{-1}$ (Segré, 1952), while the first value based on U-mica sandwich measurements and on an analysis of samples of a known age (Fleischer and Price, 1964a) was much lower: $(6.85 \pm 0.2) \times 10^{-17}\,a^{-1}$. Subsequent determinations did not resolve this discrepancy. On the contrary, they yielded scattered values more or less clustering around $7 \times 10^{-17}\,a^{-1}$ and $8.5 \times 10^{-17}\,a^{-1}$ (Bigazzi, 1981). Values obtained by the U-detector method are mainly centred around $7 \times 10^{-17}\,a^{-1}$, while direct measurements and radiochemical methods yield typical values of around $8.5 \times 10^{-17}\,a^{-1}$ (Figure 3.1). The experiments based on a fission-track analysis of samples of a known age yielded variable results which do not support either of the two above values. Of each cluster, the value quoted with the highest precision became currently used. These values are, respectively, $(7.03 \pm 0.11) \times 10^{-17}\,a^{-1}$ determined by Roberts *et al.* (1968) and $(8.46 \pm 0.06) \times 10^{-17}\,a^{-1}$ obtained by Galliker *et al.* (1970). It can be mentioned here that in 1984, Roberts and coworkers upgraded their value for the track revelation efficiency in mica by about 4%. Hence, their original value of λ_f has to be equally reduced and, in this way, comes closer to the value of Fleischer and Price (1964a).

In the past, confusion about the exact value of λ_f has created a scepticism as to the reliability of the fission-track method as an absolute geochronometer. A clear standpoint is necessary here and ours is that the direct measurements and radiochemical methods at present give the most reliable results. Methods which

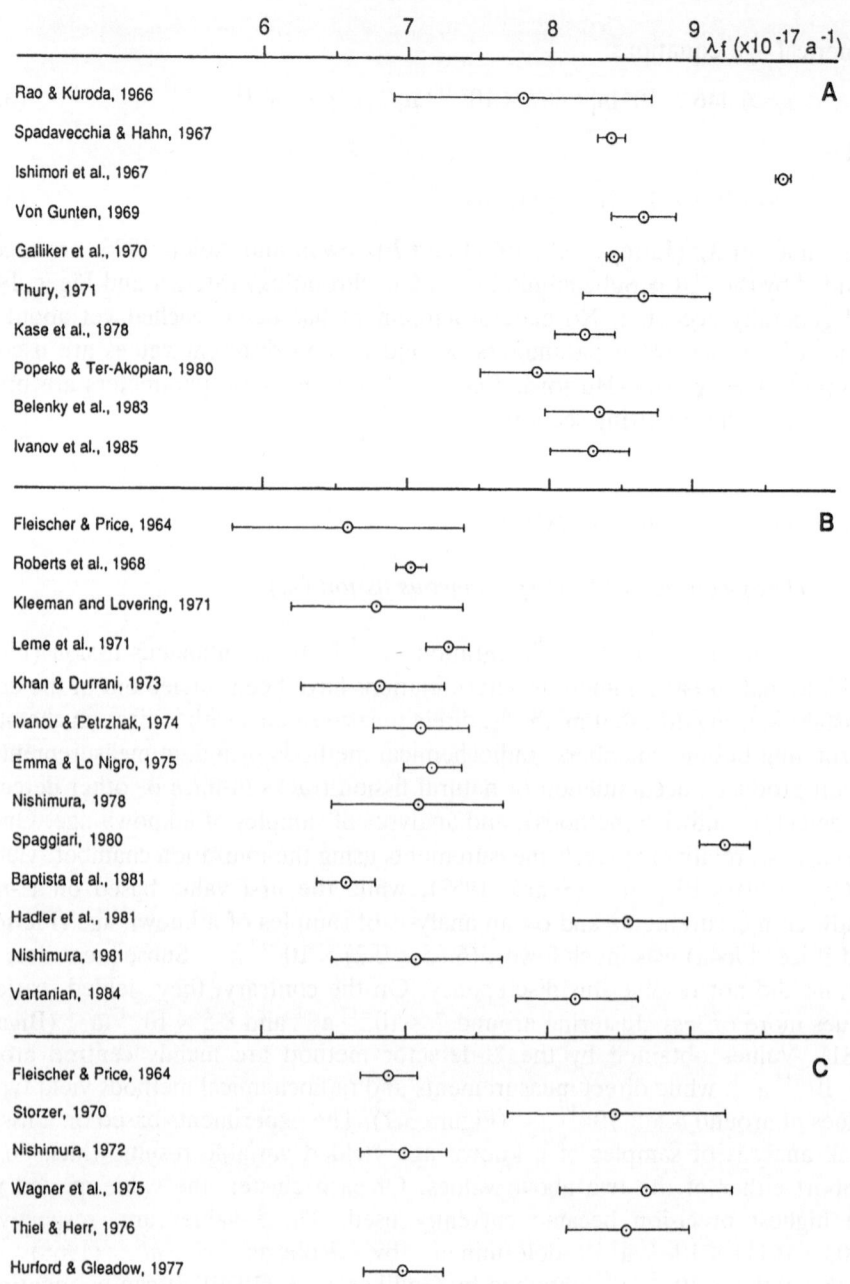

Figure 3.1. Experimental determinations of the ^{238}U spontaneous fission decay-constant (λ_f) performed since 1960, based on direct measurements or radiochemical methods (A), uranium-detector sandwich methods (B), and dating of samples of known age (C). A few determinations yielding more or less discrepant values are not included in the figure, e.g., Shukoljukov *et al.* (1968): $10.3 \pm 0.5 \times 10^{-17}\,a^{-1}$ (A-type method), De Carvalho *et al.* (1975): $11.55 \pm 0.77 \times 10^{-17}\,a^{-1}$ (B-type method), Märk *et al.* (1977), $10.1 \pm 1.2 \times 10^{-17}\,a^{-1}$ (B-type method), and de Carvalho *et al.* (1982): $11.8 \pm 0.7 \times 10^{-17}\,a^{-1}$ (B-type method).

rely on track counting are too susceptible to systematic errors. This is also stated by Holden (1981, 1989) who, for that reason when calculating his recommended value of $(8.2 \pm 0.1) \times 10^{15}$ a for the spontaneous fission half-life of ^{238}U, omits all U-detector sandwich determinations (corresponding value of $\lambda_f = 8.45 \times 10^{-17} a^{-1}$). Self-absorption of fission products in thick U-foils, overestimation of the track registration and etching efficiency of mica are possible sources of such errors. Further on, determinations which involve irradiation with thermal neutrons imply an accurate determination of the neutron fluence or, more precisely, of the amount of ^{235}U nuclei that fissioned in response to this fluence (Section 3.5.1). Systematic errors are easily comitted here and are probably also a major reason why the fission-track analysis of samples of a known age yielded varying values of λ_f (Bigazzi, 1981; Hurford and Green, 1981a). Analyses of samples of a known age finally must also take into account that spontaneous track fading may have occurred. This was not realized in the early λ_f determination of Fleischer and Price (1964a) which for that reason probably represents an underestimation.

In conclusion, the most accurate determination of $8.46 \times 10^{-17} a^{-1}$ obtained with the rotating bubble chamber (Galliker *et al.*, 1970), in our opinion, remains the best evaluation so far of the ^{238}U spontaneous fission decay constant. The corresponding value of the half-life $t_{1/2}$ is 8.19×10^{15} a. A recent analysis of apatite fission-track age standards based on a careful determination of the neutron fluence with two fluence monitors (Van den haute *et al.*, 1988) supports this value.

3.4.2. *Cross-section of ^{235}U neutron induced fission*

In order to solve the age equation, an accurate determination of the neutron fluence (ϕ) or, more precisely, of the rate of ^{235}U fission, induced by this fluence is essential. If the neutron energy distribution corresponds to an ideal distribution of purely thermal neutrons (i.e., a Maxwell–Boltzmann distribution), the reaction rate of any activation reaction is given by

$$R = \sigma_0 \phi, \tag{3.14}$$

where ϕ is the total fluence of thermal neutrons (neutr cm^{-2}) and σ_0 is the cross-section for 2200 m/sec neutrons of the specific reaction being considered. The cross-section can be understood as the ratio of the number of neutrons which actually produce the reaction per unit of time to the total flux φ of neutrons (neutr cm^{-2} s^{-1}); it has the dimension of area and is expressed in barn (1 barn = 10^{-24} cm^2). Like λ_f, σ_0 is a constant and, in contrast to the controversy about λ_f there has been little debate about the exact value of σ_0 for the ^{235}U (n, f) reaction. On the contrary, the ^{235}U$(n, f)\sigma_0$ is considered as a cross-section standard. Holden and Holden (1989) recently re-examined the measurements of σ_0 of the past 40 years and retained six values which are all very consistent. Using these six measurements, they calculated a value of 586 ± 2 barn as their recommended value. The most recent edition of the Evaluated Nuclear Data Files (ENDF-B VI, 1990) quotes a value of 584.25 barn ($\pm 0.19\%$) which is consistent with the recommendation of Holden and Holden and which we prefer to retain, taking into account the authority of these files. In fission-track dating, the value 580.2 barn quoted in the 1969 revision of Hanna *et al.* is still currently being used. The difference with the latest ENDF value is minor and the effect on a fission-track age calculation almost negligible, but there is no reason why fission-track geochronologists should neglect the advances in nuclear science by not adopting the more recent value of 584.25 barn.

Although the value of σ_0 $^{235}U(n, f)$ is accurately known, a problem rises when the rate **R** of this reaction has to be calculated. Equation (3.14) is indeed only valid if the target nuclide which is exposed to the neutron flux, behaves ideally: this means that the variation of its cross-section with neutron velocity obeys to $\sigma(v) \sim 1/v$ in the thermal neutron energy region or $\sigma(v) = \sigma_0 v_0/v$ (with $v_0 =$ 2200 m/s). This is not the case with σ $^{235}U(n, f)$ which exhibits a somewhat discrepant behaviour. The deviation from ideality can be expressed by a correction factor $g(T)$ which is neutron temperature-dependent. For σ $^{235}U(n, f)$ the deviation is small and the temperature dependence minor: $g(T) = 0.977$ at 20°C and decreases to 0.960 at 100°C (Wagemans *et al.*, 1988; Deruytter, pers. com.). If this correction is applied to the ^{235}U cross-section, a value $\sigma = 570.8$ barn is obtained. This value yields the correct $^{235}U(n, f)$ reaction rate for a perfectly thermalized neutron flux at 20°C and, for that reason, is used in the numerical age equations (3.12) and (3.13). In fission-track dating literature, it is not always clear if the σ value which is quoted corresponds to the 2200 m/s cross-section (σ_0) or to some corrected value which yields the true $^{235}U(n, f)$ reaction rate in the reactor facility being used. This may be a reason for the different σ-values that are reported by various investigators (Hurford, 1986a).

3.5. Dating Systems and their Calibration

Fission-track geochronologists who actually solve the age equation (Equation (3.10)) by performing a determination of ϕ, apply only one of the possible systems of fission-track dating. This system is known as the 'absolute approach'. Other dating systems have been developed which avoid an explicit determination of ϕ and which basically rely on a comparative analysis with age standards. Age standards are reference materials that are geologically well documented and whose ages are well established from stratigraphy and from independent age determinations with other isotopic methods (K–Ar, Rb–Sr) (Hurford and Green, 1981). Age standards are discussed in Section 3.10. The dating systems explained next represent the most commonly used approaches in fission-track dating (Hurford, 1986a).

3.5.1. *Absolute approach*

Because of its complexity which easily leads to experimental errors, the absolute approach has been criticized in the past (Hurford and Green; 1981, 1982) but it should lead to correct ages if applied properly (Van den haute *et al.*, 1988). A major concern of the absolute approach is to perform an accurate determination of the fraction of ^{235}U atoms that effectively fissioned during sample irradiation, i.e., the quantity $^{235}N\sigma\phi$. The most direct and effective way to do this is to use a uranium monitor and to measure the fission products in this monitor after irradiation (e.g., by γ-spectrometry). This procedure is rarely used in practice. Currently, it is the thermal neutron fluence ϕ_{th} which is determined with the use of another kind of metal monitor and the value of ϕ_{th} is then substituted in the age equation (Equations (3.8) or (3.10)). This implies that the $^{235}U(n,f)$ reaction rate is assumed to be given by $\mathbf{R} = 570.8\phi_{th}$ (at 20°C). The metal monitors that are to be used for this purpose are those whose nuclear parameters are very accurately known. Only few metals meet this requirement: they are Au and Co (and, to a lesser extent, Mn) which are therefore called neutron capture standards. In the past, other metals such as Cu and Fe have also been used. Although this does not necessarily imply that the fluences which are obtained in this way are incorrect, it is a practice that must be discredited because, at present, large uncertainties still exist on some of the relevant nuclear parameters of these metals (Van den haute *et al.*, 1988; De Corte *et al.*, 1991). Poor calibration and incorrect application of the techniques of neutron metrology probably were largely responsible for several inaccurate fluence determinations in the past and represent an important reason why the λ_f problem has not yet been solved.

Essentials of the approach. The fundamentals of neutron fluence determinations as applied to fission-track dating, are summarized in Green and Hurford (1984). Some aspects are also discussed by Storzer and Wagner (1982) and by Van den haute *et al.* (1988).

In principle, the determination ϕ_{th} with the use of a metal monitor, relies on the measurement of its specific activity after irradiation. This can be done by γ-spectrometry. Due to the reactor neutron bombardment, the following reactions take place in the monitors:

$$^{197}\text{Au}(n, \gamma)\,^{198}\text{Au}(t_{1/2} = 2.696\,\text{d}; \; E_\gamma = 411.8\,\text{keV}),$$

$$^{59}\text{Co}(n, \gamma)\,^{60}\text{Co}(t_{1/2} = 5.271\,\text{a}; \; E_\gamma = 1173.2\,\text{keV}, 1332.5\,\text{keV}).$$

For Co, the specific activity can thus be calculated from two γ-peaks which provides some internal control on the calibration of the measuring system.

If the absolute approach is considered, it is of major importance to perform irradiations with a well-thermalized neutron fluence, otherwise the $^{235}U(n,f)$ reaction rate $\mathbf{R}$ will not be given by $570.8\phi_{th}$ and has to be determined experimentally, which unnecessarily complicates the age determination.

In a reactor facility, the total neutron flux (φ) is built up from three components: fast neutrons (φ_f), epithermal neutrons (φ_{epi}) and thermal neutrons (φ_{th}). Fast neutrons are high-energy neutrons (0.5–10 MeV) released by the fission of the ^{235}U atoms of the reactor fuel. The reactor moderator (graphite, water) slows down these neutrons and reduces their energy to a pure thermal agitation energy. Neutrons whose kinetic energy is reduced to this level, are said to be in thermal

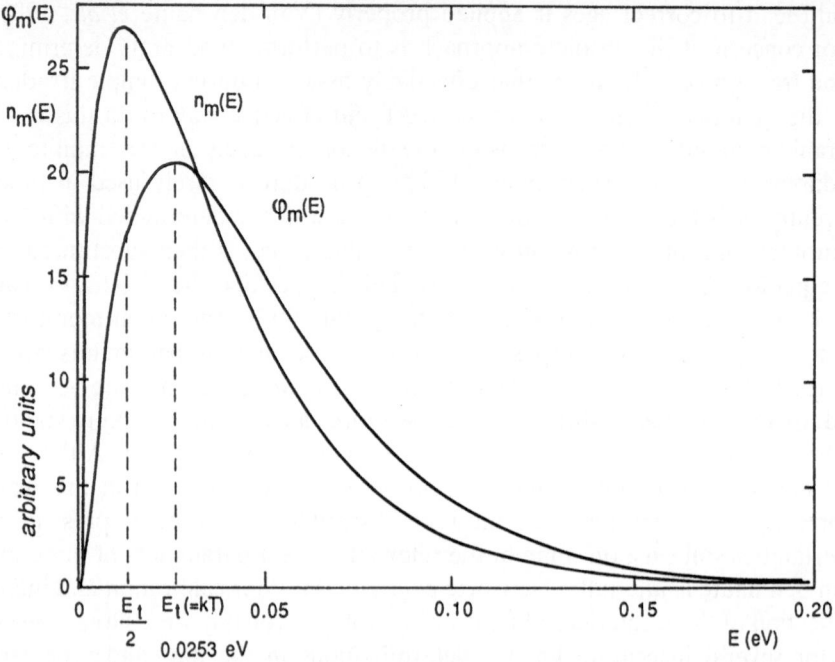

Figure 3.2. Maxwell–Boltzmann distribution of neutron density (n_m) and neutron flux (φ_m) as a function of neutron energy.

equilibrium with the moderator and are, therefore, called thermal neutrons. Their energy distribution ideally corresponds to a Maxwell–Boltzmann distribution extending from 0 to about 0.25 eV with a maximum at 2.53×10^{-2} eV (at 20.4°C) (Figure 3.2). Neutrons which are only partially slowed down are called epithermal neutrons. Their energy distribution is ideally given by $\varphi(E) \sim 1/E$ and ranges from approx. 0.1 eV to 0.5 MeV. Table 3.1 lists the cross-sections of (n, f) reactions of U and Th for thermal, epithermal and fast neutrons and helps to clarify the importance of a good thermalization. Besides thermal neutrons, epithermal neutrons also induce fission of ^{235}U. The cross-section I_0 of ^{235}U (n, f) for epithermal neutrons is about half that for thermal neutrons. The ratio $\varphi_{th}/\varphi_{epi}$ should thus

Table 3.1. Activation cross-sections in barn for thermal (σ_0), epithermal (I_0) and fast neutrons (σ_f)

Isotope	Natural abundance (%)	Reaction	σ_0	I_0	σ_f
^{232}Th	100	n, f	negl.	negl.	0.081[a]
^{235}U	0.72	n, f	586[d]	275[b]	1.203[a]
^{238}U	99.27	n, f	negl.	0.0015[b]	0.305[a]
^{59}Co	100	n, γ	37.13[d]	74[b]	negl.
^{197}Au	100	n, γ	98.65[d]	1550[c]	negl.

negl. = negligible values <1 mb ($=10^{-3}$ barn).
Sources: [a]Zijp and Baard (1979); [b]Mughabghab *et al.*, (1981); [c]Mughabghab (1984); [d]Holden and Holden (1989).

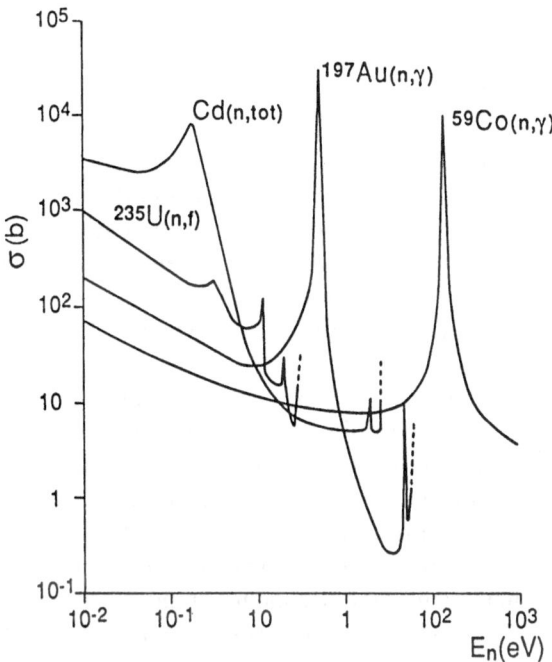

Figure 3.3. Variation of the activation cross-section of $^{59}Co(n, \gamma)$, $^{197}Au(n, \gamma)$ and $^{235}U(n, f)$ reactions and of the total absorption cross-section of Cd with neutron energy. Resonance peaks at higher energies have been omitted for clarity (after McLane *et al.*, 1988).

minimally be 50 in order to have a contribution of less than 1% of epithermal ^{235}U fission-tracks. This ratio can be regarded as the minimum value required for putting $\mathbf{R} = 570.8 \, \phi_{th}$ with an error <1%. Fast neutrons cause fission of ^{238}U and ^{232}Th, the cross-section of both isotopes being, however, small. A ratio $\varphi_{th}/\varphi_f > 7.5$ is sufficient to ensure that less than 1% of total track density results from ^{238}U fission. In the absolute approach, induced ^{238}U tracks should be avoided because they will be mistaken for induced ^{235}U tracks and, thus, lead to an underestimation of age. Fission of ^{232}Th is unwanted whatever dating system is used. This is further discussed in Section 3.7.

The γ-activity of the metal monitors, similar to the above fission reactions, may also be induced by nonthermal neutrons. If we consider the cross-sections of Au and Co (Table 3.1), a high value of I_0 is noticed for Au. This is due to the presence of high resonance peaks in the epithermal energy region (Figure 3.3). As a consequence, a ratio $\varphi_{th}/\varphi_{epi} > 1550$ is required to ensure a negligible epithermal activation in this metal. To avoid epithermal activation in Co, $\varphi_{th}/\varphi_{epi}$ may be lower but should still be >200. Such highly thermalized reactor facilities are relatively rare. Hence, for most facilities, to perform an accurate determination of ϕ_{th} with Au and Co monitors, it will be necessary to precisely determine the epithermal neutron spectrum and its contribution to the total flux and to calculate its effect on the γ-activity of both monitors.

In principle, single Au or Co monitoring is sufficient to accurately determine the neutron fluence; in practice, however, combined use of both monitors is highly advisable because it provides an efficient check on the internal consistency of the approach. Such a check is necessary, while several parameters must be accurately known such as: the calibration of the counting system, the amount of neutron self-shielding in the monitors, effects related to the geometry of the monitors, etc. Shielding effects can be avoided by using highly diluted aluminium alloy foils or wires. In this case, the exact content of Au or Co in the alloy must be known and the alloys must obviously be homogeneous over the entire foils (or wires).

If all the above considerations have been taken into account, it is possible to perform a determination of ϕ_{th} with a 2σ error of 5% or less, which is much better than reported previously (Storzer and Wagner, 1982). An analysis of one or more age standards will then provide a final test of the accuracy of the procedure. This analysis should be carried out with a technique such as the population method which ensures an identical revelation and observation of spontaneous and induced tracks ($Q = 1$; Section 3.6). An apatite age standard is the most appropriate for this purpose. All further analyses of samples of unknown age carried out with $Q = 1$ procedures, in principle do not require the addition of age standards, a fluence monitoring with Au and Co being sufficient. A check at regular intervals, e.g., with an internally calibrated secondary standard, however, always remains advisable. For $Q \neq 1$ procedures, analyses of standards together with the unknowns, always remain necessary and, in principle, allow the evaluation of Q.

The use of glass monitors. The determination of ϕ_{th} with Au and Co monitors requires an easy access to the reactor facility at every irradiation and an intimate communication and/or cooperation between the fission-track geochronologist and the reactor scientists during the setup of his dating system. For fission-track geochronologists who work at a great distance from the nearest reactor facility, this may be difficult to accomplish. To resolve this problem, the National Insitute of Standards and Technology in Gaithersburg, Washington, DC (formerly called the National Bureau of Standards) produced four series of glass wafers (SRM961 to 964) of different uranium concentrations (nominal concentration, respectively 500, 50, 1, and 0.1 ppm U) that were irradiated in the NBS reactor with a thermal neutron fluence monitored with pure Au and Cu foils (Carpenter and Reimer, 1974). These wafers were distributed as sets together with non irradiated wafers of the same type of glass. For determining the neutron fluence, the fission-track geochronologist simply has to irradiate such a nonirradiated wafer together with the samples to be dated in the reactor facility he wants to use and then to compare the track density in this wafer (ρ_u) with that observed in the pre-irradiated wafer (ρ_{NBS}) after polishing and etching both wafers under identical conditions. The neutron fluence for the specific irradiation is then calculated from:

$$\phi_u = \frac{\phi_{NBS}\rho_u}{\rho_{NBS}}.$$

In principle, the NBS–SRM glasses form an excellent solution to the problem of the determination of ϕ. Unfortunately, they possess several shortcomings which have often been criticized. The most important criticism concerns the inconsistent calibration of the fluence received by the pre-irradiated wafers, differences of about 10% being quoted on average between the Au and Cu flux values and it is not evident which of the two values is the most accurate (De Corte *et al.*, 1991). In order to set up his dating system, the fission-track geochronologist will thus have to analyze the SRM glasses in combination with age standards in order to determine which of the quoted fluxes yields the correct age together with a selected value of λ_f. In such an approach, the determination of ϕ is obviously not absolute anymore and age standards play a basic role in calibration. Hence, instead of applying such a hybrid system, it may be more efficient to use an approach which is directly based on age standards such as he ζ-method (Hurford and Green, 1982) which is outlined in the next section. Other inconveniences concern the unnatural ratio of $(2.39 \times 10^{-3}$ in SRM962) together with a considerable Th content in the 50 and 500 ppm wafers. In poorly thermalized facilities, this can be responsible for a significant amount of fission tracks other than thermal ^{235}U tracks (Crowley, 1986). Finally, the content of other elements, such as boron and REE's (SRM961, 962), is responsible for a significant thermal neutron absorption during irradiation and may cause the glasses to stay radioactive for a long time.

In 1984, new sets of SRM glasses (962a,963a) were produced replacing the 962–963 series, but still of the same composition as the previous glasses and still bearing inconsistent Au and Cu fluxes (Carpenter, 1984). They are offered together with lexan and mica flakes that were attached to them during irradiation, allowing the counting of the induced tracks in these detectors instead of in the glasses themselves. This is often preferred because track etching and counting is less problematic in mica (Hurford and Green, 1983).

In conclusion, if the absolute approach is considered in combination with a uranium doped SRM glass, it is more adequate to establish a personal calibration of the nonirradiated wafers (distributed as series SRM610-617) against Co and Au fluence determinations. Besides the SRM glasses other glasses, have been produced which are useful for this purpose, such as the U1–U7 glasses (Schreurs *et al.*, 1971) distributed by the Corning Museum (but no longer available) and the CN1 and CN2 glasses, also prepared at Corning, which contain about 38 ppm natural uranium[1] and appear to give more reproducible results (Hurford and Green, 1983). Finally, it is clear that the absolute approach, and the fission-track method as a whole, would greatly benefit from a new series of pre-irradiated glasses bearing consistent Au and Co fluences.

3.5.2. *Age standard approach: the ζ-method*

In the 1970s, in order to overcome the dispute about the value of λ_f, a dating system was proposed in which the age of a sample was determined from a comparative analysis with one or more age standards (Fleischer and Hart, 1972; Fleischer *et al.*, 1975). Later, because of the doubt that was cast on the accuracy of absolute neutron fluence determinations, this approach was elaborated and became strongly

---

[1] Recently two new glasses CN5 (12 ppmU) and CN6 (1 ppmU) have become available.

advocated by several fission-track geochronologists (Hurford and Green, 1982, 1983). The principles of such a dating system are as follows.

If we define a factor $Z = Q I \sigma / \lambda_f$ (which has a dimension of time), the age equation (3.10) can be rewritten as

$$t_u = 1/\lambda_\alpha \ln[(\lambda_\alpha)(\rho_s/\rho_i)_u GZ + 1]. \tag{3.15}$$

The value of Z is derived from an analysis of a standard of known age t_S which is irradiated together with the sample of unknown age t_u in the same reactor facility (same ϕ) and analyzed with exactly the same procedure (same Q). From the analysis of the standard, we obtain:

$$Z = \frac{(e^{\lambda_\alpha t_S} - 1)}{\lambda_\alpha (\rho_s/\rho_i)_S G}, \tag{3.16}$$

which is then substituted in Equation (3.15). If both the standard and the unknown sample are of a young age, the equations simplify to

$$t_u = (\rho_s/\rho_i)_u GZ \quad \text{with } Z = t_S/(\rho_s/\rho_i)_S G,$$

or one obtains

$$t_u = \frac{(\rho_s/\rho_i)_u}{(\rho_s/\rho_i)_S} t_S, \tag{3.17}$$

which allows the evaluation of the age of the unknown sample from a simple ratio of track-density ratios in the unknown and in the standard.

The ζ-calibration method developed by Hurford and Green (1982, 1983) represents a more practical and elaborate alternative to this approach. The method consists of two steps. In the first step, age standards are repeatedly irradiated and analzyed together with a glass monitor (SRM, U, CN) in order to establish a calibration factor ζ. The track density in the glass monitor ρ_m (or in an adjacent mica detector) is proportional to the fluence or $\phi = B\rho_m$ (B being the factor expressing this proportionality in neutrons/track), and Equation (3.16) can now be written as

$$\zeta = \frac{(e^{\lambda_\alpha t_S} - 1)}{\lambda_\alpha (\rho_s/\rho_i)_S G\rho_m} = Q I \sigma B / \lambda_f. \tag{3.18}$$

Once the ζ-calibration factor (which is expressed in year $\times$ cm^{-2} has been precisely evaluated, the age of an unknown sample can be determined after co-irradiation, with the glass monitor from the following equation:

$$t_u = 1/\lambda_\alpha \ln[(\lambda_\alpha)(\rho_s/\rho_i)_u \rho_m G\zeta + 1]. \tag{3.19}$$

The ζ-method yields calibration factors which are, to some extent, personal. For apatite, ζ-values (obtained in combination with SRM 612 glass) ranging from 310 to 360 are cited in the literature, while for different mineral species, some investigators find different and others find equal ζ-values (Green, 1985; Tagami, 1987). Hence, it is clear that the calibration must be carried out personally by every investigator who wants to apply this method and for every mineral species he wants to analyze. Taking into account the definition of ζ (Equation (3.18)), the reasons for the variation mentioned above are to be found in different B and/or Q factors between analysts and between mineral species. Different Q factors reflect differences in track revelation and observation, such as the use of dry or immersion oil objectives (Wagner, 1981; Green, 1985). The B factor, which is obtained from track counting in the glass monitor (or the adjacent mica), may vary for the same reasons but, in addition, it may also depend upon the response of the glass monitor to the neutron fluence. In poorly thermalized facilities, the total flux induces relatively more tracks other than thermal ^{235}U tracks in SRM 612 glass, which contains 37.8 ppm Th and which is depleted in ^{235}U rather than in CN glass which is practically free of Th and has a natural U isotopic ratio (Crowley, 1986; Tagami and Nishimura, 1989). This does not imply, however, that the SRM glass is unsuitable for calibration. Another reason for finding different ζ-values for different mineral species may be related to differences in thermal stability of the spontaneous tracks in these minerals and is further discussed in Section 3.9.

The ζ-calibration method at present provides an efficient way to obtain accurate results in fission-track dating but, because it depends on standards whose ages have been determined with other isotopic methods, it makes fission-track dating less dependent as a geochronological technique. It therefore remains a major challenge for fission-track geochronology to arrive at an accurate determination of the constants that make up the ζ-factor (mainly λ_f). Once this is achieved, these constants can be lifted out of ζ which will then be reduced to its essence: the procedure factor Q.

3.6. Dating Procedures and Techniques

To accomplish a fission-track age determination, several dating procedures have been elaborated, each consisting of a specific sequence of analytical steps. To some extent, these procedures have been discussed in earlier publications (Fleischer *et al.*, 1975; Naeser, 1979; Gleadow, 1981; Hurford and Green, 1982; Storzer and Wagner, 1982). The distinction made between procedures is essentially based upon the strategy that is followed for analyzing the induced tracks. If, for the induced track analysis, the same grains are used as those which served for the analysis of the spontaneous tracks, the procedure can be called a grain-by-grain procedure. If the analysis of induced tracks uses different grains, we are dealing with a grain-population procedure (Gleadow, 1981). Further distinction is made according to the techniques that are applied for revealing the induced tracks. This is outlined below and in Table 3.2 which summarizes the essentials of each method. Procedures whereby the induced tracks are etched and/or observed under different conditions than the spontaneous tracks ($Q \neq 1$ procedures), are, at present, most efficiently applied in combination with a calibration based on age standards such as the ζ-method, which avoids an explicit determination of Q. Procedures for which it can be assumed that $Q \approx 1$, allow an absolute approach.

The spontaneous tracks are generally analyzed in internal surfaces obtained by

Table 3.2. Characteristics of fission-track dating procedures

	GRAIN-BY-GRAIN METHODS			GRAIN-POPULATION METHODS	
	The same grains are used for analyzing the induced and spontaneous tracks			Induced and spontaneous tracks are analyzed in different grains	
Technique for induced track revelation	Repolishing, etching	Re-etching	Etched in external detector	Polishing, etching	Pre-irradiation annealing, polishing, etching
Counted tracks	$\rho_s + \rho_i$	ρ_i or $\rho_s + \rho_i$	ρ_i	$\rho_s + \rho_i$	ρ_i
Geometry ratio (G)	1	0.5	0.5	1	1
Procedure factor (Q)	1	$\neq 1$	$\neq 1^a$	1	1^b
Relative precision	Moderate	Moderate to high	High	Moderate to low	Moderate
Applied to[c]	Glass	Mica	Apatite, sphene zircon	Glass	Apatite

[a]May be close to unity through careful selection of crystal faces and etching conditions,
[b]If annealing does not change the track etching behaviour,
[c]Refers to the material the procedure is most commonly applied to.

cleaving or polishing of the sample. Occasionally, external surfaces have also been used for high uranium minerals such as zircon (Fleischer *et al.*, 1964; Koshimizu, 1981). Low uranium minerals are unsuitable because of a possible contamination with tracks from external sources.

3.6.1. *Grain-population methods*

In grain-population methods, an amount of several hundreds of grains (typically with a size of 70–250 μm) is normally required. This amount is split into two aliquots, a first one which serves for the analysis of the spontaneous tracks (ρ_s) and a second for the analysis of the induced tracks (ρ_i). Two alternatives can now be chosen (Table 3.2). In the first alternative, the grains for induced track analysis are simply irradiated, polished, and etched. In these grains which thus contain both spontaneous and induced tracks, the sum $\rho_s + \rho_i$ is determined and ρ_i is then obtained by subtracting the quantity ρ_s counted in the first aliquot. This procedure is known as the subtraction method. In the second alternative, the spontaneous tracks are completely annealed before the aliquot is irradiated. After irradiation, only induced tracks will thus be seen. The name 'population method' has been restricted by some authors to this last procedure but, in principle, it can be applied to every method where two populations of grains are involved. In both methods, revelation and observation of spontaneous and induced tracks are done in the same material and under identical conditions, meaning that the procedure factor Q normally $=1$. This equality always holds for the subtraction method but for the 'annealing method', an exception arises when annealing changes the etching characteristics of the material to be dated. Such a change has effectively been observed in glass where it is due to devitrification or dehydration in the case of hydrated specimens (Khan and Ahmad, 1975, Naeser *et al.*, 1980) and in minerals where it has been ascribed to the restoration of the crystal lattice (Gleadow, 1978) damaged by natural α-radiation and α-recoils. This kind of damage commonly occurs in high U and Th minerals such as zircon and leads to a complete loss of crystallinity, a process known as metamictization. Its effect on fission-track revelation has been studied most extensively in sphene (Gleadow, 1978, 1981). Coloration of sphene or zircon may be a sign of α-damage but it can also be related to other causes (Gleadow, 1978; Van den haute, 1986). A simple experiment to evaluate whether annealing significantly changes the track etching efficiency, would be to apply step etching, to watch the evolution of the shape and size of both induced and spontaneous tracks, and to verify if a clear track density plateau is reached for both types of tracks after the same etching time.

Grain-population methods are experimentally simple. The procedure, which includes annealing, is mainly used for dating apatite because the thermal treatment does not influence the track-etching characteristics of this mineral. The subtraction method is mainly applied to glass shards (Wagner, 1966; Naeser *et al.*, 1980). It becomes impractical at high spontaneous track densities.

For samples which are homogeneous in uranium and which bear isotropically etched tracks, grain-population methods yield highly reproducible results, even between laboratories (Van den haute and Chambaudet, 1990). A slight uranium heterogeneity or effects of crystal orientation are normally statistically eliminated. This goes, of course, at the expense of precision which is generally lower than that reached by grain-by-grain methods and, at strong uranium heterogeneity or strong track etching anisotropy, the grain-population methods may even become totally inappropriate.

3.6.2. *Grain-by-grain methods*

In these methods, the induced tracks are analyzed in the same grains as the spontaneous tracks. The ρ_s/ρ_i ratio and, hence, the fission-track age can thus be determined for every single grain, even at small grain size (80 μm). Grain-by-grain methods are the only ones which can be applied when the sample shows strong uranium heterogeneity among grains or is composed of grains of different age, as may be encountered in detrital rocks. Three alternative procedures have been developed: the re-etch, the repolish, and the external detector method (Table 3.2).

In the re-etch method, the induced tracks are revealed by a second etch after irradiation of the sample containing the already etched spontaneous tracks; distinction between both types of tracks is based upon track size, the spontaneous tracks being etched twice. If a clear distinction cannot be made, the induced track density has to be derived by subtraction as described above. In the repolish method, the spontaneous tracks are counted first, the sample is then repolished to remove the etched spontaneous tracks, irradiated and newly etched; the induced track density is here also derived by subtraction. In the external detector method, the induced tracks are registered and counted in an external detector of negligeable uranium content which was held in close contact with the mineral mount during irradiation. The tracks that are observed in the detector originate from U atoms which fissioned at a depth $<R_i$ under the mineral surface.

The re-etch method has been regularly used in the past (Fleischer *et al.*, 1964c; Bigazzi, 1967; Welin *et al.*, 1972) but has recently lost some favour. Although the induced and spontaneous tracks are revealed in the same material, $Q \neq 1$ because the etching conditions and the initial geometries of the investigated surfaces are different (external for the induced versus internal for the spontaneous tracks). Nevertheless, the method can be applied to micas (Bigazzi, 1967) which generally have a low spontaneous track density and a negligible etch time factor ($f(t) = 1$). The repolish method has not often been applied because the spontaneous tracks must be removed before the induced tracks can be counted and no later check on the spontaneous track density is possible (Gleadow, 1981). The method has been used on large glass specimens of a young age (Watanabe and Suzuki, 1969). The external detector method combines both the advantage of a grain-by-grain analysis with the fact that the induced and spontaneous tracks are revealed in separate mounts and is therefore frequently applied. Plastic foils (Lexan, Makrofol, Kapton) or low uranium muscovite mica are used as detectors. Often, mica is preferred because it has track registration characteristics which are more similar to those of the minerals to be dated. In the external detector method as a rule, $Q \neq 1$ because spontaneous and induced tracks are revealed and observed in different materials. Through a careful selection of crystal faces and by applying optimal etching times (Figure 2.19), the deviation from unity can be kept small. The crystal faces which exhibit narrow polishing scratches after etching, appear to be the most suitable in this respect (Gleadow, 1978, 1981; Naeser *et al.*, 1980). In apatite and zircon, they correspond to the prism faces. This, however, does not imply that these surfaces have the lowest bulk etching rates (Section 2.3.3).

In the external detector method (and, to some extent, also in the other grain-by-grain method), grains with spurious defects, anisotropically etched tracks or strongly heterogeneous track distributions can be excluded. In this way, difficult samples can still be analyzed. Care should be taken, however, not to introduce sampling bias. This may happen if the track density is used as a selection criterion, e.g., by systematically counting the grains with the highest spontaneous track density in young samples or by omitting those with no spontaneous tracks. In the population method, some selection of grains is also possible, e.g., by omitting grains with numerous spurious defects but obviously the risk of bias is greater here. Zircons, apatites, and, to a lesser extent, sphenes are the typical minerals the external detector method is applied to. It is not applied to glass.

3.7. Practical Considerations

3.7.1. *Sample preparation and irradiation*

Occasionally, the material to be dated can be collected in a pure state (large mineral or glass pieces) but, more frequently, sample preparation will have to start with the separation of the mineral grains or glass shards from a rock specimen. This starts with crushing, milling, and sieving of the rock to a grain size of about 70–250 μm. After washing, further treatment of the sand uses common separation techniques based on the physical properties of minerals such as eletromagnetic or heavy liquid separation (Bromoform, Sodiumpolytungstate, Diiodomethane). At the end of this operation, which is carried out in consecutive steps, a selection of clear and transparent grains by handpicking can be useful, especially if a grain-by-grain method is considered. For the separation of zircon out of small rock specimens, chemical methods, such as solution of the rock in HF are sometimes used (Krishnaswami *et al.*, 1974).

Samples consisting of small grains are embedded in synthetic resin. Epoxy resin is suitable in many cases because of its transparency, hardness, and resistance against most etchants at ambient temperatures. It is, however, not resistant to strong etchants at high temperatures and is better replaced by FEP (polyfluoro-ethylene) or PFA (tetrafluoroethylene-perfluoroalkoxyethylene copolymer) Teflon® (Gleadow *et al.*, 1976; Tagami *et al.*, 1988a). In order to expose a smooth internal surface, the mounted grains have to be ground and polished with corundum or diamond suspensions. Track etching is carried out with different devices depending upon the type and temperature of the etchant: in a thermostatic bath, in a furnace with good temperature control or on a hotplate, in a vessel with attached reflux-cooling jacquet if a boiling solution is used (Lal *et al.*, 1968). Identical etching of spontaneous and induced tracks, which is required for the population method, can be simply achieved by etching both mounts together.

As stated in Section 3.5.1, irradiation of the samples should be carried out in a well-thermalized reactor facility in order to avoid epithermal ^{235}U or fast ^{238}U and ^{232}Th fission. Nonthermal fission of uranium induces systematic errors in the absolute approach, but is unimportant if a dating system based on age standards is used because the same relative amount of nonthermal U tracks will be induced in the standard as in the samples to be dated. Fast ^{232}Th fission may introduce systematic errors in any dating system that is applied. The relative amount of thorium tracks that will be induced depends upon the Th/U ratio in the sample. This is illustrated in Figure 3.4 which shows the risk of Th fission for varying Th/U and φ_{th}/φ_f ratios. Taking the geochemical average Th/U = 4 as a reference, Th fission can be kept lower than 0.1% of thermal U fission if the flux ratio $\varphi_{th}/\varphi_f > 80$. This ratio together with a ratio $\varphi_{th}/\varphi_{epi} > 100$, ensures that more than 99.5% of all induced tracks will originate from ^{235}U fission, which represents an excellent thermalization.

Often, the degree of thermalization is reported as a Cadmium Ratio (CR) instead of as a ratio $\varphi_{th}/\varphi_{epi}$. The CR is determined by irradiating a target isotope

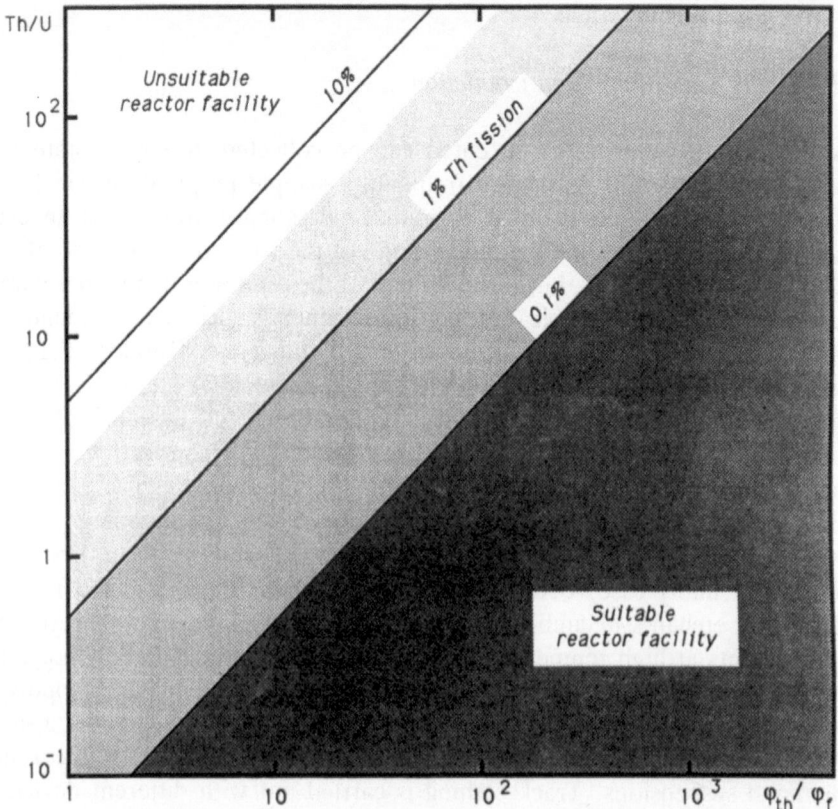

Figure 3.4. Relative amount of ^{232}Th fission compared to ^{235}U fission for varying Th/U and thermal flux/fast flux ratios. The figure allows the evaluation of the risk of creating Th fission tracks in a given reactor facility.

(preferably one of the flux monitors Au or Co), both bare and covered with a 1 mm thick Cd cover:

$$CR = \frac{\text{activity without Cd cover}}{\text{activity with Cd cover}} = \frac{\sigma_0\varphi_{th} + I_0\varphi_{epi}}{I_0\varphi_{epi}}$$

or

$$CR - 1 = \sigma_0\varphi_{th}/I_0\varphi_{epi}.$$

This procedure relies on the convention which sets the limit between thermal and epithermal neutrons at an energy of 0.55 eV, corresponding to a drastic fall in the activation cross-section of Cd (Cd-cut off energy), a metal which highly absorbs thermal neutrons (Figure 3.3).

A ratio $\varphi_{th}/\varphi_{epi} = 100$ corresponds to a CR of 6.4 for Au and of 50 for Co. A cadmium ratio, which amounts to 100 for a specific isotope, implies that epithermal activation of this isotope will represent less then 1% of its total activation. Cadmium ratios give direct information on the degree of thermalization if they have been calculated for the given isotope at infinite dilution. Often, a crude CR, e.g., measured with Au foils, is reported. Such a ratio is highly influenced by shielding effects relating to the thickness of the foils and does not reveal much about the thermalization unless the thickness of the foils is clearly specified or unless the shielding effects have been corrected for.

Another property of most reactor facilities that should be dealt with in order to accurately determine the fluence received by each sample, is the presence of a flux gradient. This gradient is typically most significant when radially moving away from the reactor core and often reaches 10%/cm. A close packing of samples and monitors may help to overcome this problem but often the best solution will be to perform an explicit determination of the gradient. This can be done by adding several U-monitor glasses at strategic places in the sample package and by counting the induced tracks in these glasses. If possible, a determination with easily activated metal foils or wires (e.g., Cu), whose activity is measured by γ-spectrometry, is to be preferred because it is much more rapid and yields more precise results (Figure 3.5).

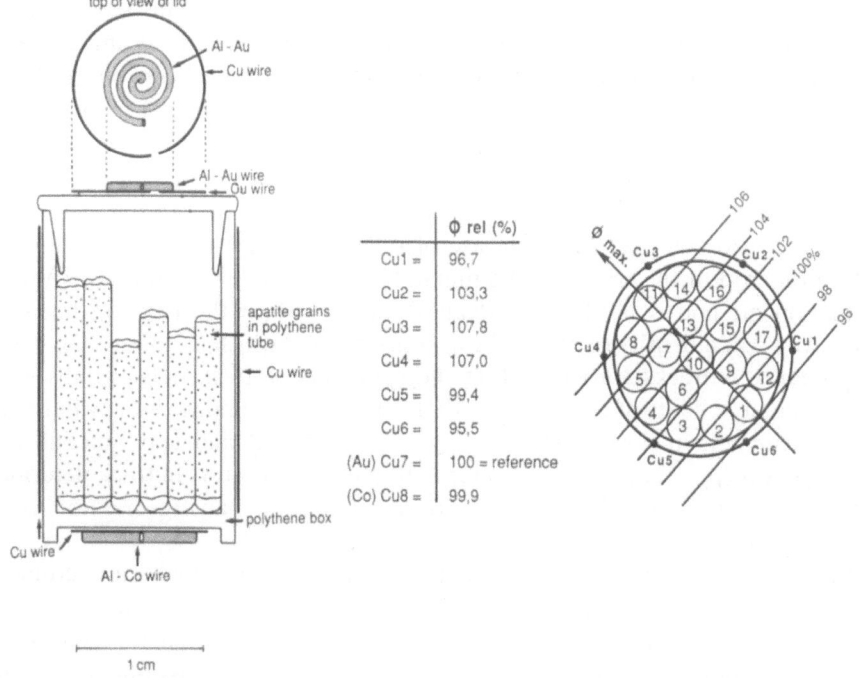

Figure 3.5. Arrangement of 17 apatite grain samples (in plastic tubes) and flux monitors prepared for irradiation (for dating with the absolute approach and the population method). The absolute flux monitors are Al–0.1% Au and Al–2% Co wires, each 1 mm thick. Six pure (99.9%) Cu wires 0.2 mm thick are regularly spaced around the sample box and two others surround the Au and Co monitors. The relative fluences registered by the Cu monitors are used for obtaining precise information on the flux gradient during irradiation and are quoted in the central column. The Cu monitor surrounding the Au monitor serves as a reference. The measurements reveal a negligible gradient in a vertical (axial) direction (compare Cu7 with Cu8), while in a radial sense the fluence gradient (derived from the Cu1–6 values) is considerable but regular, as is shown by the equally spaced 'iso-fluence' lines in the rightmost figure which shows a cross-section of the sample box with the numbered apatite tubes. The direction of the maximal gradient which amounts to 9.5%/cm is also indicated.

3.7.2. *Track counting and measuring*

As stated in Section 2.4.2, fully automatic track analysis has not yet found wide application in fission-track dating. Track counting under the microscope is thus essentially carried out by a human observer who uses a square grid mounted in one of the oculars. This grid serves as a unit-counting area and is calibrated against a stage micrometer. In grain-population methods, it is a common pratice to count the same unit area in every grain because this simplifies further statistical treatment. In the external detector method, it is of major importance to count the induced tracks in an area of the detector image which exactly matches the area of the grain that was scanned for spontaneous tracks. The motorized microscope stages (Section 2.4.3) offer an efficient way to accomplish this rapidly and accurately with the help of a dedicated computer program. Locating the grains and their detector images can also be greatly accelerated by arranging the grains in an easily recognizable pattern during embedding.

Spontaneous and induced tracks of the same sample or grain (for grain-by-grain methods) are best counted consecutively. Long interruptions between counts typically tend to deteriorate reproducibility. Also, very long counting efforts on the same mount may cause boredom. Therefore, switching from one mount to another after a certain time (e.g., from spontaneous to induced-track counting when a population method is used) can help to overcome this. Both very low ($<10^3$ tr/cm^2) and very high ($>10^7$ tr/cm^2) track densities are difficult to determine precisely. At high densities, tracks overlap each other and, in addition, care should be taken not to count TINTs. The use of reflected light which clearly shows the surface outcrop of the tracks at the mineral surface can be helpful here. When a population method is used, it is advantageous to adjust the neutron fluence in such a way that a density of induced tracks is obtained which is close to the density of spontaneous tracks (provided that the spontaneous track density falls within easily countable limits of 5×10^4–5×10^6 tr/cm^2). The counting effort will then be of approximately the same statistical size for induced and spontaneous track counting. A $\rho_s/\rho_i = 1$ ratio is also better buffered against errors in track determination and creates optimal conditions for consistent human observation (equal observation factors q_s and q_i).

Track-size measurements are very tedious if done manually with a micrometer-ocular, but they are now greatly facilitated by the use of a digitizing tablet as briefly described in Section 2.4.3. The tablet can also be used for determining angle distributions in studies of track anisotropy. In apatite fission-track dating, track-length measurements nowadays almost form an inherent part of the analysis because they enable a much more precise interpretation of the fission-track ages. In order to retrieve relevant information, a number of 50–100 tracks are necessary if confined track lengths are measured, while for projected length measurements and track diameter studies in glass, at least 10 times as many tracks are required.

3.8. Data Analysis and Error Calculation

Processes of radioactive decay, including uranium fission reactions, are so-called random processes. This implies that the distribution of their occurrences per interval of time is governed by chance alone. The statistical frequency distribution that arises in such a case is the Poisson distribution, the equation of which can be written as

$$f(x) = \frac{\mu^x}{x!} e^{-\mu}, \tag{3.20}$$

where x is the number of events and μ is the average rate of the process. Instrumental methods for measuring the activity of radionuclides, such as γ-spectrometry, α-counting, etc., which collect a number of events (counts) during a certain interval of time, typically use Poisson statistics for data analysis. If the decaying parent is homogeneously (or randomly) distributed throughout the volume of the solid, the number of events that will be registered after a certain time per unit of volume, will also vary randomly. This statement also applies to fission tracks whose spatial and areal densities can thus be expected to vary according to a Poisson distribution.

The Poisson distribution is a discrete and asymmetric distribution which is described by one parameter only: the mean μ, the variance s_P^2 being equal to the mean (hence, the standard deviation $s_P = \sqrt{\mu}$ and the coefficient of variation, i.e., the relative standard deviation of the mean $C_P = 1/\sqrt{\mu}$). As μ becomes larger (>5), the Poisson distribution tends to a normal distribution and statistical tests, which are applied to continuous normal distributions, also become applicable to Poisson distributions (Koch and Link, 1970).

3.8.1. Grain-population methods

In population methods, the spontaneous and induced track densities are determined in different grain aliquots (Section 3.6.1). Most commonly, an equal unit area is scanned in each grain and the number of spontaneous tracks N_{sj} or induced tracks N_{ij} in such an area is counted. The arithmetic mean of the track counts is then calculated as the estimate of track density or

$$\bar{\rho}_s = \frac{\Sigma_j N_{sj}}{n_s} \quad \text{and} \quad \bar{\rho}_i = \frac{\Sigma_j N_{ij}}{n_i}, \tag{3.21}$$

where n_s and n_i are the number of grains scanned for counting the spontaneous and induced tracks, respectively.

The standard deviation of the distribution of track counts, as is experimentally observed, is taken as a measure of variation and serves as a basis for error calculation (Section 3.8.3). The standard deviation is given by

$$s_{s,i} = \sqrt{\frac{\Sigma_j [N_{(s,i)j} - \bar{\rho}_{s,i}]^2}{(n_{s,i} - 1)}}. \tag{3.22}$$

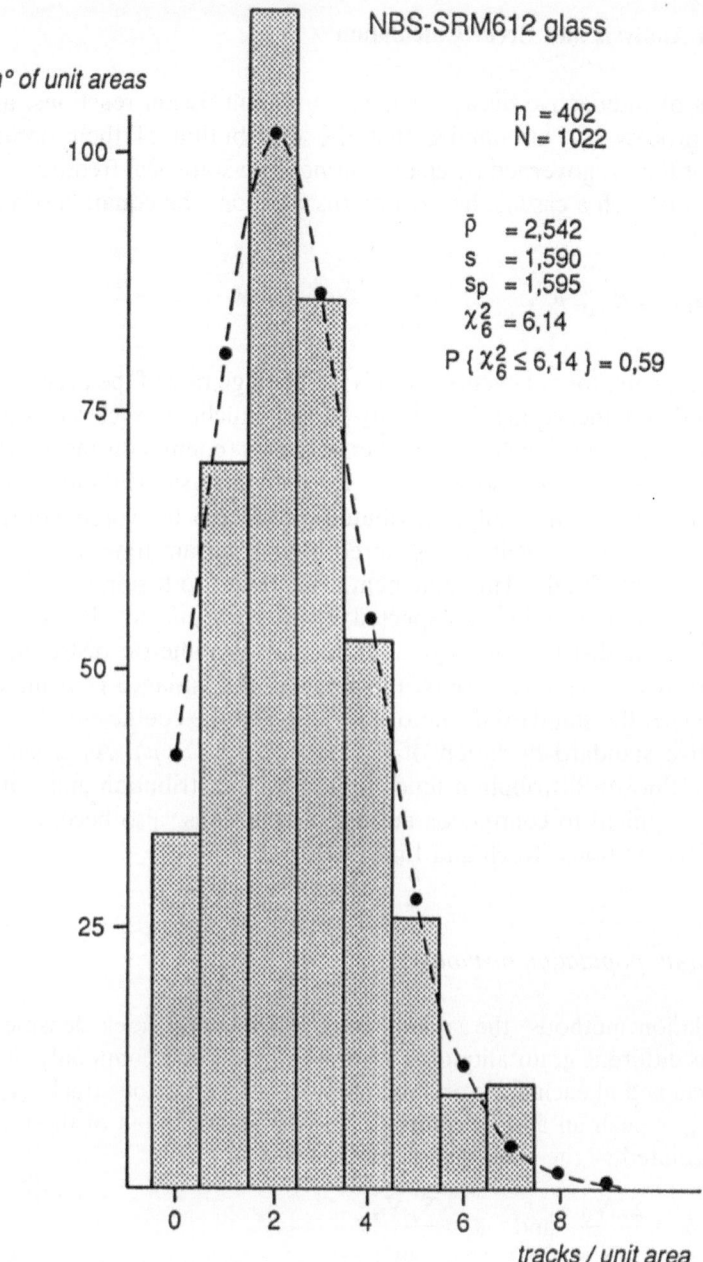

Figure 3.6. Track density histogram observed in SRM 612 glass (induced tracks). N = total number of tracks counted, n = total number of unit areas scanned, ρ = mean track density/unit area; size of unit area = $2.32 \times 10^3 \, \mu m^2$. The track density histogram corresponds well to a Poisson distribution (dashed curve). This is also illustrated by the standard deviation s which practically equals the Poisson standard deviation s_P and by the result of the χ^2 test (χ^2 value = 6.14 and corresponding probability $P(\chi^2)$ = 0.59 for 6 degrees of freedom (= number of density classes minus two)).

Track-density distributions of populations of mineral grains are seldom Poisson distributions, their standard deviations being generally larger than s_p. This is due to uranium heterogeneity between grains and also, to some extent, to differences in track revelation between the differently oriented crystal surfaces. In glasses, uranium is less heterogeneously distributed and track revelation is completely isotropic. Hence, track-density distributions approximating Poisson distributions are more commonly found (Arias *et al.*, 1981; Bigazzi *et al.*, 1986). In order to check if a track-density distribution approximates a Poisson distribution, several statistical tests are available. A classical test is the χ^2 test in which the observed frequencies in each track-density class are compared to the expected frequencies for a Poisson distribution with the same mean (Figure 3.6). Other tests imply a comparison of the experimental variances s^2 to the expected Poisson variance (s_P^2) which equals $\bar{\rho}$ (Miller and Kahn, 1962). The reader is referred to statistical handbooks for further information on this matter. Plotting the track counts on Poisson probability paper also enables a rapid visual inspection of the consistency of the counting data with a Poisson distribution, as is illustrated in Figure 3.7.

Most commonly, besides the Poisson variation, the uranium heterogeneity between grains forms a major source of variation in a track-density distribution. The total variance of a track-density distribution (spontaneous or induced) can thus be written as a statistical sum of the random Poisson variance and the variance due to U heterogeneity, or

$$s^2 = s_P^2 + s_U^2. \tag{3.23}$$

While the Poisson variance is given by $\bar{\rho}$, s_U^2 can be replaced by $\bar{\rho}^2/k$ (k being an arbitrary constant) (Galbraith, 1984). This relies on the fact that, for a grain population with a given average uranium content, the standard deviation of the track-density distribution stands in a constant proportion to the mean track density (if the Poisson variation is neglected) or $s/\bar{\rho} = $ cte ($= \sqrt{k}$). Hence, Equation (3.23) becomes

$$s^2 = \bar{\rho} + \bar{\rho}^2/k \tag{3.24a}$$

and, in terms of the coefficient of variation,

$$C^2 = 1/\bar{\rho} + 1/k. \tag{3.24b}$$

The term $1/\sqrt{k}$ represents the coefficient of variation (C_U) of the underlying uranium distribution (Galbraith, 1986a). Equation (3.24) allows calculation of this coefficient (or a value of k) for any sample that is analyzed. This can be very helpful when, due to U heterogeneity both the spontaneous and induced track-density distributions significantly deviate from a Poisson distribution and doubt exists whether the grains selected for counting both types of tracks effectively represent statistically identical samples. Calculating the value of C_U (or k) for both the spontaneous and induced track-density distributions and testing both values for their equality, offers a means to check this. Statistical tests which can be considered are the common k-test mentioned in Galbraith (1984) or an F-test (applied to C_U^2) if the track density distributions do not greatly deviate from normality. When the above tests show that the C_U values for both track distribu-

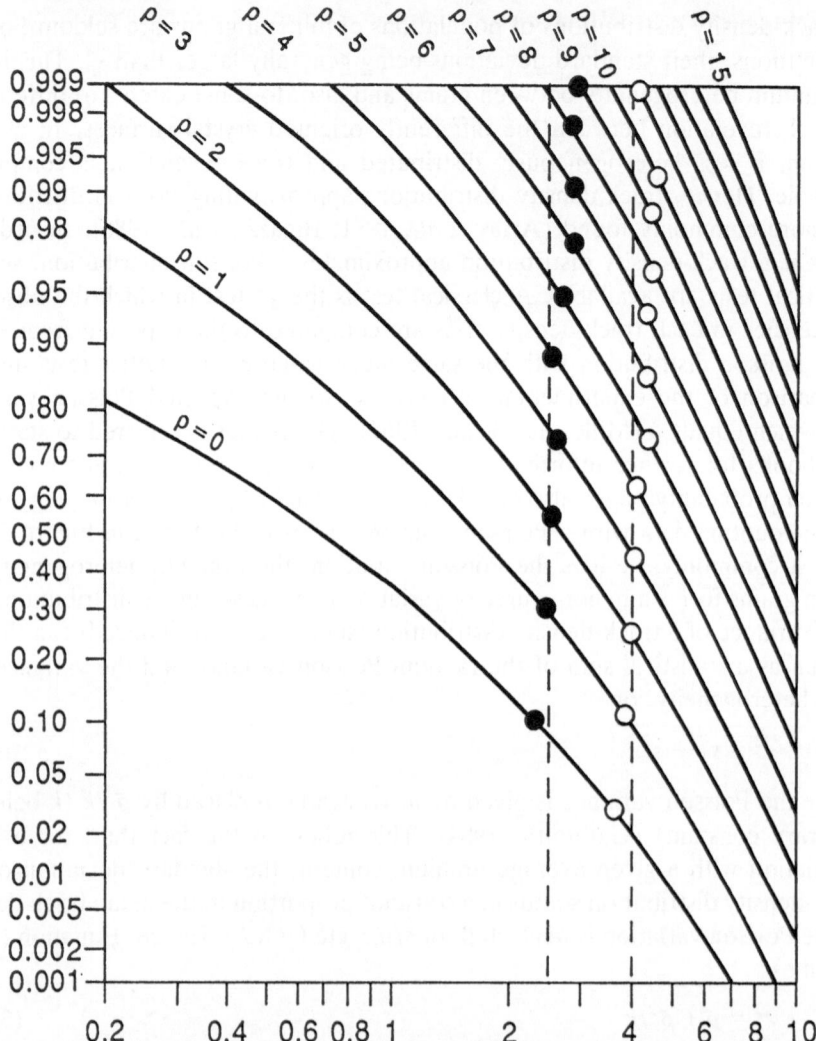

Figure 3.7. Spontaneous (black circles) and induced track counts (white circles) of Fish Canyon Tuff apatite plotted on Poisson probability paper (grain population analysis). The units on the *x*-axis correspond to track densities/unit area (logaritmic scale), the units on the *y*-axis to relative cumulative frequencies (Poisson probability scale). For each track density class (starting with $\rho = 0$) a point is plotted on the corresponding curve at an ordinate corresponding to its relative cumulative frequency. If the track density distribution is a Poisson distribution, all points will lie on a vertical line intersecting the *x*-axis at a value equal to the mean track density. As can be seen from the figure, both the spontaneous and induced track density distributions only slightly deviate from Poisson (mean track density $\bar{\rho}_s = 2.39$ tr/unit, $\bar{\rho}_i = 3.99$ tr/unit).

tions are equal, the age determination can be considered as reliable and a pooled estimate of C_U can be calculated from both separate estimates which then represents a more definite measure of the U-heterogeneity in the sample (Galbraith, 1984). If the test shows that both values are unequal, it must be concluded that the average track densities $\bar{\rho}_s$ and $\bar{\rho}_i$ were not sampled from grain populations with the same uranium distribution and, as a consequence, a systematic error may be introduced in the age calculation. In such a case, it must be considered whether to perform a new counting analysis or – which will often be more adequate – to change for a grain-by-grain dating procedure such as the external detector method.

As stated in the beginning of this section, the arithmetic mean of the spontaneous and induced tracks counts are commonly used for the calculation of ρ_s and ρ_i. For strongly asymmetric distributions, the arithmetic mean, however, becomes a rather inefficient estimator and changing to geometric means ($\bar{\rho}_g$) and lognormal statistics can be considered. In this case, the track-density ratio will be yielded by $(\bar{\rho}_g)_s - (\bar{\rho}_g)_i$. The application of lognormal statistics to geological problems is treated in statistical handbooks (e.g., Koch and Link, 1970) and also has received some attention in fission-track studies (Burchart, 1981; Galbraith, 1984).

3.8.2. *Grain-by-grain methods*

In grain-by-grain methods, the spontaneous and induced-track counts are obtained in pairs, the induced tracks being counted in an area of each grain that exactly matches the area used for spontaneous track counting (Galbraith, 1984; Galbraith and Laslett, 1985). As a consequence, each grain yields its individual ρ_s/ρ_i ratio and, hence, its individual age. Often a number (n) of 10 grains or more is analyzed in order to have a good statistical sample, the total numbers N_s and N_i of the spontaneous and induced tracks counted in all n grains is calculated and the ratio N_s/N_i taken as the estimate of ρ_s/ρ_i. There is no need to count areas of the same size in each grain, as is commonly done in the population method.

Because the uranium variation between grains bears no influence on the individual ρ_s/ρ_i ratios, no variation of these ratios is expected apart from the random variation governed by Poisson statistics only. The experimental data, however, often show that this is not the case and reveal a supplementary variation between ρ_s/ρ_i ratios (Galbraith, 1982). A χ^2-test has been developed (Galbraith, 1981; Green, 1981a) which allows the detection of this variation. The calculation (degrees of freedom = $n - 1$) uses the following formula:

$$\chi^2 = \Sigma_j \left[(N_{sj} - \hat{N}_{sj})^2 / \hat{N}_{sj} + (N_{ij} - \hat{N}_{ij})^2 / \hat{N}_{ij} \right], \qquad (3.25)$$

where $\hat{N}_{sj}$ and $\hat{N}_{ij}$ are the expected counts of spontaneous and induced tracks in the jth grain which are given by $N_s N_j / N$ and $N_i N_j / N$, respectively ($N = N_s + N_i =$ the total number of spontaneous plus induced tracks counted in all grains and $N_j = N_{sj} + N_{ij} =$ the number of spontaneous plus induced tracks counted in the jth grain). If the sample fails the test (i.e., there is less than 5% probability of finding the calculated χ^2 value), the weighted or unweighted mean of the individual ρ_s/ρ_i ratios is taken as an estimate of the overall track-density ratio instead of N_s/N_i and the standard deviation of these ratios serves as a basis for error calculations (Green, 1981a). If the ρ_s/ρ_i ratios of a grain sample are not homogeneous, the age determination should be regarded with some caution and effort should be made to find the physical reasons for this inhomogeneity.

3.8.3. *Error in the age determination*

The precision of a fission-track age is essentially determined by the precision of the track-density ratio R ($= \rho_s/\rho_i$) of spontaneous to induced tracks and the determination of a neutron fluence ϕ (directly or indirectly in the case where an age standard approach is used). As is outlined in Appendix C, the relative error on t is given by (App. C, Equation (C.3)):

$$s_t/t = K\sqrt{(s_R/R)^2 + (s_\phi/\phi)^2}, \tag{3.26}$$

where s_R/R and s_ϕ/ϕ are the relative errors on R and ϕ, respectively, and $K \approx 1$ for samples with a fission-track age <600 Ma.

The calculation of error in the track density ratio R depends upon the dating procedure which is applied. If the induced tracks are counted on separate mounts which do not contain already etched spontaneous tracks (nonsubtraction methods), the relative error on R is given by (Appendix C, Equation (C.6)):

$$s_R/R = \sqrt{(s_{\rho s}/\rho_s)^2 + (s_{\rho i}/\rho_i)^2},$$

while, in the case where a subtraction procedure is used, the error equation becomes (Appendix C; Equation (C.8)):

$$s_R/R = \frac{\rho_{s+i}}{\rho_{s+i} - \rho_s}\sqrt{(s_{\rho s}/\rho_s)^2 + (s_{\rho s+i}/\rho_{s+i})^2}.$$

In these equations, $s_{\rho s}$, $s_{\rho i}$, and $s_{\rho+i}$ are the statistical errors of the track densities ρ_s, ρ_i, and ρ_{s+i}, respectively. In grain-population procedures, these are calculated, respectively, from $s_s/\sqrt{n_s}$, $s_i/\sqrt{n_i}$ and $s_{s+i}/\sqrt{n_{s+i}}$, where $s_{s,i}$ (and s_{s+i}) are the experimentally registered standard deviations of the track counts (Equation (3.22)) and $n_{s,i}$ (and n_{s+i}) the number of grains analyzed. The relative errors on $\bar{\rho}_s$ and $\bar{\rho}_i$ (or $\bar{\rho}_{s+i}$) are given by $C_s/\sqrt{n_s}$ and $C_i/\sqrt{n_i}$ (or $C_{s+i}/\sqrt{n_{s+i}}$) and, hence, for the relative error on R it is obtained that

$$s_R/R = \sqrt{C_s^2/n_s + C_i^2/n_i} \tag{3.27a}$$
(nonsubtraction method)

and

$$s_R/R = \frac{\bar{\rho}_{s+i}}{\bar{\rho}_{s+i} - \bar{\rho}_s}\sqrt{C_s^2/n_s + C_{s+i}^2/n_{s+i}} \tag{3.27b}$$
(subtraction method).

It has been shown in simulation experiments (McGee *et al.*, 1985; Bigazzi *et al.*, 1986) that, in this way, a good approximation of true precision is obtained.

When a grain-by-grain procedure is used, the error on track density ratio R is calculated with the use of conventional Poisson statistics, meaning that if N_s and N_i are the total number of spontaneous and induced tracks counted, the relative error on these numbers is given by $1/\sqrt{N_i}$, and $1/\sqrt{N_i}$, respectively. Hence, Equations (3.27a) and (3.27b) change to

$$s_R/R = \sqrt{1/N_s + 1/N_i}$$ (3.28a)

(non-subtraction method)

and

$$s_R/R = \frac{N_{s+i}}{N_{s+i} - N_s} \sqrt{1/N_s + 1/N_{s+i}}$$ (3.28b)

(subtraction method).

The errors calculated with the above formulas include a 1σ confidence interval, meaning that there is a $\approx 68\%$ probability that the true ratio falls within the range covered by the given error. If the errors are doubled (2σ), a $\approx 95\%$ probability is obtained.

The error on ϕ is generally less important than the error on the track-density ratio and includes all uncertainties related to system calibration. In the absolute approach, s_ϕ actually corresponds to the error on the neutron-fluence determination. When ϕ is determined with metal monitors, it includes the uncertainty on the nuclear parameters of the monitors (Au, Co, U), on the calibration of the counting system (e.g., Ge–Li detector) and several experimental errors. Although these are quite large in number, a direct determination of ϕ with metal monitors can nowadays be achieved with an error of $\approx 2\%$ (1σ confidence). When the neutron fluence is derived from track counting in a glass monitor, the error of the track density in this monitor also has to be taken into account.

In the case where an age standard approach such as the ζ-method is used, s_ϕ essentially refers to the error on the track density ρ_m of the monitor glass irradiated together with the sample. The error on the calibration factor ζ is of less importance; after careful calibration with several age standards, ζ can be evaluated with an error of 1–1.5% (1σ).

When all uncertainties are accumulated, a 5% error (1σ) is commonly within reach for an age determination carried out with a normal counting effort, on the obvious condition that a sufficient number of grains are available with reasonable spontaneous track densities (Section 6.1). Grain-by-grain methods which use Poisson errors, principally require less counting effort than population methods for obtaining the same degree of precision.

The error on the age is commonly quoted as a $\pm$ error: e.g., an age, determination of 20 Ma where $s_t/t = 10\%$ is quoted as 20 ± 2 Ma. It has been argued that for a given sample, the estimates $\phi\,\rho_s/\rho_i$ (and, hence, also the age estimates) are not normally but rather lognormally distributed (Galbraith and Laslett, 1985) and that, for that reason, the upper and lower limits of the error bar are better calculated from $t(1 + s_t/t)$ and $t/(1 + s_t/t)$, respectively. In the former example, this would lead to an age interval 18.2 Ma–(20 Ma)–22 Ma. Such a scrutiny is not needed, however, if the error on the age is not too large. Moreover, the deviation from normality of the distribution of age estimates sometimes appears to be only minor (Clarke and Carter, 1987).

3.8.4. *Analysis of age groups*

When numerous samples are dated (or numerous grains in a grain-by-grain procedure, a careful comparative analysis of the obtained ages is of major importance for a correct geological interpretation. When the rocks were sampled over a large area, a distinct geographic pattern is often observed and the age interpretation is staightforward (as in several case studies discussed in Chapter 7). In other instances, however, the picture that shows up from the age data is not immediately clear and the need for a statistical analysis arises. Such a need became more urgent with the recent application of grain-by-grain dating methods to minerals of volcanic tuffs and arenaceous sediments which may contain grains from different source rocks. The presence of older detrital zircons in pyroclastic deposits has long been recognized by Japanese fission-track geochronologists and complicates the determination of the stratigraphic age of these deposits (Koshimizu, 1981).

   Several statistical methods have been applied to discriminate age groups in grain samples (mainly dated with the ED method). These methods all use rather tedious calculations which are best performed with a computer. In the approach of Hurford *et al.* (1984) applied to zircons of Lower Cretaceous sandstones from the Weald, an age probability curve is calculated for each grain as a normal distribution centered around the measured age and with a standard deviation equal to the calculated error. The age probability curves for all zircon grains from the same rock, are summed up and the resulting curve (which can be considered as a weighted age histogram) is searched for modes (Figure 3.8). These modes are interpreted to repesent the true ages of the different grain populations of which the sample is constituted. Another approach uses a χ^2-type test (Seward and Rhoades, 1986). The age estimates of the n analyzed crystals are ordered from the least to the greatest and denoted by $t_1, t_2, \ldots, t_k, \ldots, t_n$. The approach is based on the assumption that if the analyzed grains belong to two different groups with respective ages t_1 and t_2, the sum of squares of standardized residuals

$$\chi^2 = \sum_{i=1}^{n_k} (t_i - t_1)^2/s_i^2 + \sum_{i=n_k+1}^{n} (t_i - t_2)^2/s_i^2,$$

should be less than the value of the χ^2 distribution with $n - 2$ degrees of freedom for a given level of significance (e.g., 95%, or $\alpha = 0.05$). In order to work with normally distributed data, the test is applied to the logs of ages rather than to the ages themselves. Finally, Galbraith and Green (1990) introduced a method known as finite mixture modelling for the separation of age groups which they applied to artificially prepared mixtures of two and three age groups. The interested reader is referred to the original publications for more details. Mention can be made here of a recently designed graph known as the radial plot (Galbraith, 1988, 1990)

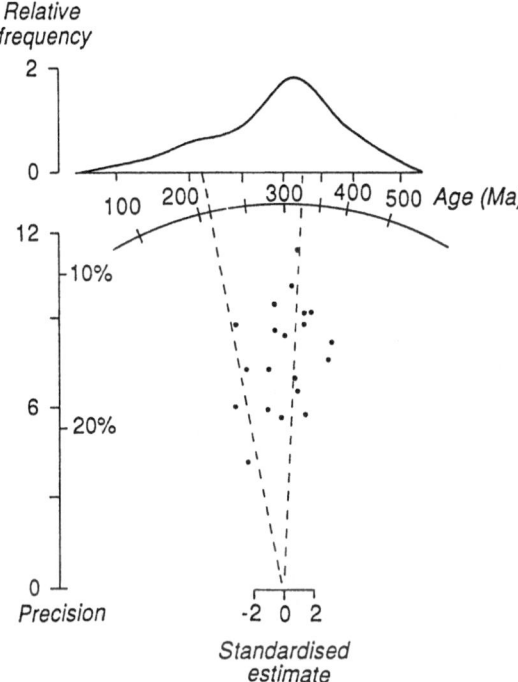

Figure 3.8. Twenty apatite grains analyzed with the external detector method and belonging to a sample composed of two age groups (respectively ≈240 Ma and ≈340 Ma) are displayed on the radial plot of Galbraith (1988, 1990). The x-axis, on which the reciprocal error is plotted on a linear scale, is drawn vertical here, while the y-axis, which displays the standardized age estimate is horizontal (see also Figure 3.9). The curve on top of the figure shows the age frequency curve of the 20 samples and gives evidence of a mixed sample. The two dashed lines show the estimation of the two component ages according to the method of finite mixture modelling of Galbraith and Green (1990). The sample shown here does not represent a natural case but an artificially prepared mixture to test the model calculations. (After Galbraith and Green, 1990).

which enables a good visual judgement of the homogeneity of a set of age determinations and a fair discrimination of outliers or age groups (Figures 3.8 and 3.9).

The radial plot is an (x, y) scatter plot which can be constructed for each set of normally distributed estimates of a given quantity z. For each estimate z_i which bears an error s_i, the x, y coordinates are respectively given by

$x = 1/s_i$,   i.e., the reciprocal error,
$y = (z_i - \bar{z})/s_i$,   i.e., the standardized residual of the (weighted) mean $\bar{z}$
         of all estimates.

In fission-track dating applications, each z_i corresponds to the log(age) and s_i to the relative error on the age. Due to the concept of the graph, all samples with the same age will plot on a line passing through the origin. A circular age-scale, at the end of the graph and centred at the mean of all ages, enables one to read off the results of the individual age determinations. Figure 3.8 illustrates how a grain sample composed of two age groups is displayed on such a plot.

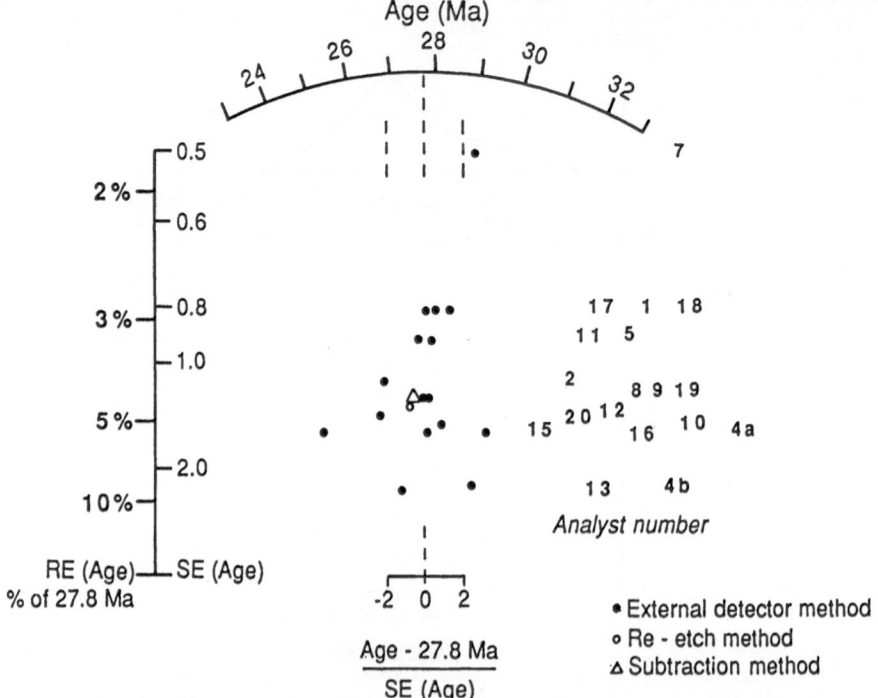

Figure 3.9. Results of the interlaboratory comparison of Fish Canyon Tuff zircon carried out in the context of the 1984 Workshop on Fission Track dating (Miller *et al.*, 1985) displayed on a radial plot (*x*-axis vertical). The circular scale is centred at the isotopic reference age of 27.8 Ma. The age which corresponds to each point can be read off on this scale by drawing a straight line through $y = 0$ and the point that is considered. Points which fall within the field limited by the dashed vertical lines at $y = \pm 2$ are age determinations which agree with the reference age within error (from Galbraith, 1990; SE = absolute standard error and RE = relative error on the age.)

3.9. Age Standards and the Accuracy of Age Determination

Until a decade ago, a lack of standardization of techniques, together with poor calibration of dating procedures, allowed doubtful results to find their way into the literature. Hence, although a considerable number of scientists already produced accurate and reliable fission-track ages, the accuracy of the fission-track dating method was still a subject of discussion. It was realized that only a general analysis of age standards would provide a solution to this problem (Hurford and Green, 1981b; Storzer and Wagner, 1982). Basic requirements that needed to be fulfilled by age standards were formulated by Wagner in 1980 (Hurford and Green, 1981b). The most important of these requirements are

- the rock formation containing the standard sample should be geologically well documented;
- if the standard sample is a mineral sample, it should be homogeneous in age, implying that the rock should contain crystals of one generation only and none derived from an older rock;

- the age of the sample should be unambiguously known both from stratigraphy and from independent isotopic dating (K–Ar, Rb–Sr), which should yield consistent results referring to the formation of the rock and not to some posterior event;
- the fission-track age should also refer to the formation age and no corrections related to spontaneous track fading should be necessary to arrive at this age.

The earliest materials that were proposed as age standards were zircons and apatites from the Fish Canyon tuff (Colorado, U.S.A.). These minerals were also the subject of the first interlaboratory comparison (Naeser *et al.*, 1981). Since then, several other age standard candidates have been investigated both by single researchers (mostly within the framework of ζ-calibration experiments) and in the context of interlaboratory comparisons (Miller *et al.*, 1985, 1990). Table 3.3 provides a list of materials which are now considered as prominent reference samples.

In 1986, the general concern about the accuracy of the method led to the constitution of an international Fission-track Working Group in order to formulate a recommendation on dating strategies and suitable age standards. Except for the Bishop tuff zircon, all materials listed in Table 3.3 have been retained as age standards by the Working Group (Hurford, 1990). Of these standards, those from the Fish Canyon tuff have been most thoroughly studied. They formed the subject of two interlaboratory comparisons (Naeser *et al.*, 1981; Miller *et al.*, 1985) of several ζ-calibration experiments (e.g., Hurford and Green, 1983; Green, 1985) and of detailed statistical analyses (Galbraith, 1982, 1986b, 1990) (Figure 3.9).

The general introduction of age standards has significantly improved consistency between laboratories and intrinsically ensures results which are more directly comparable with those from other isotopic methods. Nevertheless, some problems still remain. They are partly due to the fact that doubt may exist whether the fission-track age of the standards effectively refers to their true formation age. This especially holds for the apatite standards, all of them having a spontaneous track length which is some 5 to 10% shorter than the induced track length (Gleadow *et al.*, 1986) and which therefore can be expected to yield ages which are slightly too low. By using a ζ-type of approach based on the true formation age of these standards, an implicit age correction will be carried out. Hence, a small systematic error may be introduced for samples which have shorter lengths than the standards and the age that truly goes together with the observed spontaneous track length is not accurately assessed anymore. Different length reductions of spontaneous tracks due to natural fading may also be a possible reason why different ζ-values are found for different mineral species.

Beside age standards, a selection of the appropriate approach and dating procedure is equally important for obtaining reliable fission-track ages. In this respect, the opinion has been formulated that a uniform approach such as the ζ-calibration is to be recommended (Hurford, 1990). At present, the ζ method is certainly the most efficient if a $Q \neq 1$ procedure is used and on the obvious condition that standards and unknown samples are analyzed with the same procedure and with the same techniques. An absolute approach requires a quantitative assessment of Q which is obviously more complicated. It is, however, not inferior if a $Q = 1$

Table 3.3. Prominent reference samples used as age standards in Fission-Track dating

Reference material	Geological specification	Region, Country	Age* (Ma)	Method and reference	Relevant fission-track publications
Zircon	Bishop volc. tuff	California, U.S.A.	0.734 ± 0.024	$^{40}Ar/^{39}Ar$ (sanidine) Hurford and Hammerschmidt (1985)	Hurford and Green (1983) Green (1985)
Glass	Moldavite tektite	S. Bohemia, Czechoslovakia	15.21 ± 0.15 [15.1 ± 0.7]	K–Ar Staudacher et al. (1982) [K–Ar Gentner et al. (1967)]	Wagner (1966) Gentner et al. (1967) Naeser et al. (1980)
Zircon	Buluk Member volc. tuff Bakata Formation	Bakata Valley, N. Kenya	16.4 ± 0.2	K–Ar (alkali feldspar) McDougall and Watkins (1985)	Miller et al. (1985) Hurford and Watkins (1987)
Apatite zircon	Fish Canyon volc. tuff	Colorado, U.S.A.	27.8 ± 0.2 [27.9 ± 0.7]	$^{40}Ar/^{39}Ar$ (biotite) Hurford and Hammerschmidt (1985) [K–Ar of 4 mineral phases, Steven et al. (1967)]	Hurford and Gleadow (1977) Naeser et al. (1981) Hurford and Green (1983) Green (1985) Miller et al. (1985)
Zircon	Tardee rhyolite	Northern Ireland, U.K.	58.7 ± 1.1	K–Ar and $^{40}Ar/^{39}Ar$ (sanidine) Hurford et al. (unpubl.)	Hurford and Green (1983) Green (1985) Hurford et al. (unpubl.)
Apatite	Durango martite ore body in Carpintero volcanic group	Cerro de Mercado, Mexico	31.4 ± 0.6	K–Ar (average of Carpintero group) McDowell and Keizer (1977)	Naeser and Fleischer (1975) Hurford and Gleadow (1977) Green (1985)
Apatite sphene zircon	Mount Dromedary intrusive complex	New South Wales, Australia	98.8 ± 0.6	Rb–Sr (biotite) Williams et al. (1982)	Green (1985) Miller et al. (1985)

*Recalculated with IUGS recommended constants (Steiger and Jäger, 1977) where necessary.

Table 3.4.   Examples of fission-track data presentation

Sample reference	Mineral (No. of crystals)	Spont. tracks		Induced tracks		$P(\chi^2)$ (%)	$\overline{\rho_s/\rho_i} \pm 1\sigma$	Glass monitor		Age $\pm 1\sigma$ (Ma)
		ρ_s	N_s	ρ_i	N_i			ρ_M	N_M	
Sample 1	Zircon (15)	56.35	(1263)	32.12	(720)	35		5.683	(3210)	54.8 ± 2.7
Sample 2	Ziron (15)	7.890	(350)	25.54	(1133)	<5	0.312 ± 0.039	0.490	(1125)	0.83 ± 0.11

Sample reference	Mineral and no. of grains (spont./induced)	Spont. tracks		Induced tracks		S_{pi} (%) and (S_i/S_p)	Glass monitor		Neutr. fluence ($\pm 1\sigma$)	Age $\pm 1\sigma$ (Ma)
		ρ_s	N_s	ρ_i	N_i		ρ_M	N_M		
Sample 3	Apatite (200/200)	4.109	(1250)	4.540	1381	3.2 (1.19)	1.049	(2850)	–	31.8 ± 1.6
Sample 4	Apatite (100/150)	12.65	(1924)	8.124	1853	4.1 (1.76)	–	–	2.330 (±0.047)	175 ± 11

All track densities (ρ) are in 10^5 tracks/cm^2, while the numbers of counted tracks (N) are given in parentheses. Samples 1 and 2 were analyzed with the external detector method, using the ζ-calibration with CN1 glass ($\zeta = 110$). The second sample did not pass the χ^2-test, as can be derived from the $P(\chi^2)$ value. Hence, for this sample the mean ρ_s/ρ_i ratio and its 1σ error is quoted. Samples 3 and 4 were analyzed with the grain-population method using, respectively, the ζ-calibration with SRM612 glass ($\zeta = 335$) and the absolute approach (thermal neutron fluence in 10^{15} neutr/cm^2, average of Co and Au monitors, consistent with 3%). For the age calculation of sample 4, the values quoted in Section 3.3 have been used for the relevant nuclear parameters. The relative error S_{pi} of the mean induced track density is given as a measure of dispersion, together with the ratio S_i/S_p (between parentheses) of the experimental standard deviation of the track counts per counting grid to the Poisson standard deviation which gives more direct information on the amount of extra-Poisson variation in the sample.

procedure is used. It has to be kept in mind that factors such as neutron flux inhomogeneities, poor track revelation, determination and/or counting are other possible error sources that are not solved by an age-standard approach. Finally, recommended age standards are only available for glass and the most commonly dated minerals zircon, apatite and sphene. They do not yet exist for the other minerals discussed in Section 6.2.

3.10. Data Presentation

In the past, fission-track age determinations were often reported without technical information or counting data. Such a situation obviously impeded an evaluation of these ages or a comparison between different workers. In order to put an end to this situation, Naeser *et al.* (1979b) designed a model table for the presentation of fission-track data and formulated some items on which information should be given in every fission-track dating report. This information concerns the dating system and dating procedure used together with the values of parameters and constants needed for age calculation. The essential data which figure in the table are the number of crystals or grains analyzed, the number of spontaneous and induced tracks counted, their average densities/cm^2, and the fission-track age and its error.

The model table found wide acceptance by the fission-track community and has been newly recommended by the Fission-track Working Group in 1990 with the addition of some new elements specific to determinations with the ζ-method (Hurford, 1990). Table 3.4 gives some examples of (fictitious) data reported according to the new recommendations. A column is included with information on the statistical dispersion of the counting data. For grain-by-grain methods, the results of the χ^2-test (Section 3.8.2) serves as the information source, the probability $P(\chi^2)$ of finding the calculated χ^2-value being quoted. If it is less than 5%, the average $\overline{\rho_s/\rho_i}$ ratio, which in this case is used for the age calculation, is added. For grain-population methods, the experimental standard error of the mean of induced track counts ($s_{\rho i}$) is quoted. This was already advocated by Naeser *et al.* (1979) but does not allow a straight evaluation of dispersion as $s_{\rho i}$ also depends upon the number of grains counted. A more efficient parameter would be the standard deviation s_i of the track counts (per microscopic unit), as suggested by Hurford (1990), eventually expressed as a ratio to the Poisson standard deviation (Van den haute, 1984).

The 1990 recommendation also includes standard tables intended for newcomers to fission-track dating who wish to report on the calibration of their dating system. We refer to the original paper for further information on this matter.

CHAPTER 4

Fading of Fission Tracks

The unetched or 'latent' fission track represents a zone of intense radiation damage within a regularly built solid and which – like other kinds of radiation damage – consists of an energetically metastable, i.e., ultimately unstable solid state. The originally ordered structure of the solid which was disordered by the passage of a fission fragment, gradually becomes restored with time. This phenomenon is known as *fading*. It is manifested as the reduction of the etchable length and/or etching rate as well as the reduction of the areal density of fission tracks. The fading phenomenon is of fundamental importance for fission-track dating. Track fading violates one of the prerequisites of a radiometric clock, namely the condition, of *closed system* retention of the radiogenic daughter product, which – in the case of the fission-track clock – refers to the stability of the fission tracks. Under geological conditions, fading of fission tracks is a common phenomenon. The resulting loss of fission tracks tends to lower the apparent fission-track age. In that case, the time at which the fission-track clock was turned on and the time that is given by the clock, is no longer the same. The fact that fission-track ages are usually younger than the sample's formation age seems, at first sight, disturbing. On the other hand, apparent fission-track ages lowered by partial or complete track fading are often interesting in themselves, since they give deeper insights into thermal history. Actually, it is the track fading which qualifies the fission-track method as a unique tool for deciphering the thermo-chronology of rocks. The knowledge of the conditions under which fission tracks are stable or begin to fade, forms the vital base for the geological interpretation of fission-track ages. The relationship between geological age and apparent fission-track age will be further discussed in Chapter 5.

Owing to its importance, the phenomenon of track fading has been frequently studied for a large number of terrestrial materials. Two fundamentally different approaches have been followed. The first one involves *laboratory experiments*, in which the fading characteristics of the fission-tracks are studied under controlled conditions such as time, temperature, and pressure. In this approach, one takes samples containing freshly induced fission tracks of known areal density or size, subjects them to defined experimental conditions of track fading, and measures the density or size of the remaining fission tracks again. The degree of fading is usually expressed as the reduction of track density or size normalized to the original value prior to the fading experiment. Thereafter, the experimental data are extrapolated to geological conditions on the basis of kinetic laws. The validity of this approach remains questionable as long as the physical processes which govern track fading are not yet understood in detail. Also, additional parameters that act in nature but are omitted in the experiment, may influence the fading behaviour of the tracks.

Therefore, the second approach of observing the track stability under *geological conditions* is preferable. This can be achieved by studying tracks in samples collected from documented geological environments with still-active fading *in-situ*, particularly in deep drill-holes, or in samples of independently and well-known geological history. By relating their fission-track systems to the known age, nature, and intensity of a geological event, one might be able to assess the track-fading characteristics. These different methods of investigating the fading properties have all been applied and, for some materials, especially apatite, their results satisfactorily agree with each other. However, for other materials, such as zircon, sphene, and epidote, the experimentally derived track-stability data seem to be systematically different from the actually observed ones. Hence, although for most of the terrestrial materials which are currently used in fission-track dating, the fading behaviour is, to some extent, known, much work on this subject still needs to be done.

4.1. Causes of Track Fading

In principle, several geological parameters are capable of influencing the stability of latent fission tracks in solids. The most obvious candidates are time, temperature, hydrostatic as well as shock wave pressures, intergranular solutions, and ionizing radiation. In the experimental studies by Fleischer *et al.* (1964b, 1965b) all these parameters have been verified as potential causes affecting track stability and it appeared that temperature is by far the most influential one. In this section, these effects are discussed in some detail. The thermal fading – the *annealing* – of fission tracks is treated separately.

During their geological past, many terrestrial materials have experienced considerable *hydrostatic rock pressures* up to several kb at some depth within the Earth's crust. In order to investigate the reaction of the tracks to such conditions, Fleischer *et al.* (1965b) exposed samples of tektite glass, olivine, and zircon to high hydrostatic pressures up to 80 kb at various temperatures. The pressure was applied at a rate of about 1 kb/s. After the experiment, the samples were cleaved, etched, and the areal track density was counted. Pressures of 80 kb showed only a slight effect on the track annealing behaviour in zircon and had no measurable effect on olivine, i.e., no or little effect in addition to the annealing already observed at atmospheric pressure. Similarly, heating (250 to 350°C) under hydrostatic pressure (2 kb) was found to have no detectable effect on the annealing properties of fission tracks in apatite (Naeser and Faul, 1969). In tektite glass, a pressure of 60 kb, considerably accelerates the track annealing (Fleischer *et al.*, 1965b) – a result which is geologically irrelevant, since during their depositional history, tektites were not subjected to any noteworthy hydrostatic pressures. In

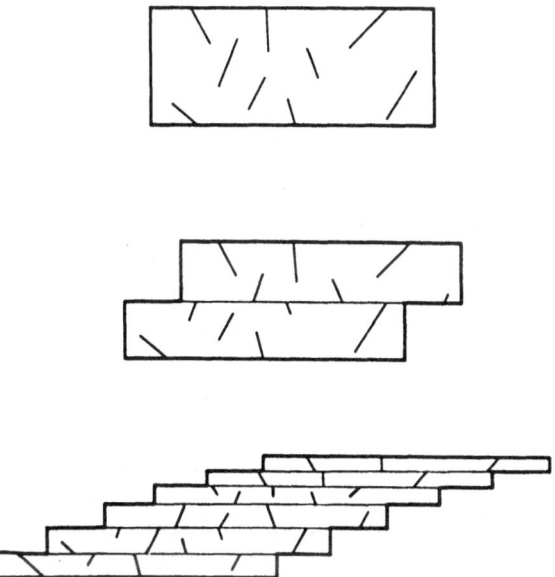

Figure 4.1. Plastic deformation breaks up fission tracks along slip planes. Such a mechanism shortens the tracks but does not reduce the areal track density. However, after etching, fragmented fission tracks with very short lengths may pass unnoticed under the optical microscope. (From Fleischer and Hart, 1972.)

muscovite, the fission tracks were reported to become increasingly unstable with increasing hydrostatic pressure between 0.25 and 2.5 kb (Lakatos and Miller, 1970).

Another possible mechanism of fission-track erasure is *plastic deformation*. Mechanical shear deformation of minerals is geologically conceivable, either as slow movement during tectonism or as shock wave deformation during meteoritic impact. After applying experimental shear stress to mica crystals and appropriate etching, Fleischer *et al.* (1965b) observed that the tracks were no longer continuous but disconnected by a slip movement into short subsegments (Figure 4.1). Although this mechanism does not qualify as track-fading *sensu stricto*, it effectively shortens the preexisting fission tracks.

The effect of experimental *shock-wave* pressure on the erasure of fission tracks was investigated by Ahrens *et al.* (1970). In tektite glass, shock pressures above 133 kb completely erased the fission tracks. On the other hand, shock pressures of 173 and 45 kb did not affect the fission tracks in biotite and apatite, respectively. Rocks exposed to underground nuclear explosions were also studied in this respect. For apatite, a decrease of fission-track density by 25% in the pressure range of 11–30 kb and by 96% at 100 kb was observed. In sphene, at least 90% of the tracks were lost at 100 kb. No fission-tracks were left in both minerals after exposure to 350–400 kb shock-wave pressure (Fleischer *et al.*, 1974). Although fission tracks do undoubtedly fade under the rigorous conditions of shock waves, the precise fading mechanism is not yet fully understood. Among other processes, thermal annealing at high peak temperatures and diaplectic deformation of the crystal structure may also be involved. If the tracks are sufficiently shortened in order to become unetchable, the fission-track clock is effectively reset and, thus, geological events involving shear stress might be datable (Fleischer *et al.*, 1972).

Rocks and minerals – in particular those with high uranium contents – are steadily exposed to a background level of *ionizing radiation* which originates from natural radioactive decay and, to a minor part, from cosmic rays. Although alpha and beta particles as well as photons themselves do not leave etchable tracks in minerals, they accumulate energy dose which may influence the stability of fission-tracks. In order to investigate this possible effect, Fleischer *et al.* (1965b) irradiated olivine, various micas, and tektite glass with electrons of 1.5 MeV energy and 10^8 R dose which is approximately equivalent to the natural dose that accumulates in a sample of 10 μg/g uranium content during 3×10^8 years. In no case was any effect on the fission-track stability observed. However, for sphene, Gleadow (1978a,b) and Watt and Durrani (1985) noted that the annealing behaviour of fossil fission tracks is different from that of induced ones and attributed this to the effect of radiation damage by α-particles and α-recoil nuclei.

Kasuya and Naeser (1988) found that fossil fission tracks in zircons with track densities ranging from 10^6 to 10^7 tracks cm^{-2} shorten significantly faster during annealing than induced tracks. They concluded that the accumulated α-dose is responsible for the lowered thermal stability of the fossil fission tracks. Since α-induced radiation damage may influence both the track retention as well as the etching behaviour, some changes in track appearance and density that were described for annealed samples, may be due to the latter effect only. Tagami *et al.* (1990) observed, for spontaneous tracks in zircon, an increase of etch-anisotropy with annealing that is caused by the erasure of the α-induced radiation damage before the track fading fully sets in. The presence of α-radiation dosage may explain why, for zircon, but also for sphene and epidote, the annealing experiments (which were always carried out on induced tracks in samples without α-damage) yield consistently higher stability than assessments from naturally annealed samples (fossil tracks in samples with more or less α-damage). These findings caution the thoughtless use of experimentally derived track-annealing data in geological application.

4.2. Track Annealing under Experimental Conditions

4.2.1. *Annealing experiments*

Temperature is by far the most dominant parameter that influences the stability of fission tracks in minerals. This has been established by Fleischer *et al.* (1965b) when investigating the effect of high temperatures on tracks in tektite, zircon, olivine, and mica in comparison to pressure and ionizing radiation. In the course of this and earlier annealing experiments (Fleischer *et al.*, 1964b; Maurette *et al.*, 1964), the tracks were observed to become gradually less numerous and shortened with increasing temperature and duration of annealing (Figure 4.2). The term *annealing* refers to the effects of both, *temperature* and *time* on the track stability, with other factors of influence kept fixed.

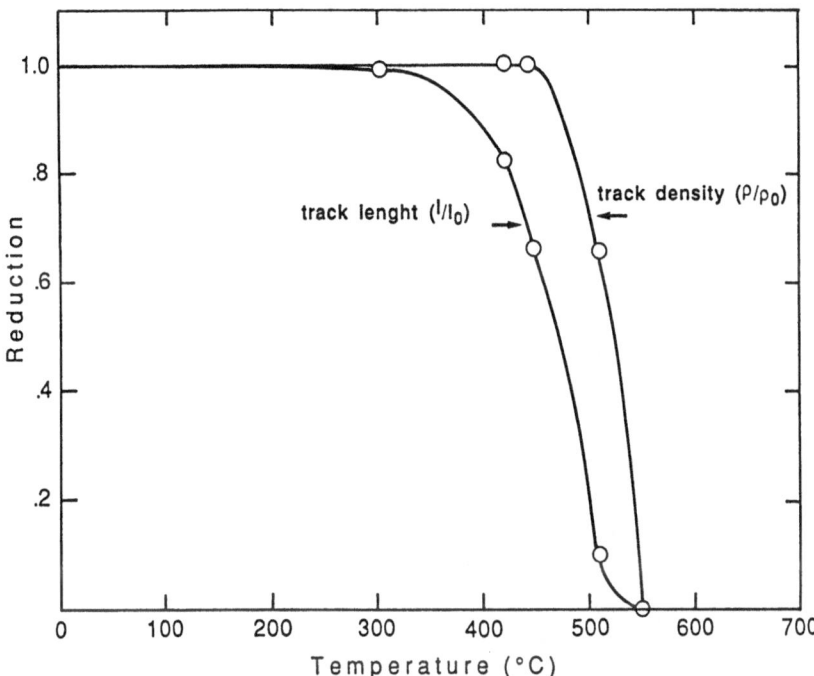

Figure 4.2. Isochronal annealing of induced fission tracks in muscovite for one hour at various temperatures. The curves show the reductions of average length and the areal density of the partially annealed fission tracks normalized to their original values without heat-treatment. (After Fleischer *et al.*, 1964b.)

In the meantime, numerous experimental studies on the annealing properties of fission tracks have been carried out in various terrestrial materials. Usually aliquots of the samples with freshly-induced, latent fission-tracks are exposed to various temperatures (up to several hundred °C) and various durations (up to about one year) and then etched. The annealing experiments are usually designed to be either isochronal with fixed annealing time but different temperatures or isothermal with fixed temperature but different durations. Continuously variable temperature annealing experiments have been reported by Duddy *et al.* (1988). The degree of annealing is commonly expressed by the reduction of track density ρ or size l – normalized to the original value ρ_0 or l_0 before annealing – under various time and temperature conditions. Typically, in the isochronal annealing modus, the original fission-track density (and size) at first remains unchanged until a certain temperature is reached and then, with further temperature increase, is gradually reduced to zero. An example of an isochronal annealing experiment is shown in Figure 4.2.

Annealing experiments have been performed on virtually all materials which have been investigated for fission-track dating. These numerous experiments, their results and extrapolation to a geological timescale are compiled in Appendix B. However, despite the impressive record of experimental annealing data in a wide variety of mineral and glasses, it seems important to emphasize that still relatively little is known about the physical nature of the annealing process itself.

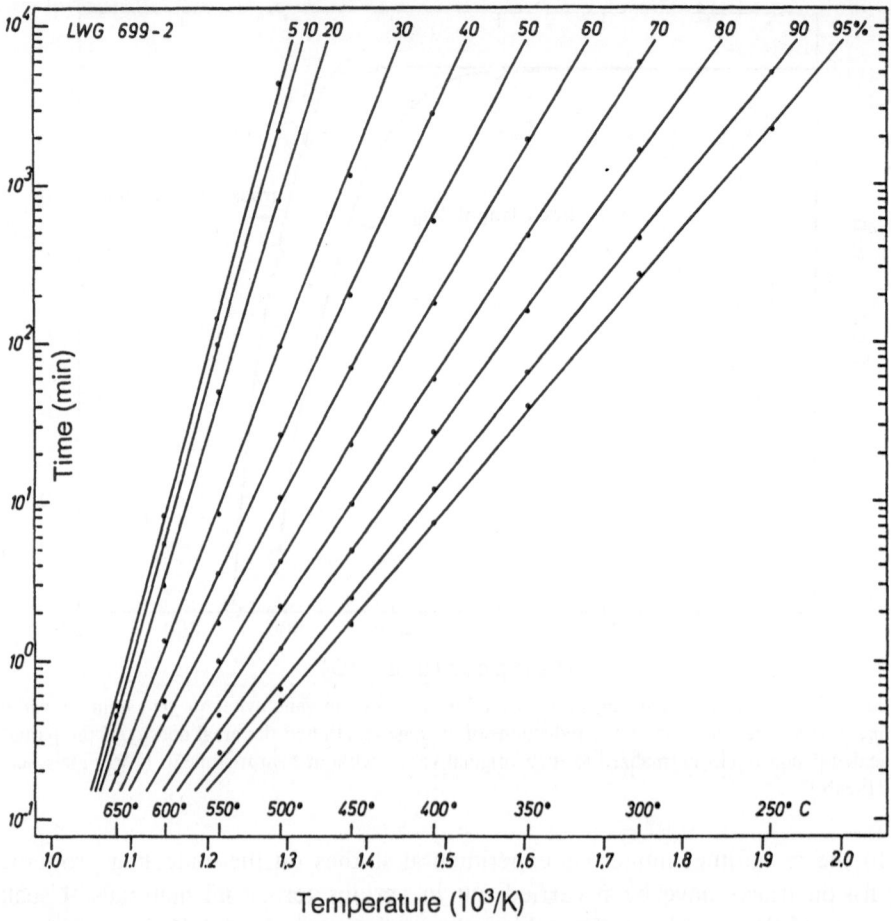

Figure 4.3. Arrhenius diagram of fission-track annealing experiments on Libyan Desert Glass. In the Arrhenius diagram, the annealing time is logarithmically plotted against the inverse temperature. All points with the same degree of track density reduction (given in %) lie on straight lines. With, increasing temperature and/or time of annealing the fission tracks become progressively erased. (After Wagner, 1972a.)

4.2.2. *Arrhenius diagram*

The experimental annealing data are traditionally presented in so-called Arrhenius plots, which are diagrams with the logarithmic annealing time t plotted against the inverse absolute temperature T. As first noticed by Fleischer *et al.* (1965b) for various materials and later confirmed by many other experimenters, the T-t-conditions of equal track-density reduction ρ/ρ_0 form straight lines in such plots (Figure 4.3) The validity of this linear relationship between logarithmic time and inverse temperature was experimentally demonstrated over time spans covering six orders of magnitude. Relying on this linearity, the geologically relevant temperature conditions for track annealing are derived simply by extrapolating the straight lines to geological durations, say to 10^6 years, and reading off the corresponding temperatures (Figure 4.4). The comparability of such experimentally derived and,

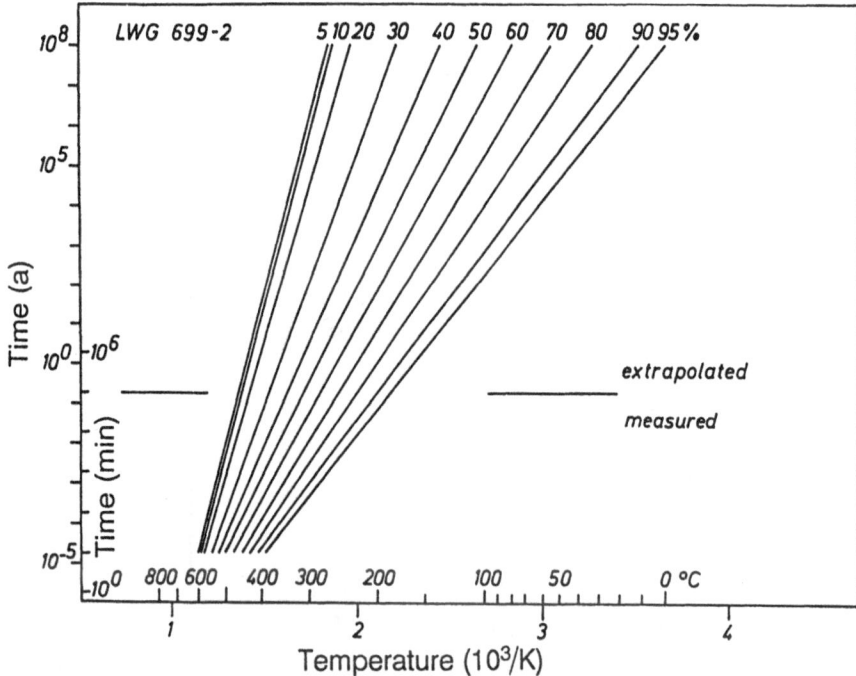

Figure 4.4. Arrhenius diagram with annealing fan of fission tracks in Libyan Desert Glass (same sample as previous figure). The experimental data are extrapolated by many orders of magnitude to geological conditions. (After Storzer and Wagner, 1971.)

subsequently, extrapolated annealing data with those directly observed under geological *in-situ* conditions will be discussed later (Section 4.3).

When the annealing data are plotted in Arrhenius diagrams, usually fan-like arrangements – such as in Figure 4.3 – of the lines marking equal track-density reduction are obtained. The slopes of these lines become steeper from the line for 100% fission-track retention to that for total track loss, with the slopes varying typically by a factor of two or three from complete retention to total erasure. The space enclosed within the fan denotes the *T–t*-conditions of partial track stability. The areas right and left outside the fan, cover the *T–t*-conditions of complete retention and erasure, respectively. The extrapolation of the diverging fan-lines from experimental annealing durations to geological times – often by six to eight orders of magnitude – results in a wide geological temperature range of partial track stability, which may even amount to a few hundred °C. However, high precision measurements of confined track lengths in annealed apatite (Durango/Mexico) by Green *et al.* (1985) imply a near-parallelism of lines for various degrees of annealing in terms of length reduction – rather than track density reduction – in the Arrhenius diagram that extrapolates to a narrower geological temperature range for partial track annealing (Figure 4.5). Since the widely fanning Arrhenius plots were derived from apatite concentrates – instead of from a single crystal with homogeneous chemical composition – Green *et al.* (1985) explained them as a kind of artifact, namely the result of superposition of a series of near parallel Arrhenius diagrams for apatites with different chemical composition and, thus,

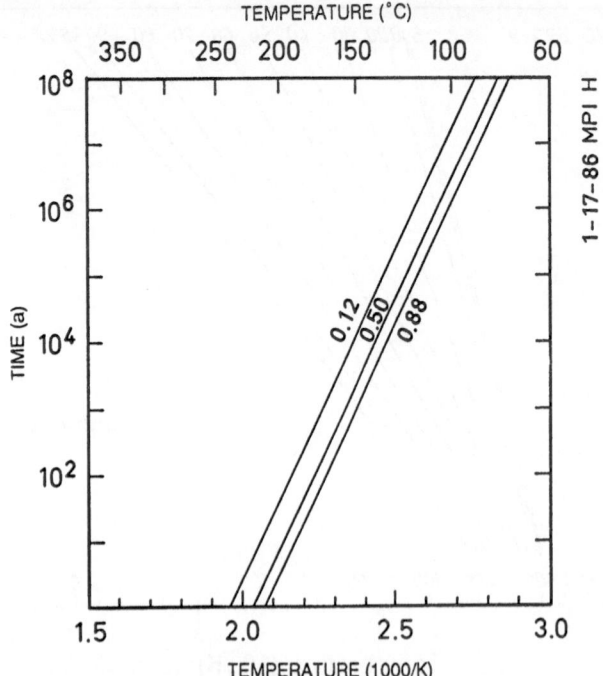

Figure 4.5. Arrhenius diagram with parallel lines of equal degrees of track-density reduction during annealing in monocompositional apatite, Sljudjanka, Siberia. (After Wagner, 1986.)

different track-annealing properties. This conclusion may also apply to annealing fans of other minerals which exhibit compositional variation.

4.2.3. *Lengths of annealed fission tracks*

The phenomenon of track size (i.e., length or diameter) reduction during annealing, is of fundamental importance in fission-track dating, because it forms the basis for the reconstruction of the sample's T–t-path from its fission-track record. As early as the very first experiments of track annealing on mica (Figure 4.2) and glass, it was noted that annealing not only results in track loss, but also in track shrinkage (Fleischer *et al.*, 1964b; Fleischer and Price, 1964c; Maurette *et al.*, 1964). In the meantime, it has been demonstrated by numerous experimental studies on various track recording materials that size reduction of fission tracks is a general phenomenon. As a consequence, the degree of fission-track annealing can also be described in terms of mean length instead of track-density reduction and lines of equal mean-length reduction can be plotted in Arrhenius diagrams (Figure 4.6).

With regard to the length of a fission track, it has already been stated in Section 2.1 that only a certain fraction of the latent fission tracks, called *the etchable length* l, is revealed by the etchant ($l < 2R$ = total combined range of the two fission fragments). The etchable length varies with annealing. In minerals, l is usually

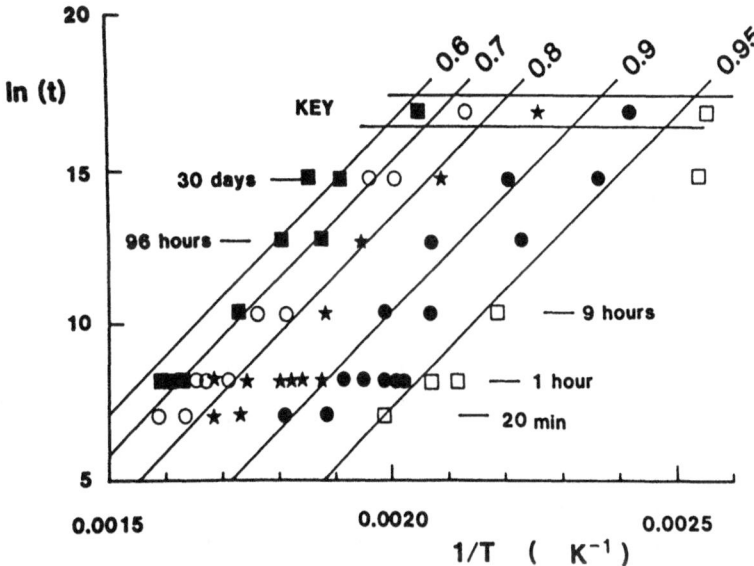

Figure 4.6. Arrhenius diagram with parallel lines of equal degrees of track length reduction during annealing in monocompositional apatite, Durango, Mexico. (After Green *et al.*, 1985.)

measured directly, whereas in glass and other materials, which are characterized by relatively large bulk etching rates with respect to the etching rate along the track, the *diameter* of the etch pit is used in order to detect the degree of annealing. The track length or diameter measurements are usually, presented as histograms which are further specified by their mean value and standard deviation (Figure 4.7).

   Three different alternatives exist for measuring track lengths in minerals: the projected length of surface tracks, the true length of surface tracks, and the full length of confined tracks. These types of length measurements show marked differences in their distribution (Section 2.3.4). For simplicity, let us consider the distribution of freshly induced fission tracks of equal etchable length l_0. The *projected lengths* of tracks intersecting an internal plane surface possess a triangular length distribution which decreases to zero at the etchable track range l_0 (Figure 2.23e). In practice, however, one also observes a fall-off at the small end below about 2 μm, since it is difficult to identify and to measure the short tracks. The mean projected track length is $l_0/3$, but is expected to be somewhat larger owing to the bias against the short tracks. For obtaining the *true length of surface tracks*, in addition to the projected length, one measures the vertical component of the track by using the fine-focus and combines both components by the Pythagoras theorem whereby the different refractive indices need to be taken into account. A rectangular distribution between zero and l_0 results with the mean length $l_0/2$, again with some fall-off at the short end (Figure 2.23d). Although both of these length-distribution types based on surface tracks are – for obvious geometric reasons – heavily biased, they have the advantage that the lengths are measured on the same tracks that are used for age determination. This allows fast, precise

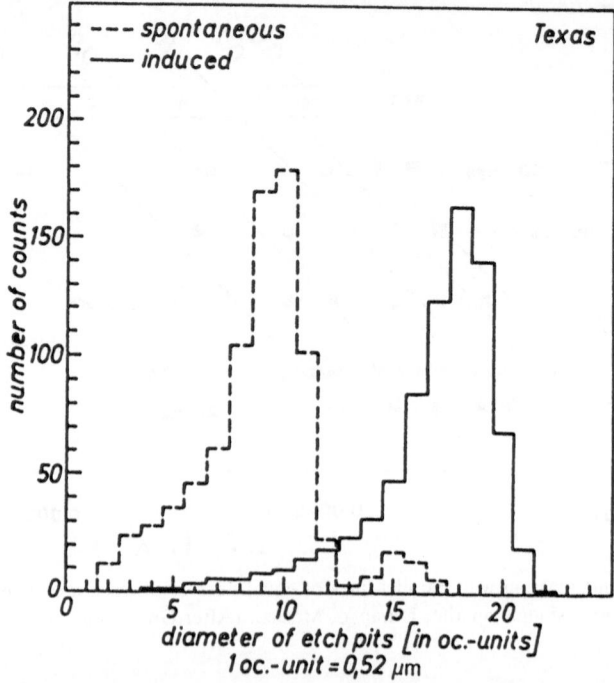

Figure 4.7. Histogram of the distribution of the etch-pit diameters (major axis) of spontaneous and induced fission tracks in tektite glass. (From Storzer and Wagner, 1971.)

measurements with image analyzing systems in the same operation as counting. The *confined lengths* (or more precisely, 'true lengths of confined tracks') are measured on tracks which are totally confined within the crystal and etched through host tracks or cleavage planes. Only tracks parallel to the horizontal surface are selected. Their length distribution might be expected to directly represent the etchable track length l_0 **(Figure 2.23b).** However, it must not be overlooked that, even for the confined tracks, the length distribution is strongly biased against short tracks (Green, 1988).

Since each fission fragment causes the highest ionization at the beginning of its path, the density of defects along a fission track decreases from the central part towards both of its ends. As annealing reduces the defect-concentration, an increasing fraction of the radiation-damaged channel becomes unetchable. The defect-concentration falls below the critical threshold which is required for etchability first at both ends of the fission-track and, thus, the etchable track length shrinks from both ends, while, in addition, the etching rate along the track is reduced.

The phenomenon of fission-track shrinking from both ends has been demonstrated for Durango apatite by Green *et al.* (1986). According to their experiments, the annealing process of fission tracks in apatite seems to consist of two steps. Shrinking of the etchable fraction from each end, whereby the rate of shortening increases with increasing angle to the crystallographic *c*-axis, is the dominant annealing process in apatite, i.e., for a given degree of annealing, there is a maximum etchable length. At high degrees of annealing, an additional process

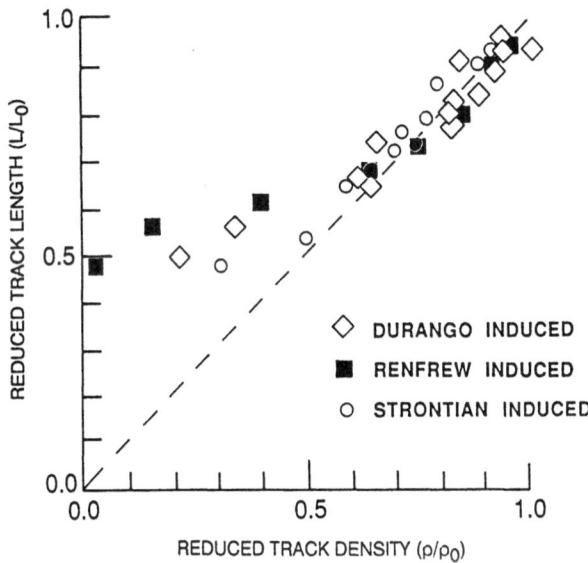

Figure 4.8. Relationship between reduction of confined length l/l_0 and reduction of areal density ρ/ρ_0 for partially annealed fission tracks in three monocompositional apatites. (After Green, 1988.)

sets in. The tracks become broken by unetchable gaps into separate segments – a phenomenon which was first observed in mica (Geguzin *et al.*, 1968; Dartyge *et al.*, 1981). At present, it is not yet clear if these two mechanisms of fission-track annealing are a general phenomenon.

As a consequence of the shortening of the etchable length of fission tracks, the areal track density is reduced. Actually, the track-density reduction observed during annealing is only a secondary effect of the primary track-length reduction. Because the areal track density (cm^{-2}) is equal to the number of tracks intersecting the surface and, for a fixed number of latent fission tracks in a unit volume (cm^{-3}), this number is proportional to the etchable length, a reduction in etchable length causes a reduction in track density (Section 1.4.3). Therefore, one should expect a one-to-one relationship between the reduction of etchable track length and that of track density. Since confined fission-track length distributions represent most closely the true etchable length distribution, the 1:1 relationship should appear when the reduction of mean confined lengths l/l_0 is plotted against the reduction of areal density ρ/ρ_0. Such relationships have been observed for apatite by Green (1988), although only for relatively weak degrees of annealing characterized by $0.7 < l/l_0 < 1$. For stronger annealed apatites, the curve diverges from the 1:1 relationship and l/l_0 becomes larger than ρ/ρ_0 (Figure 4.8). Green (1988) understands this divergence qualitatively in terms of the anisotropy of track annealing in apatite and bias against the revelation of short confined tracks. In an analogous manner, Tagami *et al.* (1990) discuss the length-vs.-density relationship of annealed fission tracks in zircon. For projected fission-track lengths, as well as track etch pit diameters, of course, one does not expect linear one-to-one relationships between track density and size (length or diameter) reductions (Van den haute, 1985). Such curves have been measured for a large number of glasses (Figure

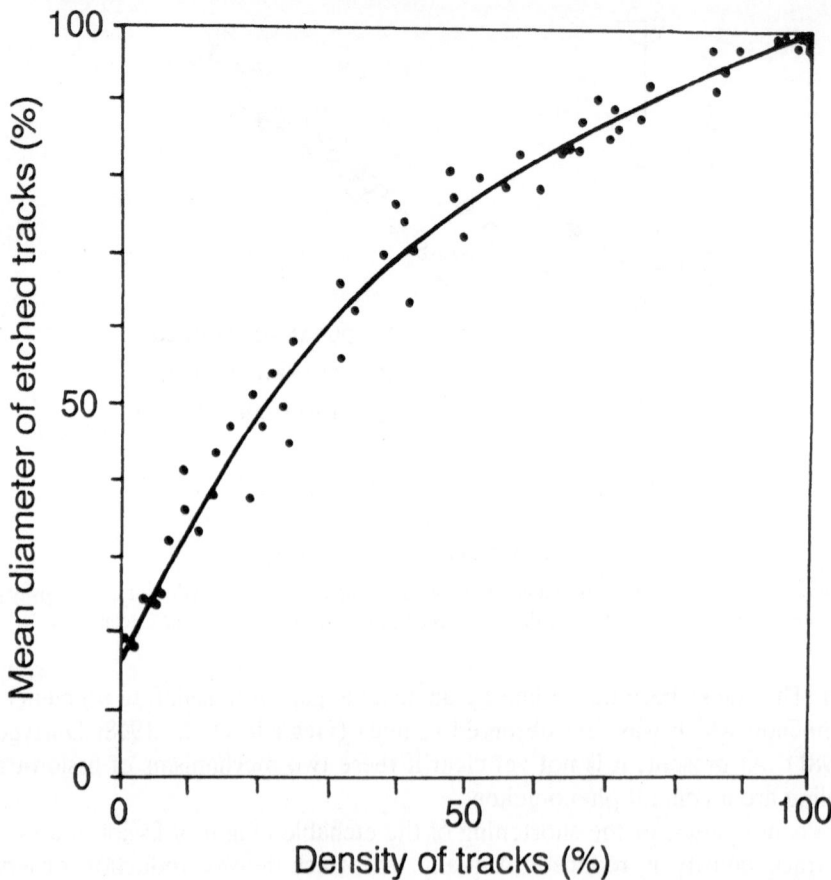

Figure 4.9. Relationship between reduction of mean etch-pit-diameter (major axis) d/d_0 and reduction of areal density ρ/ρ_0 for partially-annealed fission tracks in australite glass. (After Storzer and Wagner, 1969.)

4.9) and minerals. The shape of the curves depends, to some extent, on the etching conditions and the crystallographic orientation of the etched surface. They are useful when comparing the size of the spontaneous fission tracks with that of the induced ones in order to determine the degree of fading the spontaneous fission tracks have experienced during the geological past. They are used to correct partial track-loss due to fading and are, therefore, commonly called *correction curves* (Storzer and Wagner, 1969). They turned out to be particularly successful for correcting lowered fission-track ages of natural glasses (Section 5.7).

For apatite – even for specimens which spent all their geological past at low surface temperatures – there is ample evidence that fossil fission tracks are, on average, somewhat smaller than the induced ones (see, e.g., Green, 1988). This may also be the case for other minerals, particularly micas (Bigazzi, 1967; Green, 1980). Since the ranges of spontaneous and induced fission fragments are practically equal, one has to conclude that some annealing occurs at low ambient temperatures. This might introduce the need to correct measured fission-track ages. In the light of this conclusion, it is appropriate to mention the recent observation by Donelick *et al.* (1990) that the mean confined lengths of freshly induced fission tracks in various apatites show an almost uniform reduction, by 0.5 μm at 23°C, within several weeks after irradiation. The mechanism of this annealing is not yet understood and corresponding investigations on other minerals are highly desirable. As a consequence, mean 'initial' confined fission-track lengths, which are used as reference for annealed tracks, should be measured – at least in case of apatite – on tracks that have been aged for at least one month and that have not been heated above room temperature before or during chemical etching (Donelick *et al.*, 1990).

4.2.4. *Factors of influence on annealing*

In addition to temperature, there are other factors which may influence the annealing rate of fission tracks, notably the track's direction with respect to the crystal lattice and the sample's chemical composition. Also, the influence of intergranular hydrothermal solutions on the fission-track stability in apatite was experimentally investigated by adding NaCl in various concentrations at various annealing temperatures and some accelerating effect over simple dry-air annealing was reported for that particular setting (Burchart and Reimer, 1972; Reimer, 1972). The effects of hydrostatic rock pressure and bulk radiation damage in the mineral have already been discussed (Section 4.1).

The effect of orientation of a fission track with respect to the crystal lattice on its annealing properties, was first demonstrated for mica (Geguzin *et al.*, 1968). In muscovite, tracks with shallow dip-angles with respect to the cleavage plane, are more resistant to annealing than tracks with steep angles. In apatite, fission tracks parallel to the *c*-axis are more resistant against annealing than tracks perpendicular to the *c*-axis (Green and Durrani, 1977; Donelick, 1991) and, as the annealing proceeds, this anisotropy becomes more pronounced. Therefore, fission tracks on prismatic faces of strongly annealed apatites tend to appear parallel to the *c*-axis. Effects of anisotropy can be understood in terms of varying activation energy with varying crystallographic direction and are expected to be a general phenomenon. More experimental studies on this subject are certainly desirable. As a conseqence of this effect, the crystallographic orientations of annealed fission tracks and the crystallographic faces on which surface tracks are revealed need to be stated explicitly in reports.

The fission-track annealing rate also depends on the sample's *chemical composition*. This has been observed for glasses, where fission tracks become more resistant with increasing water (Lakatos and Miller, 1972) and silica content (compare Appendix B), and for apatite. For the latter mineral, the annealing properties are sensitive, particularly to the Cl/F ratio (Gleadow and Duddy, 1981; Green *et al.*, 1985, 1986). The apatite composition $Ca_5(PO_4)_3(F, Cl, OH)$ is predominantly fluorapatite with varying amounts of chlor- and hydroxyapatite. Electron microprobe studies show that tracks in apatite, which is rich in Cl, are more resistant to annealing than tracks in fluorapatite (Figure 4.10). For rocks with varying apatite chemistry, this effect may require single-grain dating techniques with ac-

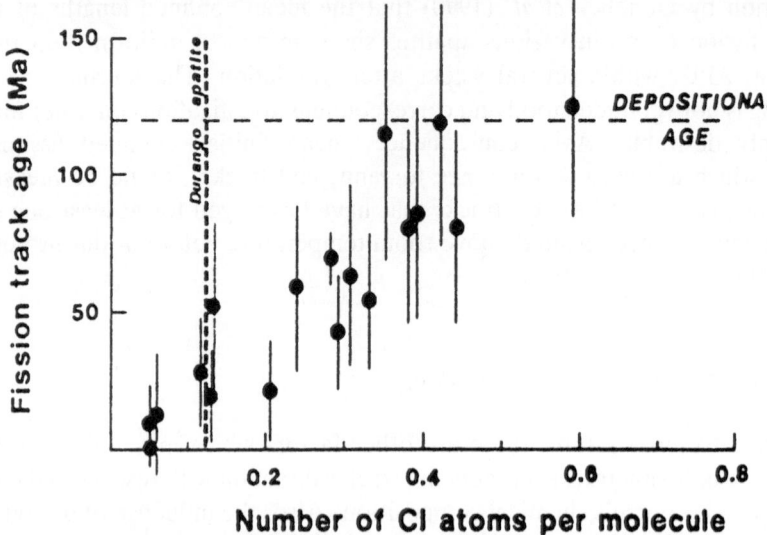

Figure 4.10. Relationship between track, retention – indicated by the apparent fission-track age – and the chlorine content of individual apatite grains from an Otway Group sandstone bore-hole sample (2595 m deep at 92°C). The composition of Durango apatite is shown for reference. (From Green *et al.*, 1986.)

companying chemical control. Studies as to this effect in other minerals are still lacking, but it seems reasonable to assume that similar effects might exist.

The etchable length of partially annealed tracks and, thus, the areal density change to some extent, with the etching conditions (Section 2.3.3). Calk and Naeser (1973) recognized that fission tracks in sphene seem to be more resistant to annealing when etched in NaOH at 130°C for 20–30 min instead of HCl at 90°C for 45 min. Similarly, Mehta and Rama (1969) were able to retrieve larger fractions of partially faded fission tracks in muscovite when they etched the mica for 3 h instead of 0.5 h in 40% HF at 25°C. Because the identification of the strongly faded, very short tracks depends on the resolution quality of the microscope and on personal criteria, even the conditions of track observation play a role. This phenomenon can be understood in terms of different levels of etchability of a given radiation damage for different chemical etchants and requires carefully defined and controlled etching conditions.

4.2.5. *Annealing kinetics*

A quantitative understanding of the track-annealing phenomenon would allow accurate prediction of the annealing under various conditions of temperature and duration of heating and, thus, would form a solid base for geological interpretation of natural fission-track systems. One approach towards understanding the quantitative behaviour of fission-track annealing in minerals is reaction kinetics.

It remains clear from Chapter 1 that, whatever the detailed nature of a latent fission-track within a solid may be, the track represents a channel-like zone of intense radiation damage. Along the fission fragment's trail, the original atomic order is strongly disordered involving a high density of displaced atoms and electrons trapped as lattice defects. Depending on their activation energy, these metastable defects may survive geological times. It is this disordered zone which leads – owing to its lowered binding energy – to etchable tracks. Since annealing results in a reduction of the etchable track length and the etching velocity along the track, it seems plausible that, during annealing, the ordered lattice structure is gradually restored. The heating supplies sufficient external energy that enables the displaced atoms and electrons to become thermally activated, to leave their trap sites, and to return by diffusive motion to their appropriate lattice sites.

Diffusion can be mathematically treated in terms of reaction kinetics. If only one kind of lattice defect exists with a specific activation energy E, the diffusion process can be described by first-order kinetics. In its general form *first-order kinetics* defines the reaction rate dn/dt of some reacting species according to

$$dn/dt = -\alpha n \tag{4.1}$$

as a function of the first order of the reagent's concentration n. The reaction probability factor $\alpha(t^{-1})$ is temperature-dependent according to

$$\alpha = \alpha_0\, e^{(-E/kT)}, \tag{4.2}$$

whereby $\alpha_0(t^{-1})$ is a specific frequency factor, $E(eV)$ is the (constant) activation energy, $T(K)$ is the absolute temperature, and k Boltzmann's constant ($= 8.616 \times 10^{-5}$ eV/K). For constant temperature T, Equation (4.1) integrates to the function

$$n = n_0\, e^{(-\alpha t)}, \tag{4.3}$$

which describes a simple exponential decay in the reagent's concentration n with time t. Equations (4.2) and (4.3) can be combined and transformed to

$$\ln t = E/(kT) + \ln[-\ln(n/n_0)] - \ln \alpha_0 \tag{4.4}$$

which, in the Arrhenius diagram, describes a straight line for a given reaction-degree n/n_0, i.e., the lines for different n/n_0 would be parallel to each other. The slope of the lines is proportional to the activation energy E.

Since its first experimental observation by Fleischer and Price (1964c), the behaviour that equal degrees of track annealing ρ/ρ_0 plot on straight lines in the Arrhenius diagram (Figure 4.3) was confirmed by many workers. The annealing process, expressed in terms of track density reduction, is then described according to Equation (4.3) by

$$\rho/\rho_0 = e^{(-\alpha t)}, \tag{4.5}$$

again with α as the temperature-dependent probability factor, as in Equation (4.2). The fan-like arrangement of the straight lines that was commonly obtained in annealing experiments, implies that the annealing process may not be controlled by a single activation energy. With increasing degrees of annealing, the activation energy seems to increase because the slope of the equal track-retention lines

becomes steeper (Storzer and Wagner, 1969) – a characteristic which is difficult
to reconcile with first-order kinetics.

However, in order to fit the laboratory-annealing data, linear and logarithmic
mathematical formulations, instead of the exponential decay of ρ/ρ_0 during isothermal annealing, have also been developed (Burchart et al., 1979; Dakowski et al.,
1974; Märk et al., 1972). These and other early formulations – essentially based
on apatite-annealing data – did not lead to a generally accepted quantitative model
on the annealing behaviour of fission tracks and left a confusing trail in the
literature.

The more recent experimental studies on fluorapatite (Durango, Mexico) by
Green et al. (1985) and Laslett et al. (1987) were focused on track length instead
of track density and revealed a rather complex time-dependent annealing function
which describes 96.7% of the variation within the experimental data set

$$\ln(l - l/l_0) = a \ln(t) - b - c/T, \tag{4.6}$$

where l/l_0 (with $0.65 < l/l_0 < 1$) is the reduction degree of the mean confined track
length and a ($= 0.22$), b ($= 3.87$), and c ($= 4220$) are all constants. Plotted in an
Arrhenius diagram, the function results in straight lines for equal l/l_0 in parallel
arrangement, indicating a single activation energy of 1.66 eV. Actually, a modified
model by Laslett et al. (1987) which allows for some variation of the activation
energy with annealing degree, gives an even better data fit than Equation (4.6).
Therefore, Green et al. (1988) conclude that the slightly fanning model is more
realistic than a parallel one. Equation (4.6) reveals a higher-order kinetic process
which is more clearly seen when it is transformed into the rate equation

$$d(l/l_0)/dt = \alpha (l - l/l_0)^n, \tag{4.7}$$

with α as in Equation (4.2) and $n = (1 - 1/\alpha)$, i.e., in the case of Durango apatite
(Green et al., 1985) the reaction order would appear to be between -3 and -4
(Green et al., 1988). The high-order models that Laslett et al. (1987) consider to
be more pertinent to fission-track annealing in apatite, ascribe far more importance
to temperature, compared to time, than in first-order kinetics and, thus, enhance
the dominant role that temperature plays in fission-track annealing under geological conditions.

The above formalism was developed on the basis of isothermal-annealing experiments on fission tracks in apatite. Since, in nature, one deals with variable temperature situations, one has to be sure before any geological prediction is made that
the formalism – apart from the problem of its extrapolation to geological timescales – can also be adapted to temperatures which vary with time. Using variable
temperature-annealing experiments and the *principle of equivalent time*, Duddy et
al. (1988) were able to demonstrate that this assumption is valid, at least for
apatite. The *equivalent time* for a certain temperature is the duration that would
be required to anneal fission tracks to the same degree already obtained at a
different temperature. According to Duddy et al. (1988), "this assumes that at
any moment, a track which has been annealed to a certain degree . . . behaves
during further annealing in a manner which is independent of the conditions which
caused the prior annealing, but which depends only on the degree of annealing

that has occurred, and the prevailing conditions of temperature and time" and that extrapolations to geological conditions "can only be made when the variable temperature behaviour has been defined".

In order to check if there is really hard experimental evidence for first-order reaction kinetics of fission-track annealing, Green *et al.* (1988) critically surveyed the literature on laboratory track-annealing for apatite, a mineral for which the largest body of experimental data are available. They found that most annealing experiments suffered from flaws in design as well as precision and concluded that there is no convincing evidence for first-order kinetics at all. Considering the processes which may take place during track-annealing, there is actually no reason why the changes of either the track density or the track length should obey first-order kinetics. First of all, the theory of reaction kinetics refers to the concentration of defects and not to the track density or length. Each latent fission track consists of many single defects with different activation energies and different densities created along the path of fission fragments in a multi-component atomic solid. Undoubtedly, the gradual annealing of each single track must be physically a rather complex process. Furthermore, the visible length and the areal density of etched tracks are related by a complex process of chemical etching to the partially annealed radiation damage along the latent tracks. Consequently, rather complex models seem more plausible than simple first-order kinetics for describing the behaviour of fission tracks during annealing (Green and Duddy, 1989).

In the future, it is desirable to achieve more accurate physical models on track-annealing kinetics that allow the quantitative extrapolation of the annealing characteristics from laboratory to geological time and temperature conditions (Carlson, 1990). As pointed out by Green *et al.* (1988), progress in this direction can only be made if better laboratory annealing data become available. For this purpose, it is necessary to have well-defined and designed experimental conditions in order to create a large body of high-precision data stretching over temperature and time ranges as widely as possible. Annealing experiments must be carried out on appropriate samples (known and homogenous chemical composition, defined crystallographic orientation) containing induced fission tracks only, rather than fossil fission tracks which have already been more or less affected during their geological history. The degree of annealing (either as reduction of track density, or mean confined track length or track diameter) should be measured with sufficient precision and should include the quotation of the measuring error.

4.3. Track Stability under Natural Conditions

Once the fission-track annealing has been satisfactorily modelled for laboratory conditions, one must seek valid geological constraints on the annealing behaviour of fission tracks in order to test the viability of the extrapolations. As will be explained in this section, this has been successfully achieved for apatite and, thus, the fission-track systematics of this mineral can be applied with reasonable confidence in order to reconstruct unknown thermal histories of rocks. It is, of course, also the aim to develop other fission-track systems, particularly those of zircon and sphene.

Geological settings which allow the 'testing' or 'calibration' of the stability of fission-track systems under geological conditions, must fulfill certain requirements, namely the nature and the T–t-path of their thermal history must be independently known. Such geological situations of well-dated thermotectonic evolution, are rarely found. One type of geological setting commonly used for assessing fission-track annealing behaviour are the cores from deep drills with *in-situ track annealing* for which at least the present temperature is known. But, as rightly pointed out by Harrison (1985), this is not sufficient because the present fission-track systems have a memory with regard to the setting's thermal history, such as thermal stability, slow cooling, slow heating, or recent episodic heating. For instance, recent episodic heating would indicate apparently higher geological retention temperatures than a regime of thermal stability. Another type of geological setting that is used to assess track retention properties, is characterized by a known *thermal history* which caused the fission-track annealing. If one knows the precise T–t-path that a rock has followed during its geological past, as in the case of steady cooling at constant rate, one may directly assign to the measured fission-track age a temperature value at which the fission tracks are effectively retained. The first approach of studying the actual *in-situ* track-annealing has been essentially restricted to apatite, whereas the second approach using palaeo-thermal events has been used, in particular, for minerals with higher track-retention properties. This is simply due to the lack of samples from sufficiently deep and hot drill holes.

4.3.1. *Bore-hole studies*

From the almost 3 km deep bore-hole Eielson, Alaska, drilled in a Mesozoic metamorphic complex undergoing slow uplift, Naeser and Forbes (1976) and Naeser (1979) reported a steady decrease in apatite fission-track age with depth from about 100 Ma near the surface to about 14 Ma at the bottom at 96°C. The apatite fission-track ages extrapolate to zero age at 105°C. A further study is the 2.9 km deep geothermal drill hole Los Alamos, New Mexico, with a bottom-hole temperature of 197°C. The upper 700 m are in Pleistocene volcanic and the lower 2200 m in Precambrian igneous and metamorphic rocks. The apatite from the Precambrian reaches a zero fission-track age at about 135°C. The sphene does not show any fission-track age reduction until the temperatures are in excess of 177°C. The rocks in the hole have only been heated very recently (within a few Ma). The higher temperature of complete track erasure in apatite from the Los Alamos hole is plausible because there the heating duration was shorter than for the apatites from the Eielson hole. However, the observed beginning of track-annealing in sphene is significantly less than expected from extrapolated laboratory data.

Gleadow and Duddy (1981) have studied the variation of apatite fission-track ages with depth in several drill holes from sedimentary basins in Victoria, Australia. The holes are between 2.5 and 3.5 km deep with bottom-hole temperatures of up to 125°C. Concordant sphene and apatite fission-track ages of about 120 Ma

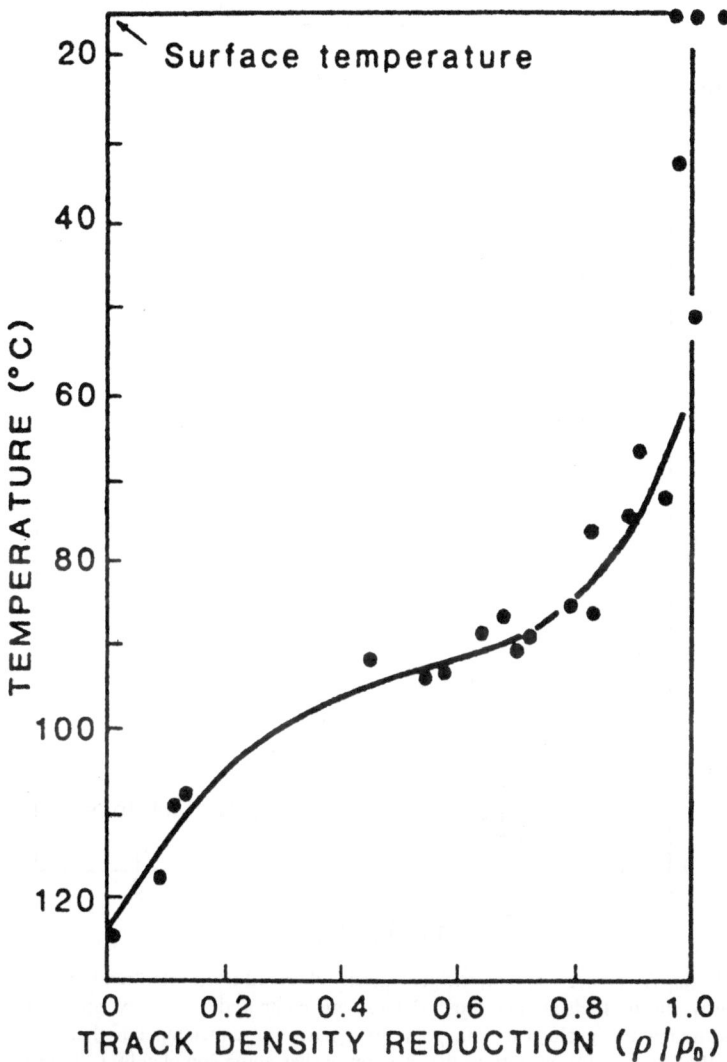

Figure 4.11. The reduction of the fossil fission-track density ρ/ρ_0 – derived from the apparent decrease of the apatite fission-track age which is 120 Ma at the surface – with down-hole temperature in drill-hole samples from the Otway Group sandstone. The zone of partial annealing is observed between 60 and 125°C. (From Gleadow and Duddy, 1981.)

from outcrop samples show that the volcanoclastic basin was formed around that time. Stratigraphic evidence indicates that the sediments reached their maximum depth of burial about 30 Ma ago and have remained at that depth ever since. The sphene fission-track ages remain unaffected in core samples, even at the highest temperature of 124°C encountered at 3.4 km depth. On the other hand, the apatite fission-track ages decrease from about 120 Ma at the surface to zero near the bottom of the deepest hole (Figure 4.11). The apatite fission-track ages begin to decrease at about 60°C and reach half their original age at 95°C defining the zone of partial track-annealing between 60 and 125°C. The mean confined fission-track

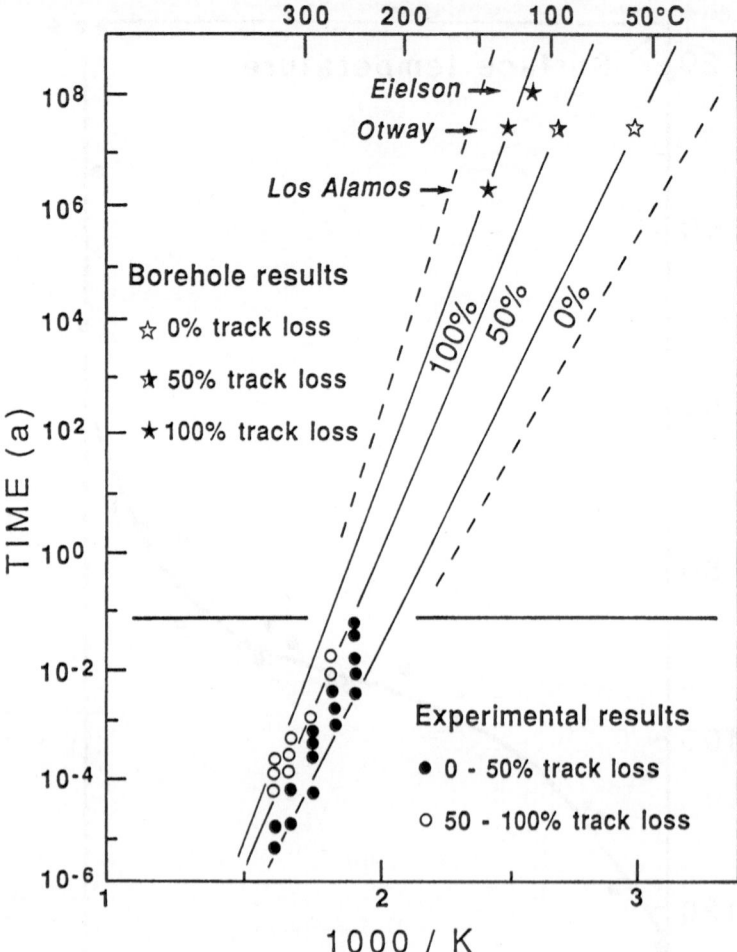

Figure 4.12. In this Arrhenius diagram, the geologically observed track-retention temperatures at the drill-holes Otway Basin, Eielson, and Los Alamos, are compared to the extrapolated laboratory predictions. Although the geological data point to a narrower annealing fan than the laboratory predictions, at the level of 50%-retention, the two approaches agree completely. (After Gleadow and Duddy, 1981.)

length also decreases systematically with increasing temperature down the holes. In Figure 4.12, the results are shown together with the extrapolation of laboratory annealing data. Both types of data sets are in excellent agreement for the line of 50% track-density reduction. However, the extrapolated temperature range over which partial annealing should occur according to the laboratory experiments, seems to be much wider than the actual observed one. In addition, the entire apatite fission-track age spectrum of 0 to 120 Ma was found for individual grains at a temperature where apatite loses, on average, half its tracks. This implies that the amount of annealing is not identical in different grains subject to the same temperature, but is controlled by their chemical composition, as was confirmed by the microprobe analysis of their F/Cl ratios (Green *et al.*, 1985).

The 3.3 km deep research drill-hole Urach-III in Germany first penetrates 900 m Mesozoic sediments and about 700 m Permian volcanoclastic sediments before

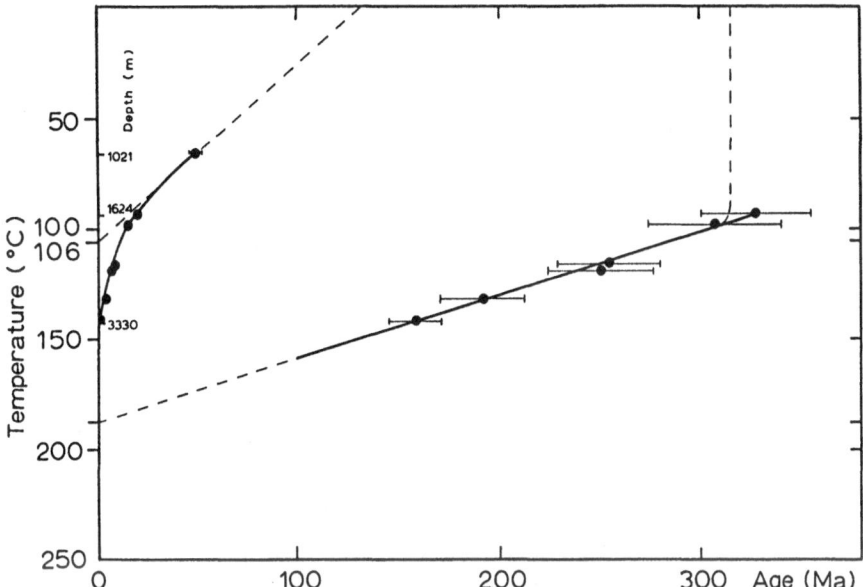

Figure 4.13. Decrease of apatite (above) and zircon (below) fission-track ages with down-hole temperature in the drill-hole Urach III. (After Zaun and Wagner, 1985.)

reaching the underlying crystalline Variscan basement. At present the area undergoes a slow uplift. For fission-track analysis, apatite and zircon have been separated from the cores below the Mesozoic cover. The apatite fission-track ages decrease from 49.5 Ma (66°C at 1021 m depth) to 0.6 Ma at the bottom of the hole (142°C at 3330 m depth). An effective closure temperature of 106°C for this slowly cooling apatite-fission-track system was evaluated from the bore-hole data (Hammerschmidt *et al.*, 1984). Also, the zircon-fission-track ages decrease with depth from 327 Ma (93°C at 1664 m depth) to 158 Ma at the bottom (Figure 4.13). Their depth-profile is believed to represent an uplifted partial-annealing zone (Section 5.2) for zircon and an effective closure temperature of 210 ± 20°C for the zircon-fission-track system under geological conditions was estimated (Zaun and Wagner, 1985).

The project of Deep Continental Drilling, KTB, in the western part of the Bohemian Massif, Germany, with a target depth of about 12 km at predicted ≈300°C, offers a unique opportunity of studying the annealing behaviour of various fission-track systems under geological conditions. The temperature range, from which drill-core samples will eventually be available, covers, in particular, the expected partial track-annealing zones of the important fission-track recording minerals zircon and sphene (Hejl and Wagner, 1990).

4.3.2. *Outcrop studies*

Fission-track systems in rocks that suffered thermal events may have become partially or completely erased. The known intensity of such events can be used in order to evaluate the annealing characteristics. A pertinent example was presented

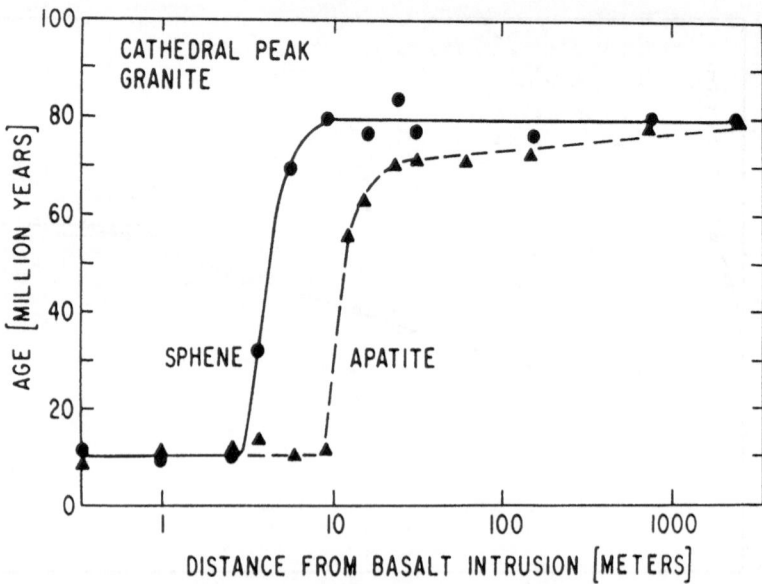

Figure 4.14. The variation of apatite and sphene fission-track ages in granite as function of distance to the contact with a younger basalt intrusion. The decrease of the fission-track ages is due to track annealing caused by the heat of the intrusion. (From Calk and Naeser, 1973.)

by Calk and Naeser (1973). They dated apatite and sphene from a granitic pluton as a function of distance from the contact with a younger basaltic intrusion. Within a few meters of the contact, both the apatite and sphene were completely reset to the age of the basaltic intrusion. Outwards follows the zone of partial resetting where the fission-track ages increase with distance and finally reach the age of the granitic intrusion (Figure 4.14). Note that, due to different retention properties of both minerals, this zone is wider for apatite than for sphene.

Cooling rocks with a known T–t-path are suitable for determining the effective track retention temperatures by linking the fission-track age to the T–t-path and reading off the corresponding temperature. This has been exploited by Reimer and Wagner (1971) for epidote. For rocks from the Central Alps, epidote fission-track ages were found to be generally younger than the biotite Rb-Sr ages but older than the apatite fission-track ages, thus the epidote-fission-track-retention temperature must be somewhere between the respective retention temperatures of 300°C (biotite Rb-Sr) and 100°C (apatite fission-track), although closer to the upper temperature limit. When working on the Late-Precambrian Damara-Orogen in South-West-Africa, Haack (1976b) observed a systematic sequence (in falling order) between biotite K/Ar, andradite-rich garnet, epidote, vesuvianite, and apatite fission-track ages. Since all ages were interpreted as cooling ages with rates in the order of 1°C/Ma, the following track-retention temperatures were assigned: ≈280°C for garnet, ≈240°C for epidote, ≈135°C for vesuvianite, and 78°C for apatite.

Gleadow and Lovering (1978b) determined sphene and apatite fission-track ages on granitic intrusions from Victoria, Australia, and related them to independent biotite K/Ar cooling ages. They derived 260 ± 20 and 80 ± 10°C as effective track

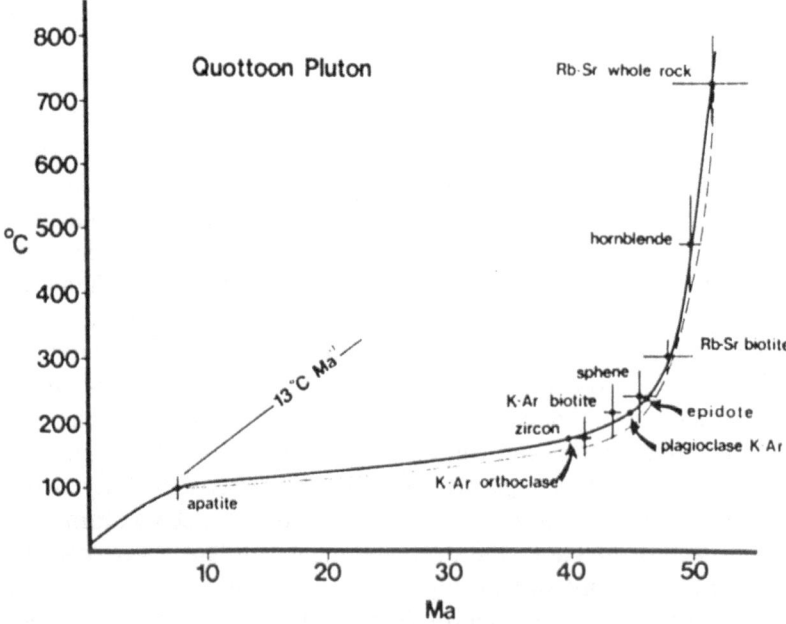

Figure 4.15. Cooling history of a sample from the Quottoon pluton, British Columbia. Individual Rb-Sr-, K-Ar-, and fission-track ages are plotted against their estimated closure temperatures. Using such known cooling curves, unknown retention temperatures of fission-track systems can be geologically 'calibrated'. (After Harrison *et al.*, 1979.)

retention temperatures for sphene and apatite, respectively. In another study, Harrison *et al.* (1979) compared fission-track ages of sphene, epidote, zircon, and apatite with K/Ar and Rb/Sr cooling ages on co-existing minerals from the Coast Plutonic Complex in British Columbia. The epidote fission-track ages indicate a partial track-annealing zone for epidote that is indistinguishable from that of sphene (200–280°C), suggesting an effective track-retention temperature of ≈240°C for, both minerals. For zircon, an effective track-retention temperature of ≈175°C was derived (Figure 4.15). Hurford (1986b) performed an extensive fission-track dating program on zircon and apatite accompanied by mica K/Ar and Rb/Sr dating in the Lepontine Alps, Switzerland. From interpolation between K/Ar and Rb/Sr mica and apatite fission-track cooling ages, 240 ± 50°C was proposed as an effective closure temperature for the retention of fission tracks in zircon. These studies have contributed much to knowledge of how the different fission-track systems behave under geological conditions and certainly more work of this kind is desirable.

4.3.3. *Temperatures of effective track retention*

Taking into account the possible variation in chemical composition for each of the investigated mineral species, the studies on fission-track retention under geological conditions reveal a remarkably consistent picture for apatite, zircon, sphene, and epidote. However, when comparing these retention temperatures to those derived from laboratory experiments extrapolated to geological timescales, the picture worsens. Hitherto, both approaches have been applied to apatite, zircon, sphene, epidote, and garnet. The comparison shows that the results agree well for apatite. In this case the methodology developed for laboratory conditions adequately explains the degree of fission-track annealing observed for a variety of geological situations, implying that, in principle, the laboratory annealing behaviour of induced fission tracks can be used with confidence to explain the geological behaviour of spontaneous fission tracks (Green *et al.*, 1989b). For zircon, sphene, epidote, and garnet, the extrapolations of the experimental data predict a significantly higher track retention than is actually observed under geological conditions. This incompatibility is not yet understood. Various reasons, such as inappropriately designed annealing experiments, nonvalidity of first-order kinetics, chemical variation, and radiation damage are conceivable and have been discussed earlier. At present, it seems best to adopt the annealing characteristics derived by direct geological observation and to improve our knowledge through additional studies. For research in this direction, the deep continental drilling programs which are under consideration or are already in progress in various countries, offer a unique opportunity. At the same time appropriate experiments should be carried out in order to model the track-annealing behaviour, as has been done for apatite.

As to the temperature zone within which partial track annealing (*partial annealing zone PAZ*) occurs under geological conditions, it may be difficult to define its top (i.e., its low temperature boundary). It is known for apatite and natural glasses that fission tracks suffer some annealing, even at a low ambient surface temperature. For other minerals, relevant studies are still lacking. Therefore, it is suggested to define the PAZ top as the temperature where the maximal gradient of stability decrease appears (Section 5.2). The definition of the PAZ base as the temperature where complete track-erasure sets in, does not seem to present a particular problem. A PAZ width of $\approx 70°C$ is found for apatite (Table 4.1). Similar ranges for other minerals – although at higher temperatures – seem reasonable. It needs to be stressed that for apatite, actually observed PAZ widths are relatively narrow with regard to most experimentally derived ones (Figure 4.12), as was demonstrated by Gleadow and Duddy (1981) and this is probably the case for other minerals, too. Green *et al.* (1985) attributed this widening of experimental annealing fans in an Arrhenius plot to the superposition of different fans for samples with variable chemical compositions.

The *closure temperature* or *effective retention temperature* refers to a discrete temperature threshold within the PAZ. At this temperature, the fission tracks are *effectively* retained and, thus, the system is closed. In principle, the concept of closure temperature is only valid for monotonously cooling systems. In accordance with the model by Dodson (1973), the effective retention temperature of cooling fission-track systems is usually assigned to the temperature at which 50% of the fission track is retained. Geologically derived track-retention temperatures are presented in Table 4.1. At present, the best assessments of the track-retention temperatures for the most important fission-track chronometers are as follows: $100 \pm 20°C$ for apatite, $210 \pm 40°C$ for zircon, $240 \pm 40°C$ for epidote, $250 \pm 40°C$

Table 4.1. Geologically derived temperatures of fission-track retention

Mineral	Eff. retention (°C)	PAZ (°C)	Reference
Apatite		<105	Naeser (1979)
		<135	Naeser (1979)
	80 ± 10		Gleadow and Lovering (1978b)
	98	60–125	Gleadow and Duddy (1981)
	106	<142	Hammerschmidt *et al.* (1984)
Zircon	±175		Harrison *et al.* (1979)
	210 ± 20		Zaun and Wagner (1985)
	240 ± 50		Hurford (1986b)
Sphene	260 ± 20		Gleadow and Lovering (1978b)
	≈240	200–280	Harrison *et al.* (1979)
Epidote	<300		Reimer and Wagner (1971)
	240	200–280	Haack (1976b)
	≈240	200–280	Harrison *et al.* (1979)
Garnet (andr)	260–280		Haack (1976b)
Vesuvianite	≈135		Haack (1976b)

for sphene, and 270 ± 30°C for andradite-rich garnet. Only for apatite does the geologically established retention temperature agree with the extrapolated experimental annealing data. For the reasons already mentioned earlier in connection with the PAZ, the experimentally predicted 50%-retention temperatures of the other investigated minerals are systematically too high compared to the geologically observed effective retention temperatures.

CHAPTER 5

Geological Interpretation

A measured, fission-track 'age' is merely a physical quantity of dimension 'time', which equals – in the first approximation – the period of fission-track accumulation in a sample. This quantity may or may not bear geological information. Its association with a geologically meaningful event is the aim of the geological interpretation. It is by this procedure that the measured 'age' becomes an age of a geological event. Therefore, it is necessary to clearly distinguish between the *physical measurement* and the *geological interpretation* of the ages. The interpretation procedure of a fission-track age is by no means straightforward and has to be as carefully performed as the analytical steps of age measurement.

Commonly, the fission-track ages are younger than the age of formation of the samples. More than two decades ago, in the early days of fission-track dating, such results were at first bewildering and disappointing. Obviously, the prerequisite that all fission tracks be retained since the sample's formation, is often not fulfilled. As early as 1964, when working on australites and indochinites, Fleischer and Price (1964c) obtained fission-track ages which were considerably younger than the expected 0.7 Ma of tektite formation. In one of the australites, they observed an "altered pit appearance" of the spontaneous tracks. Laboratory heat treatment of a sample altered the characteristics of the tracks, "such as slower etching and diminished length". Fleischer and Price (1964c) concluded that the altered appearance of the fossil fission tracks in australite yields direct evidence for natural heating some time after the tektite had fallen. Soon it was realized that fission-track fading is a common phenomenon in natural glasses and minerals (Maurette *et al.*, 1964; Fleischer *et al.*, 1965b).

The apparent disadvantage of the fission-track ages of often being 'too young', has, in the meantime, turned into an advantage. The fading of the tracks is used as a sensitive geological thermochronometer. Presently, fission-track studies are increasingly applied – apart from determining the age of rock formation – for reconstructing the temperature-time-paths of rocks in the low temperature region below 300°C. In other words, the fission-track method closes the gap between the medium-temperature chronometers, such as the Rb-Sr and K-Ar systems, and the ambient surface temperatures.

5.1. Intersecting Probability of Faded Tracks

Understanding the track fading phenomenon is crucial for the geological interpretation of fission-track ages. Although already described in Chapter 4, in the present context, it may be helpful to recall the most important properties of track fading.

(1) Various geological parameters, such as temperature, shock-wave pressure, hydrous solutions, and ionizing radiation, acting sufficiently long and strongly enough, can erase the latent fission tracks. Of these parameters, temperature is by far the most important.

(2) With increasing temperature and duration of annealing, the track density decreases in a characteristic manner, which is best described by an 'Arrhenius plot' with a logarithmic scale for the annealing duration and a reciprocal scale for the annealing temperature (Figure 4.3). All points with identical track density reduction ρ/ρ_0 lie on a straight line. There is some discussion as to whether these lines are fanning outward or are parallel (Green *et al.*, 1985).

(3) The fission tracks do not abruptly disappear when the temperature is raised, the fading process is gradual. As is evident from Figure 4.3, a transitional zone of partial track fading exists before complete track loss sets in.

(4) Generally, when geologically long annealing durations are considered, the fission tracks fade at relatively low temperatures. In some materials, such as apatites and glasses, they may not even reach full stability at ambient surface temperatures.

(5) The thermal stability characteristics of fission tracks differ from mineral to mineral. For instance, fission tracks are increasingly thermally unstable in sphene, zircon and apatite. The activation energy for track annealing appears to be related to lattice energy (Haack, 1976c). For a same mineral species the chemical composition may also bear some influence on the track retention properties. As an example, fission-tracks in chlorine-rich apatites are more stable than in fluorine-rich apatites (Green *et al.*, 1985, Figure 4.10). For glasses the track stability increases with the silica content.

(6) During the annealing process, not only is the areal track density gradually reduced, but so are the lengths of the fission tracks. (Fleischer *et al.*, 1964a; Maurette *et al.*, 1964; Berzina *et al.*, 1967). Therefore, the degree of track stability (or 'retention' and inversely 'annealing') may be alternatively expressed as a reduction of either track density (cm^{-2}) ρ/ρ_0 or mean track length l/l_0. Theoretically, a simple isotropic reduction of the etchable track length should cause the same reduction of the track density (cm^{-2}), since, for straight tracks, the probability of intersecting a plane face is proportional to their length. In reality, the linear relationship between l/l_0 and ρ/ρ_0 does not hold true either for confined (Green, 1988) or for projected track lengths (e.g., apatite in Figure 5.1). Analogous relationships between track density and length reductions have been found for other minerals. For glass, the diameter of fission-track etch pits (Figure 4.9) decreases systematically with the degree annealing (Storzer and Wagner, 1969). The diameter reduction

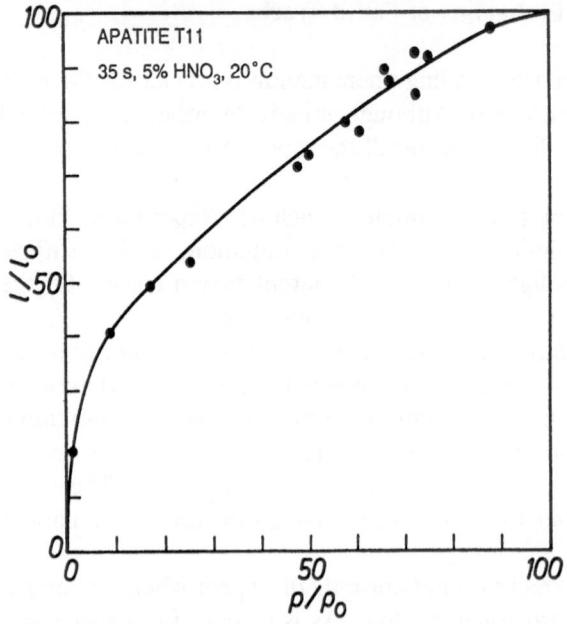

Figure 5.1. Relationship between reduction of mean projected length l/l_0 and reduction of areal density ρ/ρ_0 for partially annealed fission tracks in fluor-apatite concentrate from Ober–Flockenbach.

of etch pits in glass with increasing partial fading is probably caused by the gradual healing of the radiation damage along the track, resulting in lower etching rates and etchable track lengths (Section 2.3.5). The important implication of these curves of length or diameter versus track-density reduction is that the degree of track-density reduction can be inferred from the decrease in the size of fission tracks, i.e., from track-size measurements.

(7) Another aspect which must not be neglected is the influence which the etching conditions may have on the apparent stability behaviour of fission tracks, as noticed by Calk and Naeser (1973) for sphene and by Mehta and Rama (1969) for mica. As a consequence, one has to define the experimental conditions when studying the annealing characteristics and to apply them equally to dating procedures. *Sensu stricto*, the geological age interpretation is only valid for these conditions (Wagner, 1981).

Due to these fading characteristics, there is an important difference between induced and fossil fission tracks in the same sample. Induced fission tracks are freshly produced during neutron irradiation in the reactor. Since they are not usually subjected to any heating procedure, they are not annealed, that is, they still have their full size. Typically, such fission tracks are relatively long and exhibit a narrow length distribution (Figure 5.2). On the other hand, the fossil fission tracks have been produced under a geological regime which is commonly characterized by a lively temperature history. Only in rare geological cases does the temperature always stay cool. Since spontaneous fission occurs at a steady rate, but usually at varying temperatures throughout the geological life-time of a sample,

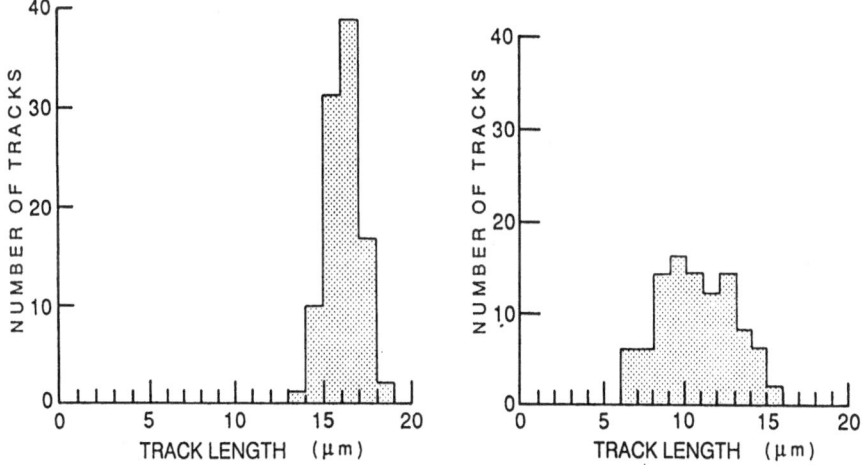

Figure 5.2. Typical narrow-length distribution of confined induced fission tracks (left) and broad-length distribution of confined fossil fission tracks (right) in apatite. (After Gleadow *et al.*, 1986.)

the resulting population of fossil fission tracks will possess a broad length distribution with its average etchable length l_s being smaller than the average induced fission-track length l_i (Figure 5.2). Neglecting the anisotropy effects of annealing, one can estimate, in principle, the age of each individual track from its size, because an older track suffers the thermal fate of all later born tracks and, hence, the older the track, the shorter its length.

If one were able to count tracks per unit volume, the track density N (cm^{-3}) would include all tracks regardless of their length. Unfortunately, a method allowing such ideal three-dimensional track measurement is not yet available. In fission-track dating, the tracks are always counted as track density ρ (cm^{-2}) per unit area. Short tracks obviously have a lower probability of intersecting a two-dimensional plane than long tracks (Section 1.4.3). Hence, the fossil fission-track population with mixed track lengths l_s $(l_s \leqslant l_i)$ is intersected with a lower probability than the induced fission-track population with uniform track length l_i. The shorter the fossil fission tracks are the lower the measured fossil-track density will be, in other words, the fission-track age depends on the length distribution of the fossil-track population which, in its turn, is a function of the sample's thermal history.

5.2. Partial Annealing and Effective Retention of Tracks

All annealing experiments on minerals and glasses have shown that thermal track annealing is a gradual process. Instead of a single 'blocking' temperature value, there exists a wide temperature range with gradually increasing track annealing. This observation led to the concept of the *partial annealing zone*, originally named *partial stability zone* (Wagner, 1972b). According to this concept, temperatures of

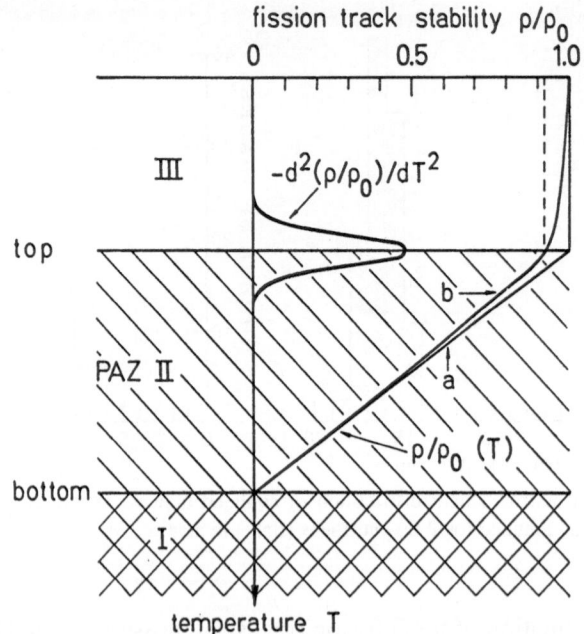

Figure 5.3. Concept of *partial annealing zone*. With regard to the track stability ρ/ρ_0, there are three temperature zones: zone (I) of instability, the partial annealing zone (II), and zone (III) of full-track stability. The idealized concept (curve a) might not always be realized in nature and, thus, the track stability-vs.-temperature function $\rho/\rho_0(T)$ may look like curve (b). In such cases, the PAZ top can be defined as the depth where the maximal gradient of the stability decrease occurs.

any geological environment are divided into three zones with respect to fission-track stability and annealing. These zones are illustrated in Figure 5.3 with the temperature increasing downward. In the *total annealing* or *instability zone* (I), at high temperatures, the latent fission tracks anneal immediately after their formation. There is no accumulation of tracks and the fission-track age of a sample residing in this zone is zero. In the *partial annealing zone* (II), at medium temperatures, the track stability ρ/ρ_0 increases steadily from bottom to top, i.e., from stability 0 to 1, respectively. A sample residing in this zone contains partially annealed fission tracks. In the *stability zone* (III), all tracks are stable at low temperatures. A sample residing in this zone may also contain, in addition to the unannealed fission-tracks which were formed and stayed throughout their life-time within zone III, annealed fission-tracks which previously underwent partial fading in zone II.

In its strict form, the PAZ concept may not be true. The bottom and top temperatures of the PAZ are inferred from annealing experiments or from direct observation of drill-core samples. Whereas the definition of the PAZ bottom with complete track-erasure should not present particular problems, the top of the PAZ is more difficult to define. As is known for apatites and some natural glasses, full track stability may not even be reached at ambient surface temperatures. Therefore, it may be more appropriate to define the PAZ **top** as the temperature corresponding to a track stability $\rho/\rho_0 = 0.95$ to 0.90, where the gradient of stability decrease $-d^2(\rho/\rho_0)/dT^2$ reaches its maximum (Figure 5.3).

As is also evident from the Arrhenius plots (Figure 4.3), the temperatures which are required for the same degree ρ/ρ_0 of track annealing, decrease with duration of heating. Consequently, the temperatures at which track annealing actually occurs not only depend on the material under consideration but also on the rate of the geological process (Wagner and Reimer, 1972; Haack, 1977b; Gleadow and Lovering, 1978b). For fast processes such as volcanism or meteorite impacts with relatively rapidly changing temperatures, track annealing requires higher temperatures than for slow processes such as the cooling of crystalline basement due to epeirogenic uplift and denudation.

Finally, if the other factors mentioned in points (5) to (7) of the previous section are taken into account, it will be understood why it is difficult to allocate, even for the same material, explicit temperature values for the bottom and top of the *partial annealing zone*. The best known in this context is the behaviour of apatite. Mostly, temperatures between 140 and 120°C are cited for the bottom and 70 to 40°C for the top of the PAZ. Such a variation may be easily explained by different cooling rates and chemical composition of the apatites. For detailed studies, microprobe analysis of the Cl/F ratio for each single apatite grain may become necessary.

In fission-track dating, discrete temperature thresholds, at which the tracks suddenly appear to change their stability behaviour, are often adapted instead of the *partial annealing zone* for age interpretaton. They are called the *closure temperature* (or 'retention', 'cooling', and 'blocking' temperature). In its simplest form, a 'closing' temperature replaces the temperature range of the PAZ by a single value. Such an oversimplification is – as discussed above – contradicted by experimental and observational evidence and is unacceptable. In its more sophisticated form, the *closure temperature* takes into account the PAZ concept (Section 4.3.3) and represents – in Dodson's (1973) sense – the temperature of the cooling system at the time given by the fission-track age. The tracks apparently become stable at this temperature. This concept makes geological sense only for rocks with more or less constant cooling rates. For such a scenario, Wagner and Reimer (1972) proposed that the closure temperature corresponds to a temperature within the PAZ, where about 50% of the tracks are stable. In this way, a closure temperature of 100°C was originally derived for the apatite fission-track system (Wagner, 1968). For geological scenarios with long-lasting thermal constancy or with reheating, the assumption of a closure temperature leads to erroneous geological interpretations and must be replaced by the full PAZ concept.

To summarize, for fission-track systems, there are no discrete 'closing' temperatures at which the tracks suddenly become retained. For more or less linearly cooling systems, the concept of an *effective retention temperature* or *closure temperature* as defined in Section 4.3.3 (100°C for apatite, presumably *ca* 210°C, for zircon and *ca* 250°C for sphene) forms a reliable base for age interpretation. Otherwise the concept of the *partial annealing zone* with defined upper and lower temperature limits may be used. Presently, these limits are well known only for apatite (140–120°C and 70–40°C, respectively).

5.3. *T–t*-Path and Fission-Track Accumulation

As long as a mineral resides in the hot zone (I) of complete track annealing, all latent fission tracks are erased immediately after their formation (Figure 5.3). There is no track accumulation, and the fission-track age is still zero. When cooling, the temperature plunges into the partial annealing zone (II) and the fission-track clock is turned on and one would expect that the fission-track age dates this moment. However, in the previous sections it has been shown that

(a) two temperature zones exist in which all latent fission tracks are recorded within the volume of the sample but with different degrees of annealing, namely partial in the medium temperature zone (II) and no annealing in the low temperature zone (III) and

(b) partially annealed tracks have reduced etchable lengths and, thus, intersect the analyzed surface of the sample with lower probability than unannealed ones (Section 1.4.3).

Because fission-track formation and partial annealing occur simultaneously in the PAZ (zone II), net accumulation rate of fission tracks (per unit area) within the PAZ depends on the temperature, i.e., the track accumulation rate increases with decreasing temperature. In zone III, the etchable fission tracks accumulate at a full, constant rate. Therefore, the areal density of accumulated fission tracks not only reflects the duration of accumulation but also the temperature (T)-time (t)-path.

   Figure 5.4 schematically illustrates the accumulation of etchable fission tracks, normalized to the uranium content, for three basic types of thermal evolution. For all three T–t-paths, it is assumed that the temperature at the mineral formation time t_0 is in the complete annealing zone (I). In the fast, steady-cooling path (A), which may represent volcanic rocks, the temperature drops quickly through the PAZ and remains constant for the rest of the time in zone III. The lower part of Figure 5.4 shows the corresponding track accumulation. One notices that the track accumulation begins soon after mineral formation. In the slow, steady-cooling path (B), which may have been experienced by rocks from an uplifted basement, the fission tracks become stable long after the mineral formation. During the passage through the PAZ, the track accumulation rate is gradually increasing as the temperature sinks. For the rest of the time, when zone III is reached, the number of fission tracks grows linearly. In this case, the measured fission-track age would be much smaller than t_0 and represents some moment during cooling within the PAZ. In the more complex T–t-path (C), with reheating, which may represent tectonic subsidence, the temperature first cools down to zones II or even III, rises again without reaching zone I, and finally drops to zone III. During the period of increasing temperature within the PAZ, previously formed fission tracks are annealed and the number of accumulated tracks is reduced. Again, the fission-track age would be much smaller than t_0. If the temperature reaches zone I, all

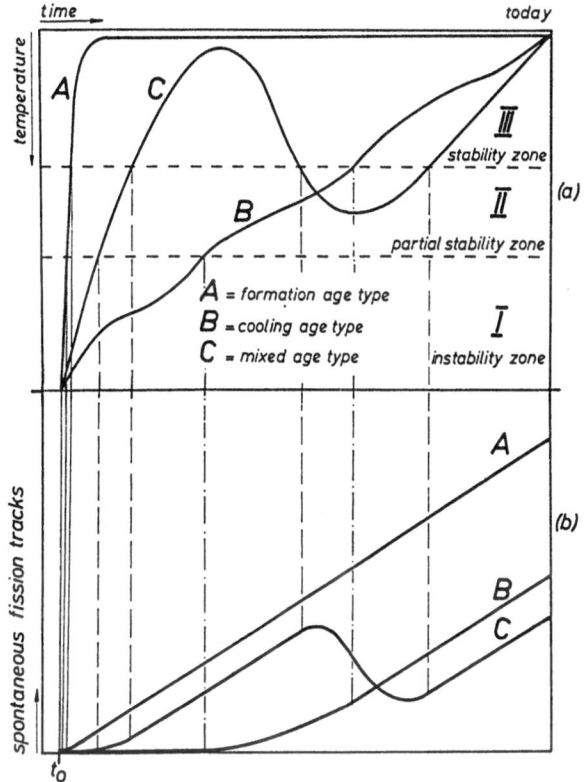

Figure 5.4. The influence of the thermal history on the accumulation of spontaneous fission tracks. Three hypothetical samples of the same age t_0 but different T–t-paths, namely rapid steady cooling (type A), slow steady cooling (type B), and complex cooling with thermal overprint (type C) are depicted with respect to the track annealing zones in the upper half of the diagram. The corresponding curves of fission-track accumulation are shown in the lower half of the diagram. The fission-track ages of the three samples are different and may be related to the geological events of sample formation (type A) and cooling (type B) or may have no direct geological meaning (type C). (From Wagner, 1972b.)

previous tracks are completely erased and, thus, case C becomes indistinguishable from cases B or A in terms of fission-track accumulation. Other possible T–t-paths with fission-track accumulation curves can be deduced from these basic types.

Attempts have been made to model the accumulation of fission tracks as a function of T and t in a quantitative way. For this purpose, the time is subdivided into equal intervals so that the T–t-path is simulated by a temperature step function. The fission-track production per interval is constant, but the degree of track erasure by annealing differs with the interval's temperature. The balance of track production and erasure of each interval is added up to the final fission-track density. In this way, the effective retention temperatures for apatites following T–t-path B (Figure 5.4) were estimated by Wagner and Reimer (1972) and Haack (1977b). Recently, computer programs have been developed which model the fission-track accumulation (i.e., age) in apatite for an assumed T–t-path (Bertagnolli *et al.*, 1983; Crowley, 1985; Green *et al.*, 1988). In the model presented by

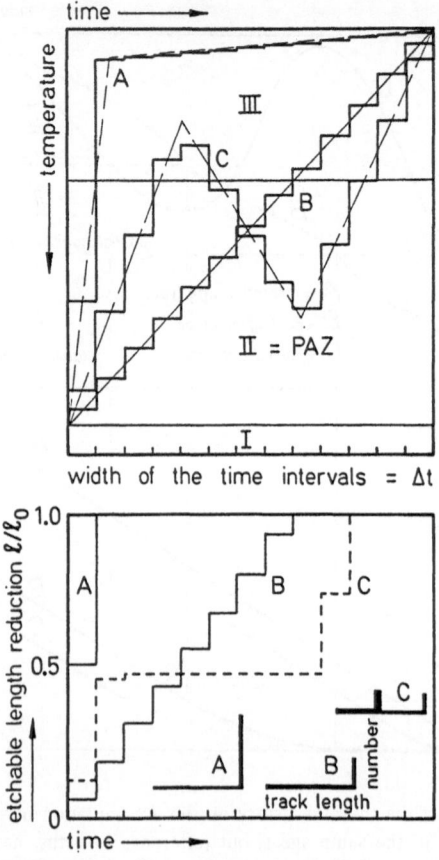

Figure 5.5. Modelling of fission-track-length distributions. Top figure: three typical *T–t*-paths (compare Figure 5.4) are approximated by step functions with equal time intervals d*t* and constant temperature within an interval. The temperature zone II is the *partial annealing zone* (compare Figure 5.3). Assuming a linear relationship between temperature and rate of shortening, the normalized length l/l_0 of the tracks is shown for the different *T–t*-paths (bottom figure). The length reduction l/l_0 of the tracks which are produced within one interval, is a measure of the interval's temperature (types A and B) or the maximum reheating temperature (type C). The resulting histograms of etchabe track lengths are given below. Full track length is retained when the temperature cools below the boundary II/III.

the latter authors, first the length of the fission tracks, produced during each interval, is calculated using laboratory annealing data of Durango apatite. These length components are then summed up in the appropriate proportions to yield the final length distribution from which the mean length and the fission track age can be calculated. One learns from such modelling that temperature dominantly controls the accumulation of fission tracks and that the geological interpretation of fission track ages crucially depends on the sample's thermal history. It also becomes evident that, for *T–t*-type C (Figure 5.4), the period with temperature increase is not preserved in the track-length record (Figure 5.5), only the cooling path is recorded and "in a single sample, a thermal pulse of short duration cannot be distinguished from slow and progressive heating" (Green *et al.*, 1988).

5.4. *T–t*-Path and Fission-Track-Length Distribution

The distribution of fission-track lengths in a sample reflects the thermal evolution which the sample has experienced during its geological past, that is, a certain *T–t*-path leaves a unique signature of track-size distribution. By analyzing the track lengths, one tries to extract information on the thermal history.

In order to illustrate the influence of the *T–t*-path on the track length distribution, let us assume a simple PAZ model in which the etchable track length l linearly increases from zero at the bottom to full length l_0 at the top of the PAZ. Three typical *T–t*-paths are approximated by step functions with equal time intervals and constant temperature per interval (Figure 5.5). Once the temperature has dropped below the I/II boundary, an equal number of latent fission-tracks (cm^{-3}) is formed per interval of time. According to the interval's temperature, the etchable range of these fission tracks is reduced to a certain degree l/l_0. Thus, the etchable length l of the fission tracks produced per interval, records the interval's temperature. This is illustrated in Figure 5.5. The resulting length distribution patterns for the three *T–t*-types clearly differ from one another: long tracks with narrow distribution for type A, a broader distribution of long tracks with a tail of small tracks for type B, and a bimodal pattern for type C. In the latter type, long tracks, which previously have been formed under cooler conditions, are shortened during reheating until the peak temperature is reached. This has the consequence that the record for the previously cooler branch of the *T–t*-path (C) is obliterated. In such distributions with various lengths, the smaller track must obviously be the older track. Note that these model distributions are valid only for the unbiased etchable track lengths in a unit volume and that annealing at ambient surface temperatures is not taken into account. Therefore, they should not be confused with actually observed length distributions of confined or projected tracks.

Nevertheless, the model predicts that the type of *T–t*-path is recorded in the track lengths and, consequently, it should be possible, at least in theory, to reconstruct the *T–t*-path (more exactly: its cooling branch) within the PAZ and to date the moments t_i and t_f when the sample for the last time cooled to and emerged from the PAZ, respectively. The age t_i, as the temperature dropped through the I/II boundary, should be represented by the total number of fission tracks (cm^{-3}) in the sample's volume, since all tracks formed in zones II and III should still be preserved regardless of their etchable length. The age t_f as temperature dropped through the II/III boundary, should be represented by the fraction of the longest, i.e., unannealed fission tracks which were formed in the track stability zone III. The *T–t*-cooling path within the PAZ should be recorded in the fraction of tracks with reduced etchable lengths $l < l_0$.

There is, however, one flaw in these model predictions: it is not possible to directly observe the true distribution of etchable track lengths in a unit volume. Fission tracks are etched and counted on an intersecting two-dimensional plane only. Presently, three practical approaches exist in order to extract information on the thermal history hidden in the track size, namely measurements of confined tracks and projected track lengths or of track diameters.

Confined tracks are still situated fully within the sample's volume but can be etched through host tracks, cracks, or cleavage planes (Section 2.3.4, Figure 2.21). They may require a longer etching duration than the tracks which are etched on the surface and used for track-density counting. Preferably, horizontal tracks are selected for length measurements (Laslett *et al.*, 1982; Gleadow *et al.*, 1986). If lengths of nonhorizontal fission tracks are to be measured, then it is also necessary to measure their dip angles. Confined tracks approximate best the true etchable length distribution of fission tracks. But if strongly faded tracks are present, the distribution and mean length of confined tracks are heavily biased against the short tracks (Section 2.3.4). The measurement of confined tracks becomes impractical for crystals with low track densities due to the small probability of intersect by host tracks and cleavage planes.

Confined fission-track-length measurements in apatite are extensively used as a diagnostic tool for thermal history analysis. According to Gleadow *et al.* (1986), who compiled fission-track length data in a wide variety of apatites from different geological environments, "the distribution of confined lengths of freshly produced induced tracks is characterized by a narrow, symmetrical distribution with a mean length of around 16.3 μm and a standard deviation of the distribution of approximately 0.9 μm". For cooling type A (Figure 5.4) "the distribution is also narrow and symmetric, but with a shorter mean of 14.5 to 15 μm, and a standard deviation of the distribution of approximately 1.0 μm". For a slow cooling pattern below 140°C, such as type B in Figure 5.4, "the distribution becomes negatively skewed, with a mean around 12 or 13 μm and a standard deviation between 1.2 and 2 μm". More complex thermal evolutions, such as type C in Figure 5.4, result in confined track-length distributions with even lower mean lengths and larger standard deviations. Such distributions may become bimodal, yielding clear evidence for a two-stage thermal history. These observations agree qualitatively with the simple model predictions of Figure 5.5. Recently, more sophisticated models have been programmed which are based on experimental annealing data and which calculate the length distributions of confined tracks in apatite for any given T–t-path (Figure 5.6.; Green *et al.*, 1989b). Also in zircons, confined track lengths have been analyzed in order to decipher the thermal history of granitic bodies (Ito *et al.*, 1989).

Projected lengths are measured on tracks intersected by and projected onto an etched surface (Section 2.3.4., Figure 2.18). The etched portions of the surface tracks (also called *semi-tracks*) only partially reside within the body of the crystal (Laslett *et al.*, 1982). Surface tracks are actually the tracks which are routinely used for age determination. The advantage of using these tracks instead of confined ones, for length studies, is their common occurrence, allowing fast and precise measurement with a digitizing tablet (Section 2.3.4), in exactly the same operation as with track counting, and so one can measure the same tracks as those counted for the age determination. On the other hand, projected track-length distributions are obviously much more biased than confined track-length distributions (Dakowski, 1978; Laslett *et al.*, 1982). It is for this reason that the mean length and standard deviation of projected track-length distributions has little meaning. Nevertheless, the projected length of fossil fission tracks in apatite remains a

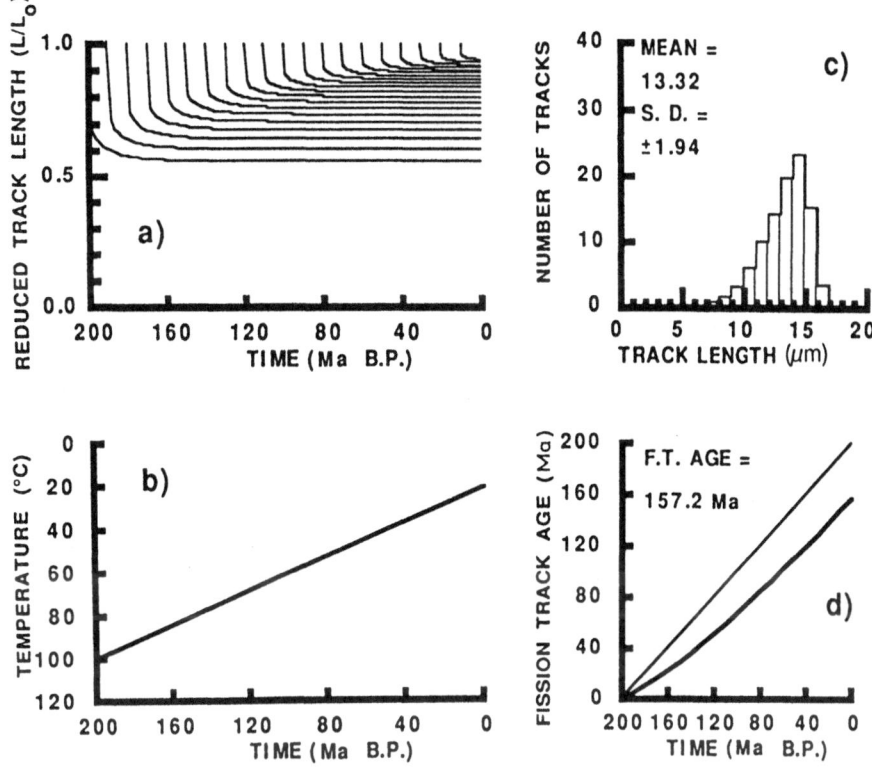

Figure 5.6. Modelling of the length distribution (a,c) of fossil fission-tracks and the fission-track age (d) of apatite for cooling at constant rate (b). (After Green *et al.*, 1989.)

potential source of information for deciphering thermal histories (Wagner and Storzer, 1975; Van den haute, 1986b).

It has been demonstrated for apatite, that the distribution of projected track lengths – expressed as fractions c_s and c_i of tracks >10 μm out of the total fossil and induced fission-track populations, respectively – bears information on the thermal history (Wagner, 1988). The ratio c_s/c_i (>10 μm) practically reflects the relative abundance of tracks effectively produced in the full stability zone (III) and is thus a sensitive diagnostic tool for deciphering the thermal evolution (see also Figure 5.9). Type B thermal histories (Figure 5.4) are indicated by c_s/c_i (>10 μm) ratios between 0.5 and 0.7; smaller ratios represent acceleration and higher ratios deceleration as a general trend for steady cooling. Recently, this concept has been further developed by Wagner and Hejl (1991). Instead of selecting a single threshold, such as 10 μm, for cutting-off shorter tracks, and calculating the fission-track age from the remaining track sub-population, the full range of cut-off values is used, namely step-wise at >0, >1, >2 μm, etc. up to the full etchable track length of ≈ 15 μm. A fission-track age is calculated from each ratio c_s/c_i (>0, >1, >2 μm, etc.). When these apparent ages are plotted versus the cut-off thresholds, an age spectrum results. Since, for each of these single ages, the respective fission-track systems effectively close at different temperatures, the spectrum carries information on the thermal history of the sample. The higher the

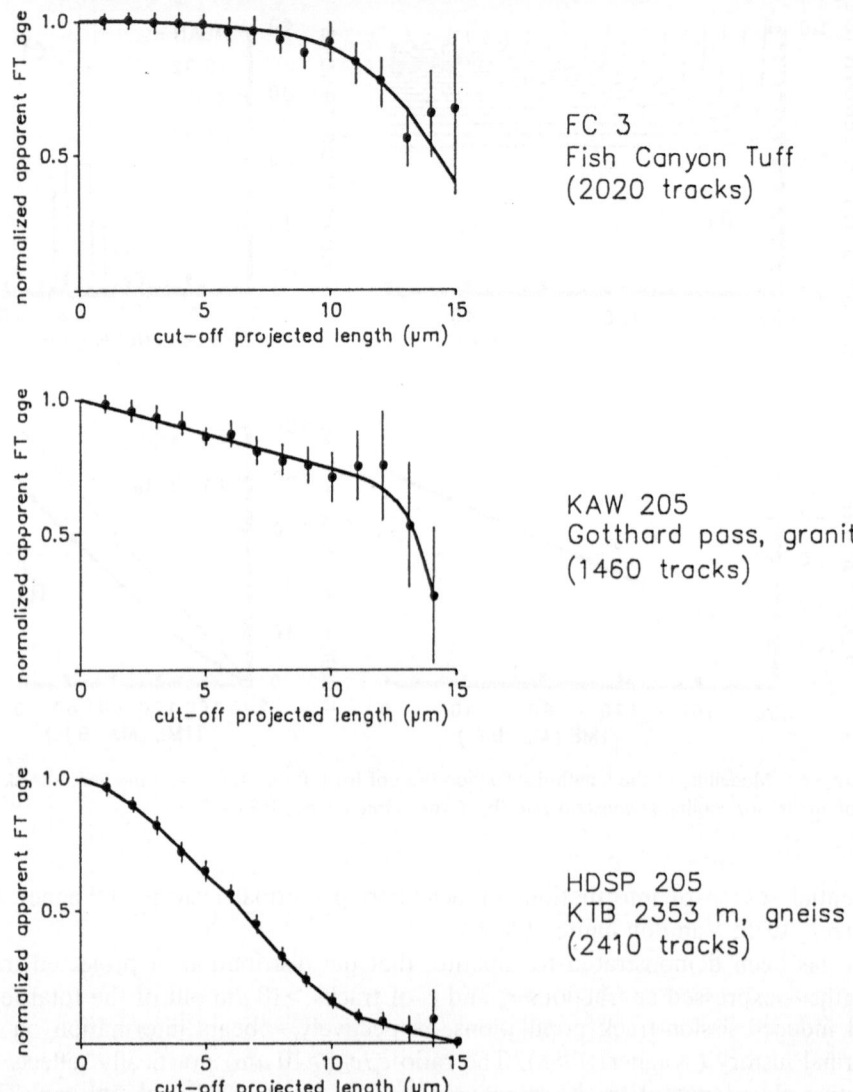

Figure 5.7. Characteristic fission-track-age spectra for three apatite samples of known thermal history including a sample from the KTB drill hole at a depth of 2353 m. The apparent ages are shown as a function of the increasing cut-off value for projected lengths of shorter surface tracks. The apparent ages are normalized to the measured ages t_m (26.7 Ma for FC3, 7.5 Ma for KAW205, 50.1 for KTB205). The vertical bars represent the 1σ precision. (From Wagner and Hejl, 1991.)

cut-off value, the shallower the temperature level which is recorded. Figure 5.7 shows such age spectra for three apatite samples with different thermal histories. The age spectrum of the Fish Canyon tuff, which probably experienced a thermal history such as type A in Figure 5.4, is characteristically different from that of the Central Alps apatite, which cooled at a constant rate such as type B in Figure 5.4 and from that of the KTB drill-core sample which resides at 76°C. Consequently, in order to extract the maximum information on the thermal evolution from track lengths in apatite, it is advisable to measure the surface track lengths as well as the confined track lengths. By measuring the true etchable lengths of surface tracks instead of their projected lengths (Section 4.2.3) age spectra can be obtained with much more precise ages for the high cut-off thresholds (>10 μm).

So far, the techniques and implications for either type of track-length measurement have been elaborated and geologically tested for apatite only (Wagner and Storzer, 1970; Bhandari et al., 1971; Wagner and Storzer, 1972; Wagner and Storzer, 1975; Bertel et al., 1977; Green, 1980; Gleadow and Duddy, 1981; Gleadow et al., 1983; Green, 1986, Hammerschmidt et al., 1984; Van den haute, 1986b). In principle, they should analogously apply to other minerals, because there various degrees of shortening of fossil tracks have also been observed, such as in micas (Maurette et al., 1964; Bigazzi, 1967; Gupta et al., 1971a; Mehta and Rama, 1969; Nagpaul et al., 1974a), sphene (Nagpaul et al., 1974b; Märk et al., 1981), zircon (Krishnaswami et al., 1974; Nishida and Takashima, 1975), epidote (Saini et al., 1978), chlorite (Sharma et al., 1977), and vermiculite (Sharma et al., 1979).

In glass, the *diameter* of the fission-track etch pits is used as an indicator of the degree of track annealing. Glass is characterized by relatively large bulk etching rates (V_g) compared to the etching rates along the track (V_g), resulting in more or less conical etch pits with circular to elliptical openings (Section 2.3). The diameter of the etch pit essentially depends on the ratio V_t/V_g. With increasing track annealing, both the etchable length and the etching rate V_t of a track decrease, causing smaller etch pit diameters. In natural glasses, fossil fission-track etch pits are commonly smaller than induced ones (Figure 4.7). This was first realized by Fleischer and Price (1964b) in australite and indochinite tektites. The correlation between the fission-track diameter (in case of elliptical etch pits, the major axis is always taken) and annealing degree was fully recognized by Storzer and Wagner, 1969). In a dating programme on australites, they found a wide spread of fission-track ages, whereby the age and the average diameter reduction were clearly correlated to each other. In nearly all australites, the fossil fission-track etch pits were smaller than the induced ones. Annealing experiments proved that the diameter reduction was caused by partial fading of fission tracks due to elevated temperatures during the geological history of the australites. The reduction ratio d/d_0 (d and d_0 being the average diameters for the fossil and induced fission tracks, respectively) gives the degree of track annealing (Figure 4.9). There have also been attempts to use the distribution of track diameters for deciphering more complex thermal histories of natural glasses (Storzer, 1970; Wagner and Storzer, 1970; Storzer and Wagner, 1971; Storzer et al., 1973), but this approach is problematic, since the relation between the etch-pit-diameter distribution and the thermal history of a sample is not straightforward (Van den haute, 1985). Other observations of fossil track fading in natural glasses have been reported for tektites (Gentner et al., 1969a; Durrani and Khan, 1970; Glass et al., 1973; Komarov and Raichlin, 1976), impact crater glasses (Gentner et al., 1969b; Gentner and Wagner, 1969; Gentner et al., 1973), and for volcanic glasses (Komarov et al., 1972; Suzuki, 1973; Wagner et al., 1976; Miller and Wagner, 1981; Bigazzi et al., 1990).

When the track length or size measurements have been achieved, one tries to interpret them in terms of a sample's thermal history. However, a given, observed track length or size distribution does not necessarily result from a unique T–t-path. In principle, an indefinite number of possible T–t-paths exist, which can produce the same length distribution. Fortunately, in most cases, this number can be reduced to few or only one solution by independent geological information.

As already mentioned (Section 5.3), computer models have been developed which, for an assumed T–t-path, predict the track length distribution and the fission-track age of apatite (Bertagnolli *et al.*, 1983; Crowley, 1985; Green *et al.*, 1988). One selects, by trial and error, an appropriate T–t-path that is compatible with the analytical track data and geological observations. The present methodo-logical developments aim not only at the crude recognition of the T–t-type, which forms the basis for appropriate age interpretation, but also at the precise recon-struction of the geological T–t-path.

5.5. Types of Fission-Track Age

In the previous sections, it has been pointed out that the geological interpretation of a fission-track age depends on the thermal history which is reflected in track size. Depending on the sample's T–t-evolution, the fission-track age may date quite different geological processes, such as rock formation, cooling, uplift, ero-sion, thermal overprint, or even none at all (Wagner, 1972b). In the following, this is discussed in detail for the three typical T–t-paths shown in Figure 5.8. Since these T–t-types result in characteristic fractions of partially annealed tracks, it is convenient for this discussion to define them in terms of the time t with respect to the track annealing zones I (full annealing), II (partial annealing), and III (no annealing), whereby t_0 = sample formation (Figure 5.4), t_i = last cooling to the PAZ bottom, and t_f = last cooling from the PAZ top (Figure 5.8). An obvious condition for the measured fission-track age t_m is: $t_i > t_m > t_f$. The results of the following discussion are summarized in Table 5.1.

5.5.1. *A-type ages (formation and early cooling)*

The A-type T–t-path (Figure 5.8) is characterized by $(t_i - t_f) \ll t_f$, i.e., the sample cools through temperature zone II within a period $(t_i - t_f)$ which is negligible compared to the period t_f over which it stays in the temperature zone III (*insignifi-cant partial track annealing*). As to the track size, no significant fraction of spon-taneous fission tracks with reduced sizes is expected due to the relatively short duration the rock has spent in zone II compared to zone III. Gleadow *et al.* (1986) introduced the term *undisturbed 'volcanic'* for this type. Since it is not restricted to volcanic rocks, this term should be dropped.

With respect to geological interpretation, one has to distinguish two sub-cases: if the sample's formation is followed by steady, rapid cooling (AA), then $t_0 \approx t_i \approx t_f$ and the measured fission-track age t_m dates the sample's formation (*formation*

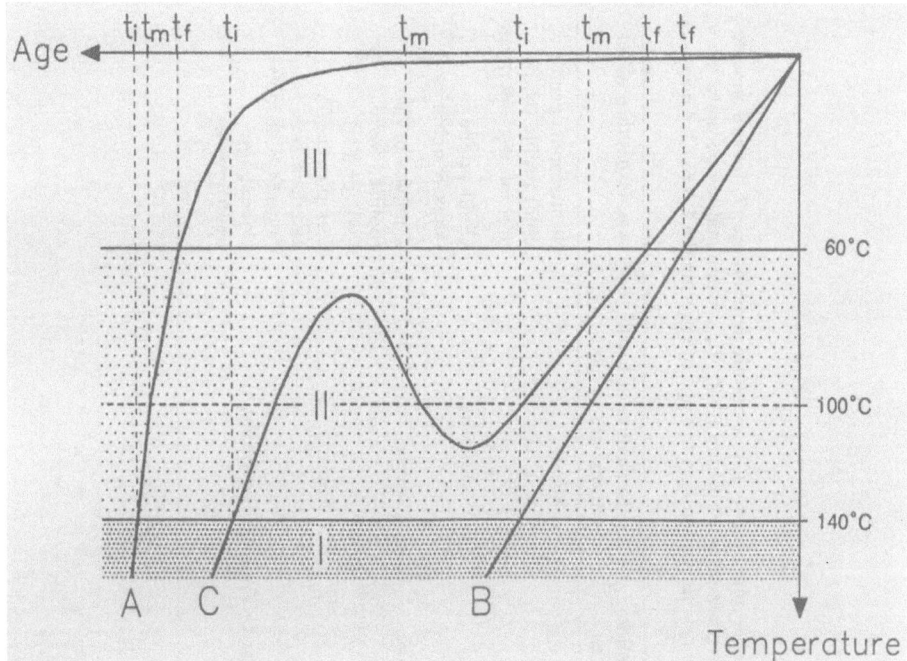

Figure 5.8. For three typical *T–t*-paths A, B, and C (compare Figure 5.4), the measured fission-track age t_m, the time t_i of cooling to and the time t_f of cooling from the partial annealing zone (II) are indicated. (From Wagner, 1990.)

age). If the sample's formation is followed by slow cooling (AB), then the times $t_i \approx t_f$ are significantly later than time t_0 and the measured fission-track age t_m dates an early event of cooling (*early cooling age*).

Formation ages (AA-type) are commonly observed for volcanic rocks and im-pactites. Concordant fission-track ages – as well as other radiometric ages – should be found on co-existing minerals. In the literature, there exist numerous examples of applying fission-track dating to rock formation. Mostly, the minerals zircon and apatite, separated from volcanic ash layers, are used. Extensive chronostrati-graphic fission-track work on tephra horizons in the south-western United States has been reported by Naeser and co-workers (e.g., Naeser and McKee, 1970; Izett and Naeser, 1976; Izett *et al.*, 1981; Naeser and Maldonado, 1981). However, there is mounting evidence that fission tracks in apatite may fade to some extent, even at ambient surface temperatures, as indicated by the length reduction of the fossil tracks in volcanic apatites such as Durango and Fish Canyon (Bertel *et al.*, 1977; Green, 1988). This has far-reaching consequences since these apatites are commonly used as age standards in fission-track dating (Section 3.5.2). Fission-track dating is also frequently applied to volcanic glasses, such as pumice (Fleischer *et al.*, 1965f), obsidian (Fleischer *et al.*, 1965g, Suzuki, 1973; Miller and Wagner, 1981), glass shards from tephra layers (Seward, 1975; Seward *et al.*, 1980) and basaltic deep-sea glass (Fleischer *et al.*, 1968b; Storzer and Selo, 1978). Besides volcanic rocks, impactites from meteorite craters have been dated with the fission-track technique; examples include impact glasses (Fleischer *et al.*, 1965c; Wagner,

Table 5.1. Geological interpretation of fission-track ages of (apatite) samples with different thermal history. (Temperature symbols are: T_I for full annealing zone, T_{II} for partial annealing zone PAZ, $T_{I/II}$ for the top of PAZ, T_{IIeff} effective retention temperature within PAZ, T_{min} = minimal temperature during thermal overprint, T_{max} = maximal temperature during thermal overprint; time symbols are: t_0 = formation age, t_i = age of entering the PAZ, t_m = measured fission track age, t_f = age of leaving the PAZ)

Thermal history (T–t-path)	Geological environment	Geological interpretation of measured FT age t_m	Reduction of fossil track size	Age-vs-depth profiles
Early cooling (type A)	Young volcanism early uplift	Formation age early uplift age $(t_0 \gtrsim) t_i \approx t_m \approx t_f$	Minor	Stratigraphic profile uplift profile (with unresolvable uplift rate)
Steady cooling (type B)	Plutonism steady uplift	Cooling age T_{IIeff} $(t_0 >) t_i > t_m > t_f$	Significant	Uplift profile (slope = uplift rate)
Complex history with thermal overprint (type C) — Weak $(T \approx T_{II})$	Subsidence increased geothermal gradient	Meaningless mixed age $(t_0 >) t_i > t_m > t_f$ $+ T_{max}$ of event	Strong	Complex profile (uplifted fossil PAZ profile)
Strong $(T \approx T_I)$ but short	Meteorite impact, contact heating (flows, dykes)	Heating event age $(t_0 >) t_i \approx t_m \approx t_f$ $+ T_{min}$ of event	Minor	Complex profile age decrease towards heat-source
Strong $(T \approx T_I)$ but long	Metamorphism, contact heating (batholiths)	Post-event cooling age to T_{IIeff} $(t_0 >) t_i > t_m > t_f$ $+ T_{min}$ of event	Significant	Complex or uplift profile

1966; Storzer and Wagner, 1979), tektites (Fleischer and Price, 1964b; Gentner *et al.*, 1969a) and impact breccias (Miller and Wagner, 1979). However, most volcanic glasses, as well as impact glasses, have suffered some fossil-track fading at ambient surface temperatures. Such samples are recognized by reduced etch-pit sizes of the fossil tracks (Wagner *et al.*, 1976).

The common occurrence of track fading at ambient surface temperatures introduces a systematic error which more or less lowers the fission-track age, although the sample may have cooled very rapidly. This error, in addition to the error of the analytical age measurement, may seriously limit the use of fission tracks for chronostratigraphic application, at least in the case of apatite and glass. There have been attempts to correct this systematic error (Section 5.7). For zircon, it is believed that such effects of low temperature fading need not to be taken into account. Nevertheless, the discussion of fission-track ages in terms of rock and mineral formation and chronostratigraphy requires careful and critical evaluation (Storzer and Wagner, 1982).

5.5.2. *B-type ages (cooling and uplift)*

In the B-type T–t-path (Figure 5.8), the sample cools steadily through temperature zones II and III whereby the period $(t_i - t_f)$ is significant compared to the period t_f (*significant partial track annealing*). Consequently, a significant fraction of spontaneous fission tracks with reduced sizes is expected due to the comparably long duration the rock has spent in zone II. During B-type cooling, the sample reaches the temperatures of the PAZ, long after its formation time t_0. The measured fission-track age t_m dates a moment of cooling to a temperature somewhere within the PAZ where the fission tracks are effectively retained, with $t_0 > t_i > t_m > t_f$. Therefore, a fission-track age of a sample with B-type T–t-path is geologically a *cooling age*. Compared to the A2-type (Section 5.5.1), the B-type cooling age refers to a more recent event, probably in geologically still active regions. The different types are identified from track-size distribution and, of course, gradual transitions between them are possible.

For constant cooling rates, the accumulation of spontaneous fission tracks first follows – under the assumption of a linear model of track stability increase within the PAZ – a quadratic function and then turns into a straight growth line when the PAZ top is reached (Figure 5.9). In such a case, the effective retention temperature lies exactly in the middle of the PAZ and t_m becomes $t_m = (t_i + t_f)/2$.

Since samples of B-type thermal history spent a significantly long period within zone II, the fossil fission-track population should consist of tracks of different sizes, namely shortened tracks from zone II and full-sized tracks from zone III. The latter track fraction should bear information on time t_f (t_1 in Figure 5.9) when the sample emerged from zone II. In the case of apatite, this potential has been exploited for fission-track dating. Wagner (1988) proposed the use of fractions of tracks with projected lengths >10 μm (spontaneous tracks c_s normalized to induced ones c_i) to date the moment t_f when the cooling path crossed the $\approx 60°C$ according

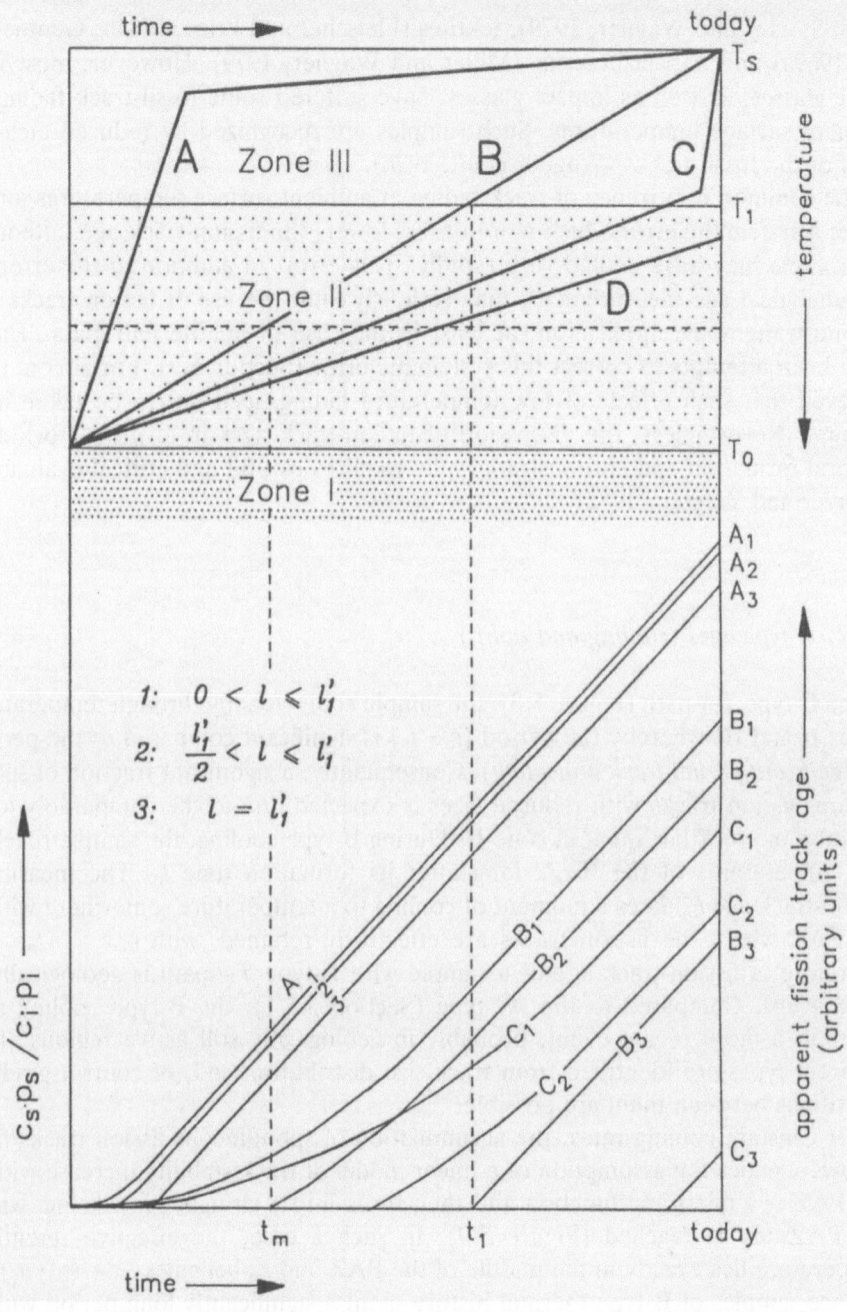

Figure 5.9. The accumulation of spontaneous fission tracks is schematically shown for various T–t-paths (A–C) of apatite with respect to three different annealing zones (I: full annealing, II: partial annealing, III: no annealing). For modelling the track accumulation, sharply defined upper (T_0) and lower (T_1) temperature limits, for and linear increase within the PAZ (=zone II) are assumed (T_s = surface temperature within zone III). The track density increase grows in zone II as a quadratic and in zone III as a linear function with time. The track density increase of spontaneous fission-track densities $c_s \times \rho s/c_i$ (when normalized to ρ_i, this corresponds to the fission-track age) is modelled for tracks with various ranges of projected lengths (1: total population of all lengths, 2: sub-population of tracks longer than 1/2 of the full length, 3: sub-population of tracks with full lengths). (From Wagner and Hejl, 1991.)

to equation

$$t_f = t_m \times (c_s/c_i).\tag{5.1}$$

This allows one to date the cooling to $\approx 60°C$ in addition to the 'conventional' cooling age t_m at the effective track-retention temperature around $100°C$. Note that, analogous to the t_m-age, the t_f-age is a cooling age. It also corresponds to an *effective* retention temperature because the accumulation rate of the track population $c_s \times \rho_s$ ($>10\ \mu m$ projected length) increases over a certain temperature range. For constant cooling rates, the accumulation of $c_s \times \rho_s/c_i \times \rho_i$ follows a similar mathematical function as the total number of spontaneous fission tracks ρ_s/ρ_i but starting later at a lower temperature level (Figure 5.9). The first tracks with etchable track length $>10\ \mu m$ appear during cooling at $\approx 85°C$ (Gleadow and Duddy, 1981). As the temperature gradually drops, the etchable length of the new tracks increases. Due to the linear relationship between the etchable track length and its probability of being intersected by a plane surface, the production rate of surface tracks grows until the top of zone II is reached. Clearly, according to this model, the cooling temperature at time t_f is not identical with the temperature at the the PAZ top but rather corresponds to a temperature when the track stability reaches $\rho/\rho_0 = 0.72$. Owing to the problem of defining the PAZ top for apatite (Section 5.2, Figure 5.3), the age t_f probably dates the effective cooling to a temperature of $\approx 60°C$.

When co-existing minerals are dated, discordant fission-track ages should be found, whereby the apparent ages should decrease with decreasing track-retention properties. The pairs of retention-temperature versus age for different minerals, trace the T–t-path taken by the sample.

Slow, steady cooling is typical for crystalline basement rocks, where it may be caused by a decaying heat source or by uplift/denudation but it may also occur in sedimentary rocks. In case of post-metamorphic cooling, it is important that the peak temperature of the thermal event is higher than the upper temperature limit of the PAZ and, hence, the fission-track system in the mineral under consideration was completely reset. Also, sedimentary rocks may undergo slow cooling during uplift after a phase of subsidence with sufficient heating for resetting the apatite fission-track system. One then determines, with fission-tracks, the post-event cooling age. Early examples of fission-track cooling ages have been presented by Wagner (1968) using apatite from the Central European Variscan basement, by Naeser and Faul (1969) as well as Naeser and Dodge (1969) dating apatite and sphene from Sierra Nevada granitic batholiths, and by Reimer and Wagner (1971) using apatite and epidote from the Central Alps. Cooling ages of co-existing minerals with different retention temperatures, allow the reconstruction of the T–t-path. Wagner *et al.* (1977) combined mica Rb-Sr and K-Ar ages with apatite fission-track ages in order to decipher the post-metamorphic cooling history of the Central Alps (Section 7.2.2). Analogously, Gleadow and Brooks (1979) traced the cooling history of intrusive rocks in Greenland with sphene, zircon, and apatite fission-track ages. Another example of this kind was reported by Harrison *et al.* (1979) for a British Columbian plutonic complex.

In order to decide if a fission-track age can be interpreted as a cooling age, several criteria can be used. Often, one has *independent geological information on the thermal history* of the region from where the sample was taken. Fission-track as well as other radiometric ages of co-existing minerals, can support the interpretation in terms of cooling ages; in this case, the ages systematically decrease in the order of decreasing closure temperatures. In addition, there exists a methodical criterion inherent in the fission-track system itself in the form of the *track-length distribution*. So far, this criterion has only been satisfactorily corroborated for apatite. Mean lengths of between 12 and 13.5 μm and standard deviations of between 1.2 and 2 μm for confined fossil tracks in apatite are characteristic for cooling ages (Gleadow *et al.*, 1986). As to the projected track lengths, c_s/c_i ratios between 0.5 to 0.7 are indicative for steady cooling with a constant rate (Wagner, 1988). As to other minerals, there is actually no reason why similar length criteria should not work for age interpretation. Methodical investigations in this direction are, however, still lacking.

Once fission-track ages are established as cooling ages, interesting geological information on the thermo-tectonic history can be drawn from them. The fission-track age t_m of a sample, which is now outcropping at the surface, directly reveals the mean cooling rate $(T_m - T_{surf})/t_m$ between the effective retention temperature T_m at cooling age t_m and the ambient surface temperature T_{surf}. In the low temperature regime of fission-track retention, tectonic uplift is the most important geological process causing rock cooling. As the rock column is lifted towards the surface, it cools continuously, *provided the surface layers are equally removed by denudation and the geothermal gradient stays constant* (i.e., the track-retention isotherm stays at constant depth). Then the mean uplift rate can be calculated from the cooling rate according to

$$\text{uplift rate} = \text{cooling rate/geothermal gradient} \qquad (5.2)$$

and the fission-track age becomes an *uplift age* dating the moment when the sample, now at the surface, passed the depth of the retention isotherm. Note that the cooling is not caused by the uplift itself, but by the *uplift-induced denudation*. Under geological circumstances, there might be nonnegligible time lags between the onset of uplift and its compensation by denudation, as well as between the lowering of the surface by denudation and the depression of the isotherms down in the Earth's crust (Figure 5.10).

If the cooling was not purely due to uplift but also to a decaying heat source, i.e., the geothermal gradient has not been constant with time, then one may determine the geothermal palaeogradient at time t_m if the uplift rate is independently known according to

$$\text{geothermal palaeogradient} = \text{cooling rate/uplift rate.} \qquad (5.3)$$

This discussion indicates that fission-track ages not only contain information on the thermal but also on the tectonic evolution of rocks. However, in order to separate the thermal from the tectonic factors, it is necessary – besides taking into account independent geological information – to date samples from horizontal and vertical profiles.

In thermo-tectonic applications of fission-track dating, mostly of apatite, one

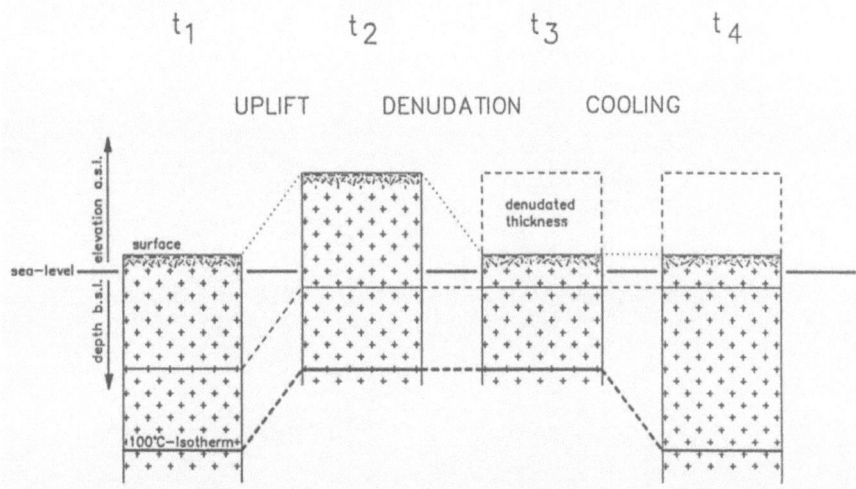

Figure 5.10. Schematic illustration of the causal relationship between uplift, denudation, and the depression of the isotherms in a crustal block. Initially, at time t_1, strong tectonic uplift raises the elevation of the morphological surface since the uplift movement is not simultaneously compensated by denudation. The isotherms (shown is the 100°C isotherm for effective track retention in apatite) are passively uplifted with the rock column. After the uplift movement comes to a rest (t_2), denudation sets in until, at time t_3, the surface is lowered to the original elevation. The denudation compresses the isotherms and increases the geothermal gradient causing higher heat transport to the surface until the original geothermal regime is installed (t_4). Only during this latter stage do the isotherms move downward and finally turn on the respective fission-track clocks (for apatite when the 100°C isotherm drops below the depth level of the sample). Although the various processes occur more or less simultaneously in geological reality, there may exist a nonnegligible time lag between uplift and its recording by the fission-track system. The fission-track age may represent the time of denudation (constant geothermal gradient) or uplift (constant geothermal gradient + denudation compensates uplift). (After Hejl and Wagner, 1990.)

usually samples a larger region. Let us assume a variation with low and high fission-track ages of apatite, t_{m1} and t_{m2}, respectively, along a horizontal profile (Figure 5.11). Such an age variation tells us that the mean cooling rate between 100°C ($=T_m$) and surface temperature T_{surf} in the region with the lower fission-track ages was faster than in the other region. In principle, this may be due to two different geological scenarios which both produce uplift ages with the horizontal age variation, as illustrated in Figure 5.11.

(a) *Tectonic model*: If a regionally equal geothermal gradient is assumed, at least from time t_{m2} till time t_{m1}, then the different fission-track ages indicates a differential uplift. The low age indicates a high uplift rate and *vice-versa*. The uplift rates may be calculated according to Equation (5.2). The low fission-track age t_{m1} sets an upper limit for the age t_d of the last vertical displacement between both blocks ($t_{m1} > t_d$).

(b) *Thermic model*: If the uplift rate was regionally equal since time t_{m2}, then the different fission-track ages indicate differential geothermal gradients prevailing at least until t_{m1}. The low age indicates a high thermogradient and *vice-versa*. Note that this requires a horizontal thermogradient within the crust. The different geothermal palaeogradients may be calculated according to Equation (5.3). If this

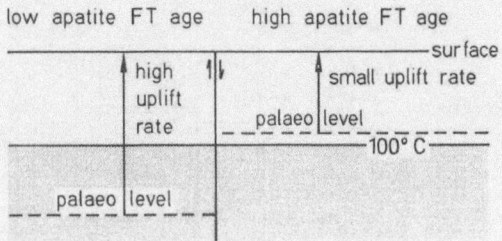

Geological Interpretation of Apatite FT ages

a) tectonic model
assumption: equal thermogradient
result　　 : differential uplift

low apatite FT age　　　high apatite FT age

————————————————————————surface
high uplift rate ↑ ↓　　↑ small uplift rate
　　　　　　　　　palaeo level
————————————————————— 100° C —————
palaeo level

b) thermic model
assumption: equal uplift
result　　 : differential thermogradient

low apatite FT age　　　high apatite FT age

high thermogradient　uplift　small thermogradient
　　　　　　　　　　　　　palaeolevel
———————————— 100° C ————————

intenstity of the thermal overprint in zone II, a dominant fraction of spontaneous fission tracks with reduced size can be expected.

The measured fission-track ages represent the net balance of track accumulation against erasure and the age interpretation may become very complex. Of course, the measured fission-track age t_m is always smaller than the time t_i of the last cooling through the base and larger than the time t_f of the last cooling through the top of zone II. The shallower the temperature path dips into zone II during the thermal overprint, the less pre-existing tracks are annealed and the closer t_m stays to the original age, which may have been either a formation or a cooling age. The deeper it dips into zone II, the more of the pre-existing tracks are annealed and the closer t_m becomes to the age of the overprint event. In other words, t_m is a *complex* or *mixed age*, which may have no geological meaning. Only in the case of very strong events reaching temperatures of zone I, can it approximate the *overprint age*. In principle, the fission-track analysis of samples with complex thermal history may reveal information on both the *age* and *degree* of the secondary thermal event.

The size distribution of fossil fission tracks in samples with mixed ages, is equally complex due to the more or less strong fading effects. One expects a distribution composed of two fractions, one with annealed pre-existing tracks and the other one with unannealed post-event tracks. With increasing overprint, the pre-existing tracks are increasingly reduced in size and number, thus their fraction becomes less important. From confined fission tracks in apatite as well as from etch pit diameters in glasses, bimodal distributions have been observed which give direct evidence for the two-staged thermal history (Laslett *et al.*, 1982; Storzer, 1970b). For confined fission tracks in apatite, Green (1986) reported a relationship between age and length distribution. The age variation is believed to represent a gradual transition from partial to complete track fading during subsidence. Samples with the youngest ages have been completely thermally reset and possess long narrow length distributions (Figure 5.12). As the (mixed) ages increase, the component of the annealed tracks becomes dominant but less shortened. Thus, the changing length distribution reflects, in a characteristic manner, the various degrees of thermal overprint.

As to co-existing minerals, discordant fission-track ages are to be expected. Similar to T–t-type B, the ages should decrease with decreasing track stability characteristics, but their retention-temperature-vs.-age pairs do not necessarily mark the sample's T–t-path. Depending on their track retention properties, co-existing minerals may require a quite different geological interpretation. For example, the fission-track age of zircon may still be an unaffected cooling age, whereas the more sensitive apatite from the same rock sample represents a mixed age.

Complex cooling histories may be caused by heating due to intrusive and extrusive contacts, meteorite impact, metamorphism, and tectonic subsidence. According to the intensity and duration of the event, one has to distinguish between three subcases which are connected by gradual transitions. If the thermal overprint is relatively weak and only reaches temperatures within zone II, then the pre-event fission tracks are not completely erased and a *mixed age* results. The mixed age is younger than the original fission-track age and older than the age of the

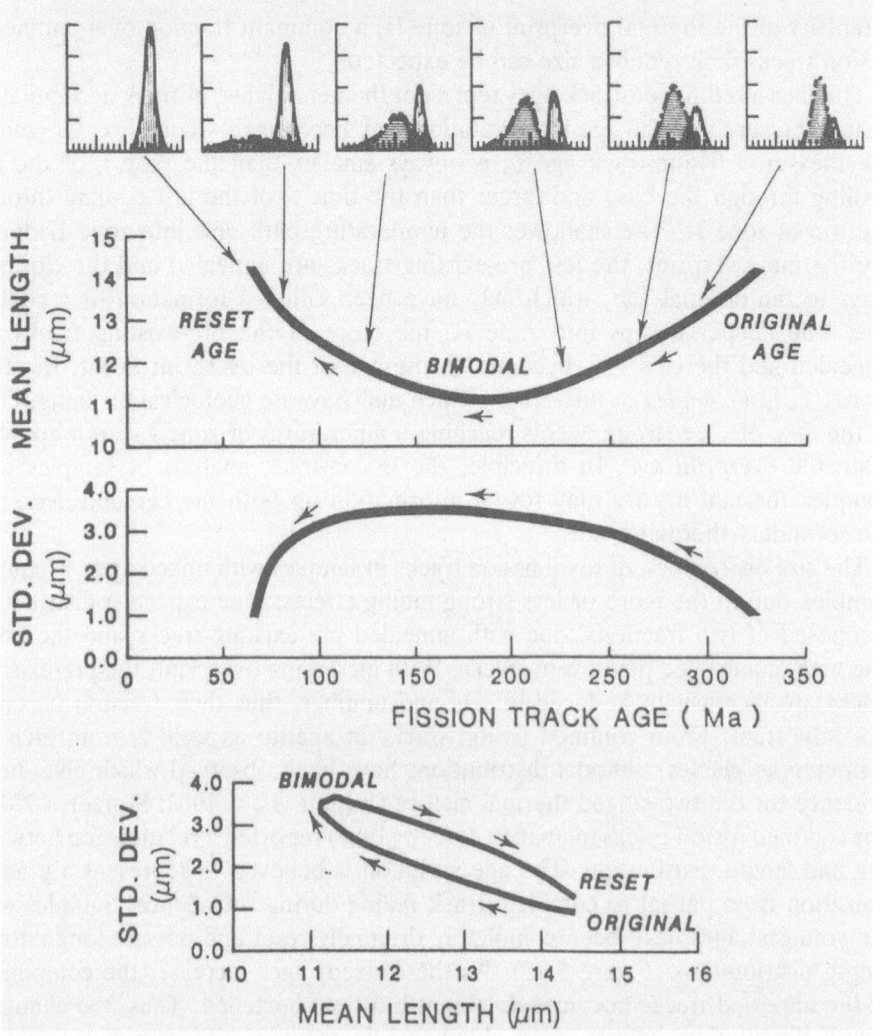

Figure 5.12. Relationship between length distribution of confined tracks and geological interpretation of complex apatite fission-track ages from Northern England (after Green, 1986). Transitions between the original age and the reset age due to increasing intensity of a thermal overprint *ca.* 60 Ma ago are visible. The length distributions are schematically separated into their two components produced before and after the overprint event.

event. The population of fossil tracks should consist of two components, one of them more or less shortened according to the event's intensity and the other one of unshortened post-event tracks, typically yielding bimodal size distributions. An early example of such partially reset fission-track ages has been presented by Storzer (1970b) for Permian volcanic glasses. Although mixed ages are often geologically meaningless, one may derive from them the intensity of the thermal event if its age is independently known. This approach has been applied to several rocks which were thermally affected by the Ries, Germany, meteorite impact (Miller and Wagner, 1979). Of special interest in this context are sedimentary rocks which contain detrital apatites. During subsidence, the previously accumulated fission tracks are annealed according to the prevailing geothermal gradient and, therefore, the thermal history can be reconstructed from fission-track analysis. Since fission tracks in apatite anneal in the same temperature-time regime as hydrocarbons maturate, fission-track analysis is an important tool for the evaluation of the hydrocarbon potential of sedimentary basins (Gleadow *et al.*, 1983) (Section 7.5).

If the thermal overprint is strong enough to reach temperatures of zone I $(T > T_{I/II})$, all pre-event tracks are erased. For short events such as extrusive contact and impact heating, which are followed by rapid cooling, the fission-track age dates these events, i.e., it is an *overprint age*. The post-event $T–t$ history and the track size distribution are analogous with $T–t$-path A (Section 5.5.1). An example is the Ries meteorite crater, where apatite and sphene fission-track ages of the suevite breccias yielded an age of 15 Ma for the impact (Miller and Wagner, 1979). For longer-lasting events, such as contact heating by igneous intrusion, metamorphism, and deep subsidence, the fission-track ages post-date the event and date the time of subsequent cooling of the rocks through the track retention temperatures according to the $T–t$-path (B). Actually, all the apatite fission-track ages measured by Wagner *et al.* (1977) in the Central Alps are post-metamorphic cooling ages. As to the intensity of the thermal event, in the case of complete track erasure, a lower limit of its temperature can be evaluated from the track-annealing characteristics of the mineral under consideration.

5.6. Age-Depth Profiles

Age-vs.-depth (or age-vs.-elevation) profiles are familiar to geologists. Usually, the geological age increases with depth according to the principle of superposition. Fission-track ages, however, especially from crystalline terranes, often behave conversely; they may decrease with depth even if the rocks become stratigraphically older. The trend and the shape of the age-vs.-depth profiles is closely connected with the interpretation of the fission-track ages and yields additional clues for deciphering the thermal evolution of the rocks. Therefore, in fission-track analysis, it is always rewarding to follow a double strategy, namely to sample vertical profiles in addition to horizontal profiles with regionally distributed samples. Several types of age-depth profiles can be distinguished. Also, composed profiles may exist which consist of different types in different depth-levels and, within the same section, different minerals may exhibit different types of age-depth profiles.

5.6.1. *Stratigraphic profile*

In the stratigraphic profile, the fission-track ages are formation ages and, hence, increase with depth according to the stratigraphic ages of the rocks. This presumes that they are not affected by any track fading due to secondary contact heating or ambient surface temperatures, that is, they must comply with the criteria defined for formation ages (AA-type). Stratigraphic profiles of fission-track ages (Figure 5.13) have been observed for glasses, apatites, and zircons from young volcanic beds (Seward *et al.*, 1980; Naeser *et al.*, 1987). An increase of fission-track ages with depth does not necessarily imply that they can be interpreted as formation ages. Applications of fission-track dating to stratigraphic sections, as the study by Ross *et al.* (1982) on Ordovician and Silurian stratotypes, are not reliable without support by rigorous track-length analyses, although the ages increase reasonably with depth.

5.6.2. *Uplift profile*

In the uplift profile, the fission-track ages are uplift ages and increase with elevation, i.e., reversed with respect to the stratigraphic profile. This is commonly observed in mountainous regions, when samples from surface outcrops at different elevation but short horizontal distance are compared. In the case of drill-hole samples, the ages decrease with depth. The slope of the curve in which the sampling elevation is plotted against the fission-track age, directly gives the uplift rate (Figure 5.14). This was first realized by Wagner and Reimer (1972), when finding apatite fission-track ages increasing with sampling elevation in the Central Alps. Hitherto, the approach is mostly applied to apatite but also to garnet and zircon in order to study the uplift evolution of mountain ranges (Benjamin *et al.*, 1987; Haack, 1983; Schaer *et al.*, 1975; Wagner *et al.*, 1977; Zeitler *et al.*, 1982a,b; Kohn *et al.*, 1984).

Uplift profiles within a vertical rock column develop during steady uplift. The fission-track clock is turned on when the rocks move through the track-retention isotherm whereby this isotherm is fixed to a constant depth. Then the fission-track ages which are all cooling ages, successively become larger with elevation and the slope of the profile reveals the uplift rate. However, due to two complications, such a reversed vertical age trend does not necessarily imply that it represents an uplift profile. The first complication is caused by the existence of a *partial annealing zone* instead of a single temperature threshold at which the fission-track clock is suddenly turned on (Section 5.6.4). Secondly, the assumption of a fixed depth of the closing temperature with respect to sea level is often not fulfilled. Unless the uplift is immediately compensated by denudation, the retention temperature stays in a static position relative to the surface but not to the sea level (Figure 5.10). This means that uplift profiles are essentially uplift-induced *denudation profiles*, provided the geothermal gradient stays constant. Further complications may be caused by disturbed geothermal gradients due to the surface morphology. For very fast and very slow uplift the age profiles become increasingly complex. Hence, very steep (>1000 m/Ma) and very flat (≈10 m/Ma) age profiles probably do not represent pure uplift profiles (Parrish, 1985; Van den haute, 1986b).

Owing to the partial annealing zone, sudden changes in the uplift rate cause curved sections in the profile (Section 5.6.4). Pure uplift profiles exhibit straight slopes which result from constant uplift rates but, conversely, not all straight slopes are uplift profiles. *Sensu stricto*, the term *uplift profile* is only applicable to

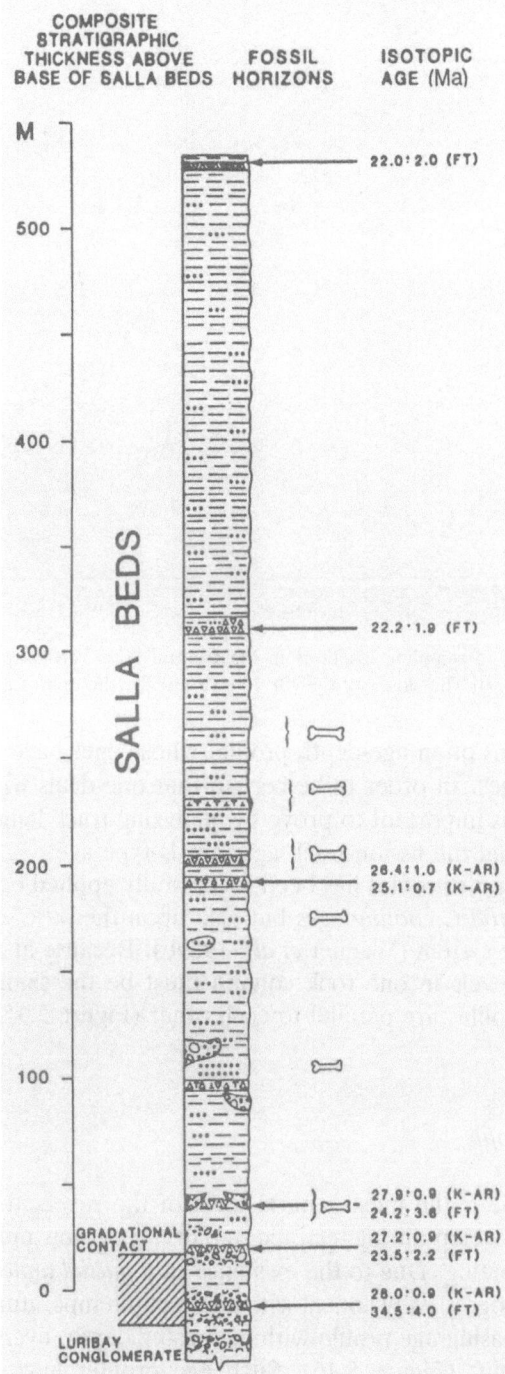

Figure 5.13.   Stratigraphic profile of the Late Oligocene-Early Miocene Deseadan Salla Beds, Bolivia (after Naeser *et al.*, 1987). The zircon fission-track ages of the volcanic ashes are formation ages and are consistent with the chronostratigraphy.

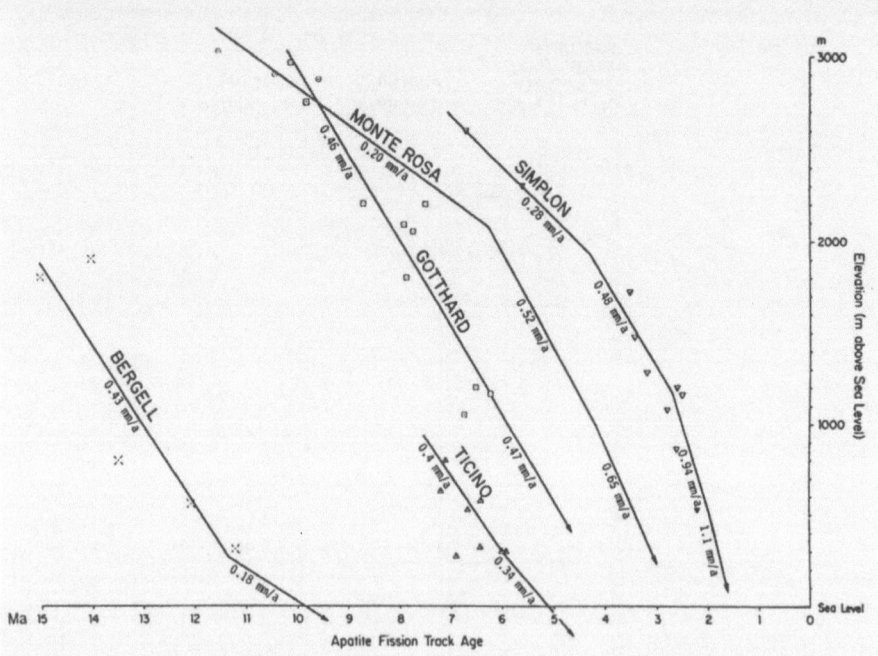

Figure 5.14. Elevation of sampling locations in the Central Alps versus apatite fission-track ages (from Wagner *et al.*, 1977). The slopes of the curves give directly the uplift rates.

such straight sections of an age-depth profile, whose ages have been proved to be B-type (cooling) ages. In order to be certain that one deals with uplift (or denudation) profiles, it is important to prove by analyzing track lengths (Gleadow and Fitzgerald, 1987) that the fission-track ages are B-type ages.

The concept of uplift profiles has been successfully applied based not only upon the '*conventional*' *100°C-cooling ages* but also upon the *60°C cooling ages* of the apatite fission-track system (Wagner *et al.*, 1989b). Because at any time the uplift rate for different levels in one rock column must be the same, the '100°C' and '60°C' uplift age profiles are parallel to each other (Figure 5.15).

5.6.3. *Complex profile*

The complex profile is the most general form of the reversed age-depth profile. Complex profiles develop during tectonic stability, very slow uplift, uplift changes, subsidence, or reheating. Due to the existence of *a partial annealing zone (PAZ)*, the fission tracks progressively anneal with depth and temperature within the PAZ, resulting in a decreasing age profile with increasing depth, even in the case of the absence of any uplift (Figure 5.16). Such age profiles have been observed in apatites from deep drill-hole samples under *in-situ* annealing conditions (Naeser and Forbes, 1976; Naeser, 1979; Gleadow and Duddy, 1981; Gleadow *et al.*, 1983; Hammerschmidt *et al.*, 1984). Their shape essentially corresponds to the present temperature profile as well as the previous thermal history (Gleadow *et al.* 1983;

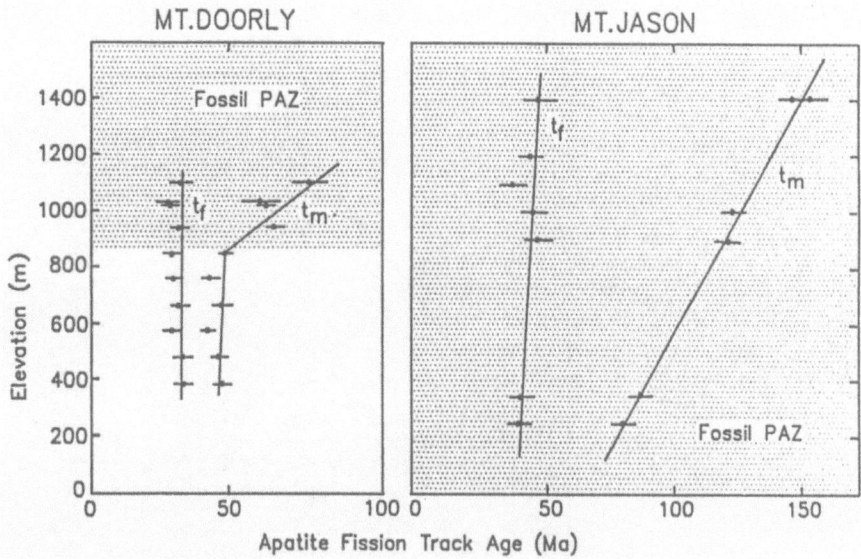

Figure 5.15. Elevation of sampling locations at Mt. Doorly and Mt. Jason, Transantarctic Mountains versus apatite fission-track ages (from Wagner *et al.*, 1989b). The measured ages t_m are shown as well as the ages t_f for the same samples. All t_f-ages are cooling ages to 60°C. In the case of t_m-ages, only those below the break-in-slope at Mt. Doorly are cooling ages to 100°C. The t_m ages above the break-in-slope at Mt. Doorly and all t_m-ages at Mt. Jason are complex ages of an uplifted PAZ and their slope does not represent an uplift rate. The uplift rate is given by the slope of the t_f-age profile and by the slope of the t_m-age profile below the break-in-slope at Mt. Doorly. The break-in-slope defines the bottom of the uplifted fossil PAZ. At Mt. Jason, the break-in-slope can be expected around sea level, indicating 800 m less uplift during the last 50 Ma.

N. D. Naeser *et al.*, 1990). When such PAZ profiles are uplifted to the surface, their original slope is more or less modified by uplift and cooling. In the case of uplift with a constant rate for a sufficiently long time (at least the duration a sample needs to move through the whole PAZ), an uplift profile develops. *Complex profiles* result in all other thermo-tectonic scenarios. The fission-track ages are C-type ages and the slope of the age-elevation profile no longer represents the uplift rate. The fossil uplift rate may have been smaller or larger than the one apparently represented by the curve's slope.

This discussion illustrates the complexity of the age-vs. elevation-profiles. Since all transitions between tectonic stability and fast uplift exist, no clear distinction between uplift and complex profiles is possible. A recent example was presented by Gleadow *et al.* (1984) and Gleadow and Fitzgerald (1987). They discovered a *break-in-slope* (Figure 5.15) in the elevation-vs.-fission-track-age curve for apatites from the Transantarctic Mountains, originally interpreted as the time when uplift acceleration set in. The profile is now believed to be composed of an uplift and a complex section with the break-in-slope representing the base of the uplifted fossil PAZ, i.e., the section of the curve below the break is an uplift profile with B-type ages and the section above the break is a complex profile with C-type ages. After long tectonic stability, the uplift set in about 10 Ma before the time (about 50 Ma) marked by the break. This revised interpretation is supported by length

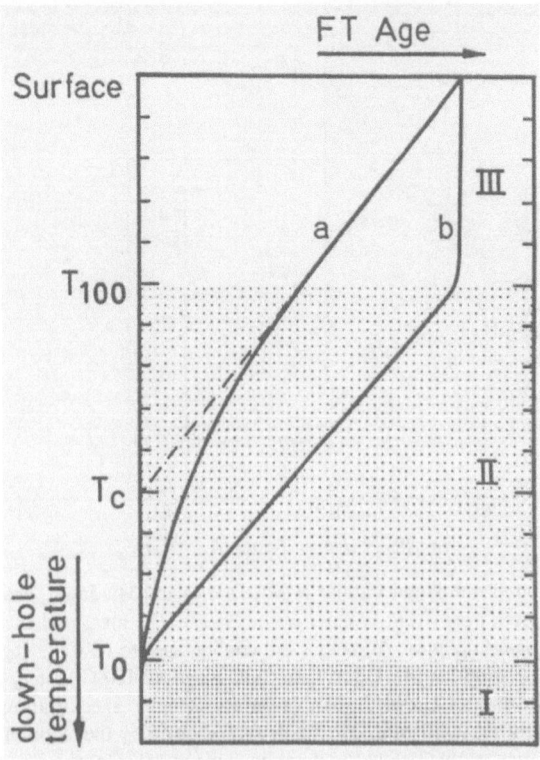

Figure 5.16. Model profiles of fission-track age versus down-hole temperature (=depth) for two tectonic settings: (a) constant uplift and (b) static condition (from Hammerschmidt *et al.*, 1984). Note that within the PAZ for neither setting, the slope of the age-vs.-depth-profile (II) represents an uplift rate. Only above the PAZ in the zone (III) of full track stability can the slope give the uplift rate.

criteria on confined and projected fission tracks (Gleadow and Fitzgerald, 1987; Wagner *et al.*, 1989b).

Complex profiles are regularly observed for apatites from crystalline terranes. Only a few age-depth profiles have been published for other minerals, such as zircon and sphene (Fitzgerald and Gleadow, 1988; Zaun and Wagner, 1985). The discrimination between uplift profiles and complex profiles is rather difficult. When one finds increasing fission-track ages with elevation (or decreasing with depth), it is safe to first assume that one deals with complex instead of uplift profiles. Probably, many of the age-vs.-elevation profiles reported in the fission-track literature are not as claimed, pure uplift profiles. Only after careful consideration of other parameters, in particular the track lengths and the profile shape, can one confidently interpret linear sections as uplift profiles. Once a complex profile is identified to represent an uplifted fossil PAZ, it allows the evaluation of the amount and rate of uplift (Gleadow and Fitzgerald, 1987; Wagner *et al.*, 1989b).

In the foregoing discussion, the heat source, which anneals the tracks, is situated below in the Earth's depth. However, in the rare cases of very large meteorite impacts, the heat source acts from the Earth's surface downward. The impact-induced transient temperature gradient is inverse to the normal geothermal gradient. The shock wave dissipates its energy as heat to the surrounding ground leading – apart from evaporation and impact metamorphism – to thermal track fading. Provided the impact was sufficiently strong, the fission-track systems are partially or even completely reset. The apparent fission-track ages increase with depth, forming a characteristic impact profile, which is essentially an overturned complex profile. Such a profile has actually been observed for apatites selected from a drill core penetrating the crystalline floor of the Ries crater (Miller and Wagner, 1979). Depending on the degree of the resetting of the fission-track systems, resulting in complex or formation ages, the track analysis reveals information on the age and mechanism of crater formation (Section 7.7). Another example of an inverted geothermal palaeogradient was reported by Carpena (1985) for the Gran Paradiso/Alps basement. During the nappe overthrust 40 Ma ago, the underlying basement was heated causing partial track annealing in zircon. The inverted gradient of this thermal overprint is reflected by the zircon fission-track ages which increase towards lower elevations, but this interpretation has been called to question by Hurford and Hunziker (1989).

5.6.4. *Modelled uplift age-depth profiles*

Fission-track age-depth profiles contain important geological information on the thermal history of crustal sections. Age-depth profiles can be studied on rocks from vertical surface outcrops in mountainous regions. Since outcrops of sufficient vertical extension are rare in nature, one has to compromise and to include samples taken within short horizontal distances from each other. This practice is acceptable as long as the samples belong to the same tectonic block. More convenient are vertical sampling profiles from drill cores.

Apatite with its low thermal track stability is the most important material for investigating age-depth profiles. During uplift the apatite-containing rocks gradually cool through the track-retention temperature. Apatites which are situated higher in the rock column cool earlier through this temperature than those of a lower position. Consequently, the fission-track age decreases from surface to depth and finally reaches zero. The elevation-vs.-age curve should record the uplift movement with its slope representing the uplift rate. However, this simple model requires several assumptions: (a) The amount of uplift is immediately compensated by denudation, i.e., the depth of the isotherms is absolutely fixed; (b) the geothermal gradient stays constant with time; (c) the retention is a discrete threshold at which the tracks become stable and at which the fission-track clock is suddenly turned on. In geological reality, all of these assumptions are easily invalidated. In the following, the significance of the partial annealing zone (PAZ) is discussed. For the discussion of the first two assumptions, see Figure 5.10.

Although the PAZ concept is not strictly valid for apatite because, even at ambient surface temperatures, no complete track stability exists, it is nevertheless useful (Wagner *et al.*, 1989b). For this discussion, a simple PAZ model (Figure 5.3) is selected with well-defined lower and upper temperature boundaries and a linear increase of track stability from 0% at the bottom z_0, 50% at the middle z_{50} and 100% at the top z_{100} of the PAZ. The model approximates reasonably the actual observations in bore-holes as, for example, the Otway (Figure 4.11). The model undoubtedly represents an oversimplification but the conclusions that can be drawn from it are basically correct. Based on this model, the age-depth profiles for a given tectonic setting can be predicted and compared with those actually observed. Four types of tectonic settings are discussed: constant uplift, sudden acceleration and deceleration of uplift, as well as sudden uplift after a long period of tectonic stability. These types are illustrated in Figures 5.17 to 5.20 by the elevation-vs.-time path (shaded) of the original PAZ. The following symbols are used:

Dz = thickness of the PAZ,
u = constant rate of uplift,
t_c = time of sudden change in uplift rate from u_1 to u_2,
m = slope of straight sections of elevation-vs.-age profiles,
t_1 and t_2 = ages of upper and lower limit of the bend (break) in the age-depth profile, respectively,
z_1 and z_2 = depths of upper and lower limit of the bend (break) in the age-depth profile.

(a) *Constant uplift.* When a sample within a rock column which is uplifted at a constant rate u enters the PAZ-bottom z_0, the fission-track accumulation starts. As time t passes, the sample moves steadily upward through the PAZ. Simultaneously, its temperature drops and the rate of track accumulation linearly increases from 0 to 100%. Consequently, within the PAZ, the track density (=fission-track age) increases as a quadratic function of time and elevation until the PAZ top z_{100} is reached. Afterwards, within the zone of full track stability, the tracks linearly accumulate with time and elevation. The resulting age-depth profile is shown in Figure 5.17. The extrapolation of the straight section of the age-depth profile intersects the depth axis in the PAZ middle z_{50}. Thus, the fission-track ages t_m of the straight section give the time elapsed since the various levels of the rock column were uplifted through depth z_{50} and the slope m is the uplift rate u at that time. An apatite fission-track age-depth profile of this kind has been observed at the Urach-III drill core (Hammerschmidt et al., 1984).

(b) *Sudden uplift increase.* In this tectonic setting, the constant uplift rate u_1 accelerates suddenly at time t_c to uplift rate u_2, which, again, is constant (Figure 5.18). Before the change of uplift occurs, an age-depth profile develops, such as in case (a). Starting at time t_c, the track accumulation evolves along steeper profile lines according to the faster uplift rate, since the slope of the age-depth profile increases with the uplift rate. The resulting age-depth profile is plotted in Figure 5.18. It appears that the uplift rate u_1 is still preserved in the upper straight section with the slope m_1. However, below the point (t_1, z_1) the profile is bent (convex if seen from above). Below the point (t_2, z_2), the profile again becomes straight with the slope m_2 representing the new uplift rate u_2. The curved section between the

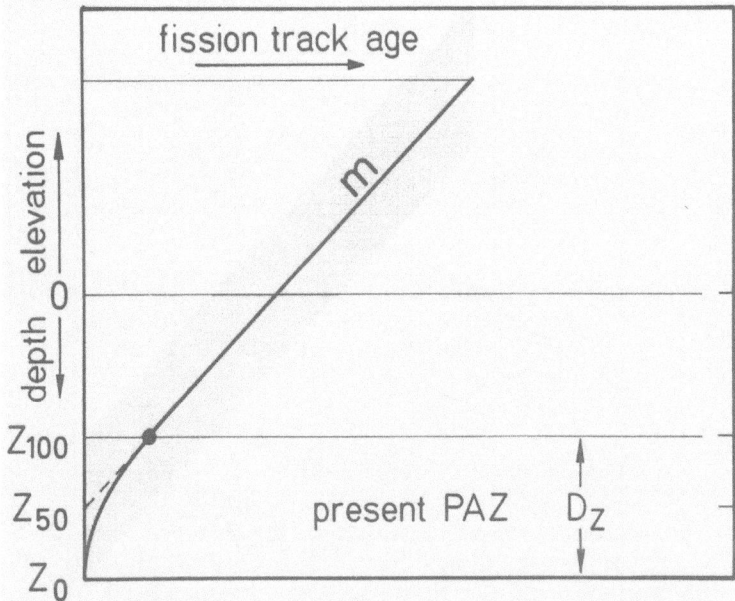

Figure 5.17. Model profile of fission-track age versus depth (or inversely elevation) for a tectonic scenario with constant uplift u. The present partial annealing zone (PAZ) is situated between depths z_0 and z_{100}. The slope m of the profile above the PAZ top z_{100} gives the uplift rate u. The trace of the uplifted fossil PAZ is shaded.

two straight sections of the profile represents the uplifted palaeo-PAZ of age t_1. Note that a sudden change in uplift does not cause a break-in-slope of the age-depth-profile, but a bend-in-slope. The time t_c, when the uplift rate changed, lies within this curved section between t_1 and t_2, quantitatively described by the following equations

$$(z_1 - z_2) = Dz, \tag{5.4}$$

$$(t_1 - t_2) = Dz/2 \times (1/m_1 + 1/m_2), \tag{5.5}$$

$$t_c = t_1 - Dz/(2 \times m_1), \tag{5.6}$$

$$t_c = t_2 + Dz/(2 \times m_2). \tag{5.7}$$

Of both points (t_1, z_1) and (t_2, z_2) of the bend-in-slope, the lower one (t_2, z_2) has a more pronounced curvature. It postdates the change-in-uplift t_c by $Dz/(2 \times m_2)$ (Equation (5.7)), which is the time required for passing half the PAZ with the new uplift rate u_2. If the bend at point (t_2, z_2) is recognized in an observed age-depth-profile, the age t_c of uplift change can be calculated according to Equation (5.7). Obviously, another bend occurs further down at the top of the present PAZ, as in case (a).

(c) *Sudden uplift decrease.* In this tectonic setting, the constant uplift rate u_1 suddenly declerates at time t_c to the uplift rate u_2, which, again, is constant. The resulting age-depth profile is plotted in Figure 5.19. As in case (b), the uplift rate

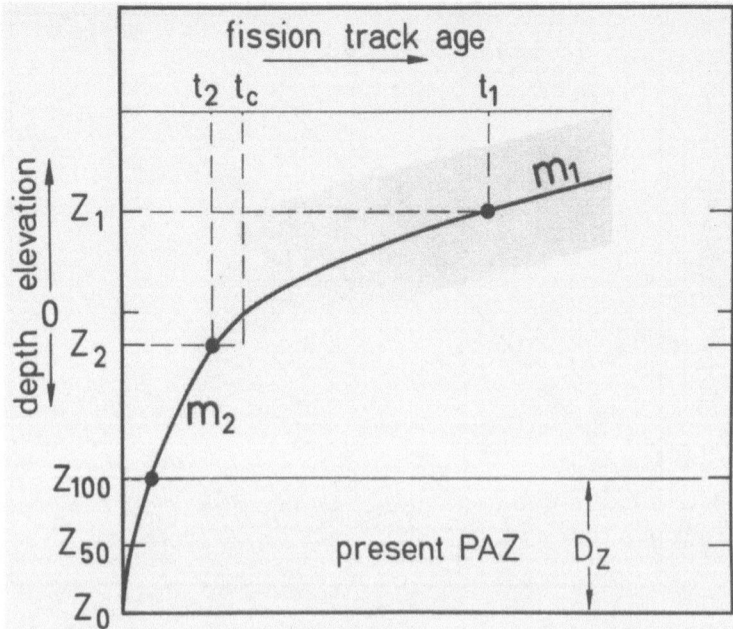

Figure 5.18. Model profile of fission-track age versus depth (or inversely elevation) for a tectonic scenario with sudden uplift increase at time t_c between constant uplift rates u_1 and u_2. The present partial annealing zone (PAZ) is situated between depths z_0 and z_{100}. Note that the sudden change of uplift causes a curved section instead of a break in the slope of the profile. The straight slope m_1 of the profile above the curved section (z_1) gives the uplift rate u_1 and the straight slope m_2 below the curved section (z_2) but, above z_{100}, gives the uplift rate u_2. The width $z_1 - z_2$ of the curved section represents the thickness of the uplifted fossil PAZ. The age of sudden uplift change t_c lies within the curved section between t_1 and t_2. The trace of the uplifted fossil PAZ is shaded.

u_1 is still preserved in the upper straight section with the slope m_1, but below the point (t_1, z_1), the profile is bent concavely (if seen from above). Below the point (t_2, z_2), the profile again becomes straight with the slope m_2 representing the new uplift rate u_2. The curved section between these two straight sections of the profile represents again the uplifted palaeo-PAZ of age t_1 and Equation (5.4) to (5.7), derived for the acceleration case, apply equally for the deceleration case. Note that in this case, the curvature of the bend-in-slope is most clearly pronounced close to its higher limit (t_1, z_1). It predates the change-in-uplift t_c by $Dz/(2 \times m_2)$. Accelerating and decelerating uplift scenarios have been observed on apatites from the Central Alps (Section 7.2.2).

(d) *Tectonic stability followed by uplift.* In this tectonic setting, a constant uplift rate u sets in after a long-lasting tectonic standstill. During the static stage, a linear age-depth profile has developed within the PAZ. Its slope m_{PAZ} depends on Dz and the duration of the static stage (Figure 5.20). Starting at time t_c, the track accumulation evolves along steeper profile lines according to uplift rate u. After the bottom of the original PAZ is uplifted through z_{100}, a pronounced convex break-in-slope appears at point (t_2, z_2) in the age-depth profile. The straight

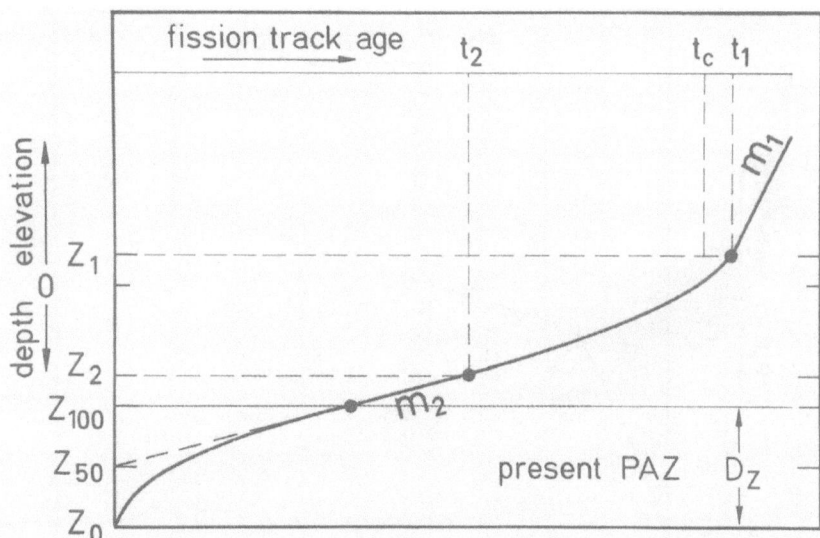

Figure 5.19. Model profile of fission-track age versus depth (or inversely elevation) for a tectonic scenario with sudden uplift decrease at time t_c from constant uplift rates u_1 to u_2. The present partial annealing zone (PAZ) is situated between depths z_0 and z_{100}. Note that the sudden change of uplift causes a curved section instead of a break in the slope of the profile. The straight slope m_1 of the profile above the curved section (z_1) gives the uplift rate u_1 and the straight slope m_2 below the curved section (z_2) but, above z_{100}, gives the uplift rate u_2. The width $z_1 - z_2$ of the curved section represents the thickness of the uplifted fossil PAZ. The age of sudden uplift change t_c lies within the curved section between t_1 and t_2. The trace of the uplifted fossil PAZ is shaded.

slope m_2 below the break corresponds to the uplift rate u and the straight slope m_1 above the break is the uplifted fossil PAZ profile. Note that the slope m_1 approaches m_{PAZ} but is somewhat flatter and it does not represent an uplift rate. In order to distinguish such fossil PAZ slopes from genuine uplift slopes, length measurements on tracks are required (Gleadow and Fitzgerald, 1987; Wagner *et al.*, 1989b). The base of the uplifted fossil PAZ is marked by z_2 and the top by the upper concave break-in-slope at z_1. For the quantitative description of the uplifted fossil PAZ and the age t_c of uplift-initiation Equations (5.4) and (5.7) are valid, Equation (5.6) is not valid, since m_1 is not an uplift rate. Equation (5.5) is replaced by

$$(t_1 - t_2) = Dz/m_1. \tag{5.8}$$

For apatites from the Transantarctic Mountains, a vertical section bracketing the lower, convex break-in-slope has actually been observed (Gleadow and Fitzgerald, 1987). Age-depth profiles for the 60°C cooling ages have also been observed and modelled in the same samples (Wagner *et al.*, 1989b). Since these, cooling ages refer to a narrower temperature zone at the top of the PAZ – compared to the wider temperature range of the PAZ for the 'conventional' apatite ages – the complications discussed here are probably less severe for them. The age-depth profiles of the 60°C cooling ages should therefore resemble the uplift trace of the fossil PAZ top in Figures 5.17 to 5.20.

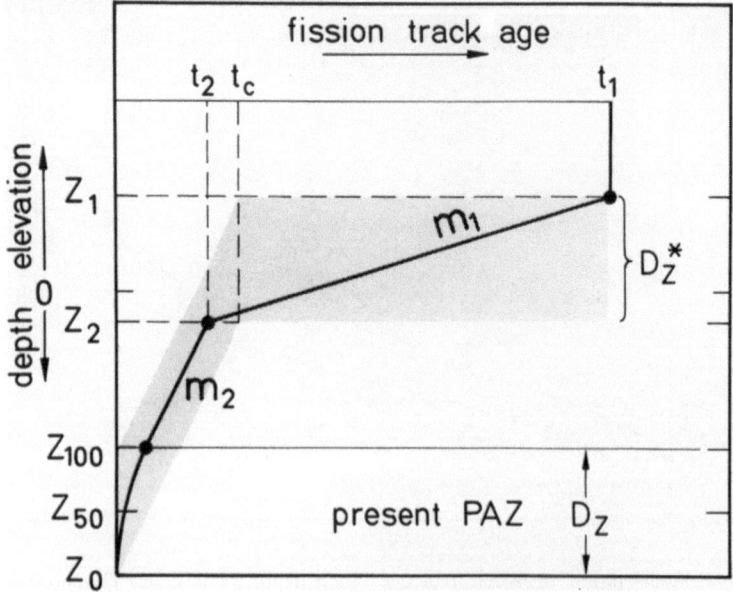

Figure 5.20. Model profile of fission-track age versus depth (or inversely elevation) for a tectonic scenario with sudden uplift beginning at time t_c with constant uplift rate u_2 after a long tectonic stand-still. The present partial annealing zone (PAZ) is situated between depths z_0 and z_{100}. Note that the sudden change of uplift causes a break in the slope of the profile. The section with slope m_1 above the break (z_2) is the uplifted fossil PAZ and does not give an uplift rate. The straight slope m_2 below the break (z_2) but, above z_{100}, gives the uplift rate u_2. The width $z_1 - z_2$ between the two breaks represents the thickness of the uplifted fossil PAZ. The age of sudden uplift change t_c predicates the break t_1. The trace of the uplifted fossil PAZ is shaded.

For assumed temperature histories, the apatite fission-track ages – as well as the mean confined track-lengths – versus depth, have been computer-modelled. The model-predicted depth-dependent fission-track parameters *age* and *track length* can then act as rigorous constraints and help to exclude unrealistic geological thermal histories for an actual observed profile (Green *et al.*, 1989; Fitzgerald and Gleadow, 1990).

5.7. Correction of Track Fading

During the early days of fission-track dating it was noticed that fossil fission tracks are commonly smaller in size in minerals and glasses than the induced fission tracks (Maurette *et al.*, 1964; Fleischer *et al.*, 1964b; Bigazzi, 1967). This is now known to be a general phenomenon. Fossil tracks have a size distribution shifted towards smaller sizes than induced ones, yielding a lower average size. This difference is not due to nuclear physical reasons. A freshly formed and etched spontaneous fission track has the same size – that is, length in the case of minerals and diameter in the case of glasses – as an induced fission track. The difference is caused by natural annealing of the fossil tracks within the sample's geological lifetime. The annealing degree of fossil tracks may vary widely from sample to sample. In some samples, the difference in track size is minor and hardly detectable, in others it is strong and obvious.

As track fading does not only affect the length, but also the number of tracks intersecting a surface, the areal density of fossil tracks and, thus, the age, is also reduced. Early experimental proof that annealing reduces both track length and density, has been reported for mica and glass by Maurette *et al.* (1964) and Berzina *et al.* (1967). Due to the correlation between track size and track-density reduction, it was a logical step to use length measurements for correcting ages lowered by fading. This was first accomplished by Storzer and Wagner (1969). Working on australite tektites, they found apparent fission-track ages between 0.13 and 0.7 Ma with a clear correlation between fossil track size and age. From annealing experiments, a curve relating the track density and track-size reduction was established. When using this curve, corrected ages of 0.7 Ma for all specimens were obtained, which is also in good agreement with the K-Ar-age-determination. This procedure became known as the track-size correction technique. In the meantime, another correction technique, known as the 'plateau technique', has been developed. Both techniques are discussed below.

5.7.1. *Track-size correction technique*

In the track-size correction technique (Storzer and Wagner, 1969), the smaller size of partially faded fission tracks is compared to freshly induced tracks. From the degree of track-size reduction, the original prefading track density can be recalculated by using the experimentally established correlation between track density and size reduction due to partial fading. Such correlation curves have been called correction curves (Figure 4.9). Correction curves have been determined for many materials which are of interest in fission-track dating, such as tektites, impact glasses, and lunar glass spherules (Durrani and Khan, 1970; Gentner *et al.*, 1969a 1973; Komarov and Raichlin, 1976; Storzer and Wagner, 1971), volcanic glasses (Arias *et al.*, 1981; Komarov *et al.*, 1972; Storzer, 1970b; Suzuki, 1973), apatites (Green, 1988; Märk *et al.*, 1973; Nagpaul *et al.*, 1974b; Sharma *et al.*, 1978; Wagner and Storzer, 1970, 1972; Watt and Durrani, 1985), sphene (Märk *et al.*, 1981; Nagpaul *et al.*, 1974b), epidote (Saini *et al.*, 1978), tourmaline (Nand Lal *et al.*, 1977b), biotite (Nagpaul *et al.*, 1974b; Nand Lal *et al.*, 1976b), muscovite (Saini *et al.*, 1977), chlorite (Sharma *et al.*, 1977), phlogopite (Parshad *et al.*, 1978), and vermiculite (Sharma *et al.*, 1979). Note that in some minerals, different kinds of track lengths, i.e., on confined and projected tracks, have been determined. These do not necessarily produce identical correction curves. In glasses, the major axis of the elliptical etch pits is usually measured. Fleischer and Hart (1972), Somogyi and Nagy (1972), Dakowski (1978), Hashemi-Nezhad and Durrani (1981a) and Van den haute (1985) have calculated theoretical correction curves, which closely resemble those actually observed.

The size-correction technique is very time-consuming, because it necessitates tedious size measurements. Also, correction curves must be newly determined for each type of material and experimental condition. The *track-size correction technique* has been applied to tektites (Durrani and Khan, 1970; Glass *et al.*, 1973; Storzer and Wagner, 1969; Storzer and Wagner, 1971; Storzer *et al.*, 1973), volcanic glasses (Selo and Storzer, 1981; Storzer, 1970b; Suzuki, 1973), muscovite (Mehta and Nagpaul, 1971), and apatite (Wagner and Storzer, 1970, 1975). Most of the corrected ages obtained for glasses, which had suffered their track fading during a relatively recent event, turned out to be concordant with independently known ages. If the thermal event occurred some time far back in the geological past, then the present track population consists of two populations, the pre-event one with faded tracks and the post-event one with unfaded tracks. In such a case, the correction procedure must be applied only to the pre-event population, which may be difficult to recognize and separate. For such complex thermal histories as well as for other materials than glass, the track-size correction technique is not yet sufficiently elaborated.

5.7.2. *Plateau-correction technique*

The plateau-correction technique (Storzer and Popeau, 1973; Galazka and Burchart, 1976) uses the higher thermal stability of partially faded fission tracks compared to freshly induced tracks. The apparently higher resistance against further annealing, which fossil tracks with previous fading show in comparison to freshly induced tracks, can be ascribed to a gradual increase of the activation energy of the annealing process with increasing degree of annealing. Two portions of each sample, one with fossil and the other with induced fission tracks, are simultaneously annealed in the laboratory in several steps of increasing temperature (*isochronal plateau*) or duration (*isothermal plateau*, Burchart *et al.*, 1975) under controlled conditions. The ratio of fossil to induced track densities is plotted as a function of temperature (or time, respectively) for each annealing step. The ratio increases for a sample with partial loss of fossil tracks, as long as the fading rate of the induced tracks is larger than that of the fossil tracks. Finally, when both fading rates become equal, a plateau value is reached (Figure 5.21). The age calculated from the ratio within the plateau region corrects for the partial track loss.

The *plateau-correction technique* is even more time-consuming than the track-size technique, because, after each annealing step, the spontaneous and induced fission-track densities have to be counted. For identifying the plateau region, at least a minimum of three data points are required. Therefore, a modified, speeded-up version was developed by Miller and Wagner (1981) and further improved by Bigazzi *et al.* (1990) and Westgate (1989). After counting the tracks and measuring their size without any laboratory pre-annealing, the degree of fossil-track fading is estimated. Then both portions of the sample are annealed to such an extent that the plateau region is safely reached in one step. Reaching the plateau is characterized by identical, although diminished, average size of both the fossil and induced fission tracks. Plateau-corrected fission-track ages have been obtained for

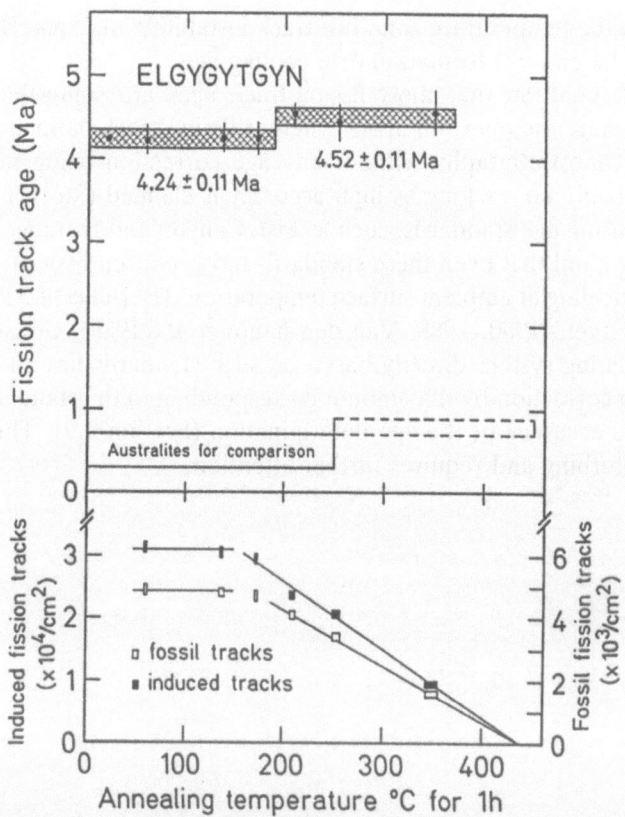

Figure 5.21. Isochronal annealing of fossil (ρ_s) and induced (ρ_i) fission-track densities in impact glass Elgygytgyn, Siberia (Storzer and Wagner, 1979). With progressive annealing temperature the fossil (ρ_s) and induced (ρ_i) fission-track densities decrease but their ratio ρ_s/ρ_i increases and reaches a plateau value from which the corrected age is calculated. The much younger fission-track plateau age of australites, shown for comparison, excludes Elgygytgyn as the parental crater for the australites.

tektites and obsidians (Arias *et al.*, 1981; Naeser *et al.*, 1980; Seward *et al.*, 1980; Storzer and Poupeau, 1973; Storzer and Wagner, 1980; Westgate *et al.*, 1987; Westgate, 1988) as well as apatites (Naeser and Fleischer, 1975; Poupeau *et al.*, 1980; Storzer and Poupeau, 1973) were corrected. Most of the plateau-corrected fission-track ages of natural glasses agree satisfactorily with independently known ages. For apatite and other minerals, however, the validity of this correction procedure has not been convincingly demonstrated.

When both correction techniques were applied to the same samples, mostly concordant corrected ages were determined for volcanic glasses (Bigazzi *et al.*, 1976; Macdougall, 1976; Wagner *et al.*, 1976; Arias *et al.*, 1981; Miller and Wagner, 1981) as well as tektites and crater glasses (Storzer and Wagner, 1977, 1979). For the interpretation of corrected fission-track ages, it must be kept in mind that only partial track fading can be corrected. Periods of complete track fading in the sample's geological history obviously cannot be retrieved. Consequently, the corrected fission-track age should date the time t_i when the sample passed for the last time from the temperature zone I of track instability into zone II (Figure 5.8). This age may be either a formation or a cooling age.

With the recognition that most fission-track ages are somewhat reduced by thermal annealing, the question arises whether fission-track dating can be reliably applied to chronostratigraphy without any age correction being necessary. The answer is probably no, as long as high accuracy is claimed (Storzer and Wagner, 1982). When using age standards, such as Fish Canyon and Durango apatites, one has to keep in mind that even these standards have reduced fission-track ages due to thermal annealing at ambient surface temperatures (Bertel *et al.*, 1977; Gleadow *et al.*, 1986; Green, 1980, 1988; Van den haute *et al.*, 1988). Consequently, any fission-track-dating system directly based on such standards has an automatically built-in fading correction by the amount corresponding to the standard used which may affect the accuracy of the age determination (Section 3.9). This problem is, of course, disturbing and requires further attention.

CHAPTER 6

Applicability

6.1. Time Span

The time span which is covered by fission-track dating surpasses that of any other radiometric dating method. Man-made objects only several tens of years old, as well as of rocks and meteorites several billions of years old, have been successfully dated with fission tracks. This wide range of applicability is surprising with regard to the large half-life of 8×10^{15} years for the spontaneous fission of uranium. The time span covered by the fission-track dating method is essentially determined by the areal density of fission tracks (tracks/cm^2) in the sample. In order to calculate a precise age, hundreds or, preferably, thousands of fission tracks have to be counted. Consequently, the fission-track dating presupposes materials of sufficient age or of sufficient uranium content, or both. Finally, it comes to the product of age and uranium content which must fall between specific limits, which is depicted in Figure 6.1. If large counting areas are available, such as in the case of obsidians, tektites, or micas, track densities as low as a few tracks/cm^2 might be sufficient. However, when dating 100 μm sized grains of apatite, zircon, or sphene, track densities of more than 10^5 tracks/cm^2 are required. With the optical microscope, the upper limit for track resolution and counting is reached around 10^7 tracks/cm^2. For higher track densities, the scanning electron microscope or the transmission electron microsope can be used. In the latter case, track densities of several 10^9 tracks/cm^2 have been counted on replicas of an etched surface (Weiland et al., 1980).

6.2. Geological Materials

As uranium contents in the order of μg/g, which yield sufficient fission events in geological time, are common in nature, and as fission fragments leave etchable tracks in all di-electric solids, one might expect that many geological materials should, at least in principle, be suitable for fission-track dating. Also, the amount of material required is comparatively little, since single radioactive decay events, i.e., the fission tracks, are counted. During the past 25 years, quite a lot of terrestrial materials have been investigated for the occurrence of fossil fission tracks. Only in relatively few of them have fossil fission tracks been identified with

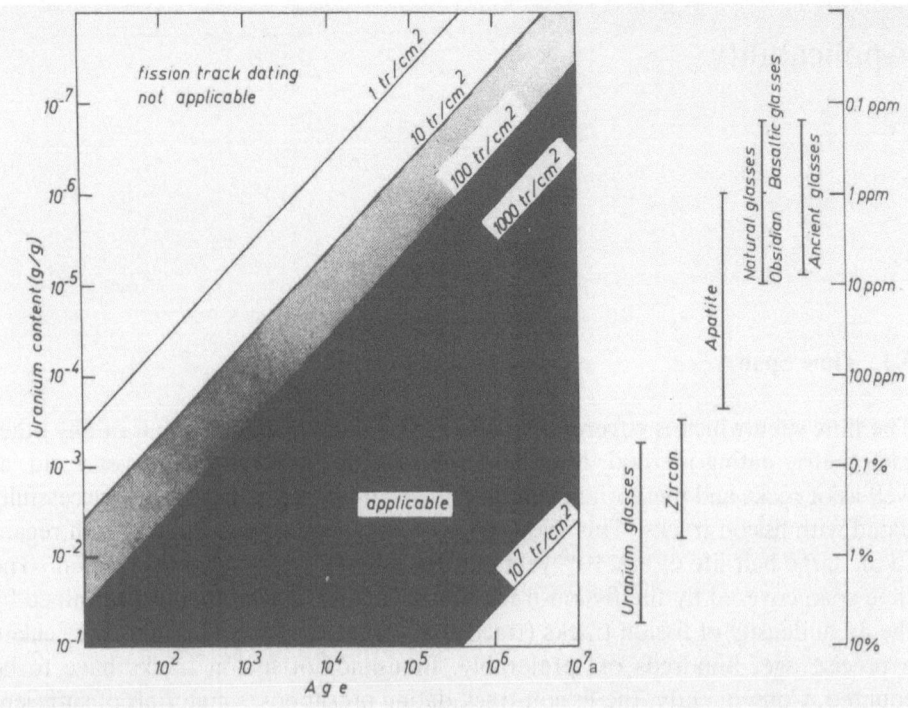

Figure 6.1. Applicable time-span of the fission-track clock and its relationship to the sample's uranium content (after Wagner, 1978). Counting track densities below 10^2 tracks/cm^2 is very time-consuming. Track densities above 10^7 tracks/cm^2 require scanning or transmission electron instead of optical microscopy techniques.

certainty and, in fact, only four or five of them are routinely used in fission-track dating. This limitation is due to various reasons such as inappropiate etching conditions, very low uranium content, sensitive thermal annealing of tracks, high background of spurious etch pits, and, last but not least, incomplete investigation of other materials. Apatite is by far the most favoured mineral for fission-track dating, followed by zircon, natural glass, mica and sphene. In this section, all terrestrial minerals and glasses, in which fossil fission tracks have been observed and more or less successfully counted, are briefly discussed together with some of their properties relevant to fission-track dating. The etching and annealing properties are listed in Appendices A and B, respectively. General information concerning dating techniques and sample preparation is given in Sections 3.6 and 3.7.

6.2.1. *Allanite*

Allanite (= orthite) $(Ca,Ce)_2(Al,Fe,Mg)_3Si_3O_{12}(OH)$ is a uranium- (up to 700 $\mu g/g$) and thorium-rich accessory mineral in igneous and metamorphic rocks. So far, only a few allanite fission-track ages have been reported (Naeser *et al.*, 1968, 1970). The samples were collected from various granodiorites in the U.S.A. The allanite, as well as the epidote fission-track ages, agree well with radiometric ages on co-existing minerals. Sample preparation and etching is similar to epidote. This promising mineral deserves more attention by the fission-track community. Due to high thorium concentrations, care must be taken that allanites are irradiated only in well-thermalized reactor positions in order to prevent fast neutron-induced fission of thorium (Section 3.5).

6.2.2. *Amazonite*

Amazonite is a variety of the potassium feldspar microcline $KAlSi_3O_8$. A crystal from Broken Hill, Australia, was polished and etched for several minutes in hydrofluoric acid. Scanning an area of 0.125 cm^2 revealed eight fossil fission tracks (Haack, 1973). A uranium determination and age calculation has not been reported. For feldspar, one expects rather stable thermal track retention (Haack, 1972) and low uranium content. More studies on the fission-track systematics of feldspars are desirable.

6.2.3. *Apatite*

Apatite $Ca_5(PO_4)_3(F,Cl,OH)$ is a common accessory mineral in various kinds of rocks. It is found in plutonic and volcanic igneous rocks, ore veins and metamorphic shists as well as in sediments. Its wide occurrence in many geological environments is one of the reasons for its frequent use in fission-track dating. Another favourable prerequisite is its relatively high uranium content in the order of 1 to 100 μg/g.

For polishing, the crystals may be mounted in random orientation or, what sometimes may be preferable due to annealing and etching anisotropy effects (Green and Durrani, 1977; Green, 1981b; Laslett *et al.*, 1984), parallel to the prism faces. Apatite is routinely etched in diluted or concentrated nitric acid at room temperature (Figures 2.15, 2.17, 2.22, 2.25, 2.26, 2.27, 6.2 and 6.3). The population or the external-detector technique may be applied as dating procedures. Experimental details are described elsewhere (Wagner, 1969a; Naeser, 1979; Gleadow and Lovering, 1975). However, a technique of single-grain analysis has to be used if the uranium distribution varies either within (Figure 6.3) or among the grains, which might be the case for detrital apatites (Gleadow, 1981). Also, variations in the chemical composition, essentially the Cl/F ratio, among the grains influences the annealing characteristics and, thus, the apparent fission-track age of the single apatite grains (Green *et al.*, 1985). Therefore, it is recommended to use the external-detector technique wherever there is a danger that apatites are chemically different, which can be expected for detrital grains in sediments, for paragneisses and tephra layers. The simpler population technique can be used for most igneous rocks which contain apatite of one generation.

The first apatite ages were reported in the very early years of fission-track dating (Fleischer and Price, 1964d; Naeser, 1967; Wagner, 1968). In the meantime, the number of apatite fission-track ages, published in more than 100 scientific

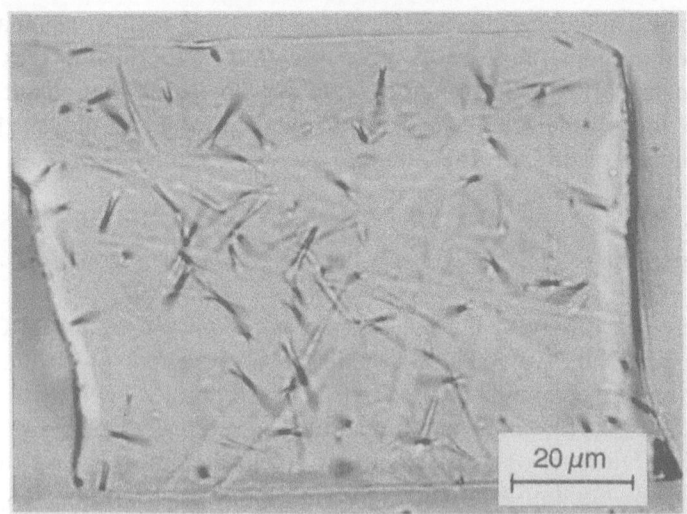

Figure 6.2.   The prism face of an apatite crystal from Oberflockenbach, Odenwald was etched for 35 s in 5% nitric acid at 20°C in order to reveal induced fission tracks. (From Wagner, 1972a.)

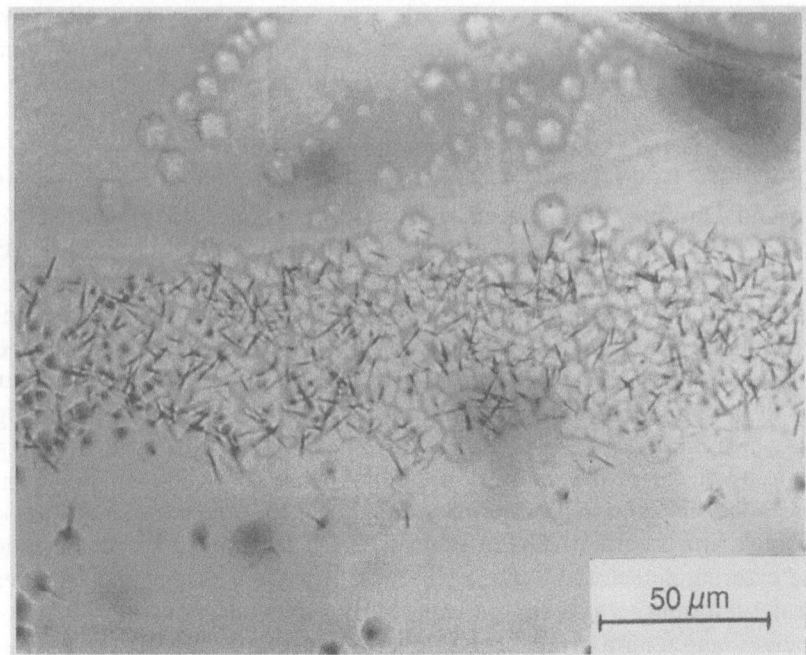

Figure 6.3.   The basal face of an apatite crystal from Waldstein, Fichtelgebirge was etched for 45 s in 5% nitric acid at 20°C in order to reveal induced fission tracks. The tracks reflect the zonal distribution of uranium with 450 μg/g U (high track density) against 7 μg/g U (low track density). (From Wagner, 1973.)

contributions, amounts to at least 1500 dates. The dated apatites originate from virtually all types of rocks except calcareous sediments. The ages range from several hundred thousand to more than one billion years. At present, apatite fission-track dating is predominantly used for deciphering the thermo-tectonic history of rocks. This application as a geothermo-chronometer is based on the relatively low thermal stability of tracks around 100°C (Wagner, 1968; Naeser and Faul, 1969). The potential of apatite fission tracks as a geological low temperature thermo-chronometer is supplemented by the information contained in the length distribution of the fossil tracks (Gleadow *et al.*, 1986; Wagner, 1988). Therefore, it is advisable to also report, together with any apatite fission-track age, the length distribution of the confined as well as the surface tracks. Another important source of information are vertical age profiles. Hence, when collecting rocks for apatite fission-track studies, one should ideally follow a double strategy, namely collecting along horizontal as well as vertical profiles, which can be achieved in areas with sufficient relief or by analysing drill-hole samples. Compared to the thermo-tectonic applications, the chrono-stratigraphic use of the apatite fission-track system is – as reflected by the literature – of less importance. The accuracy of absolute apatite fission-track ages is, to some extent, lowered due to minor, but not negligible track fading in apatite at ambient surface temperatures.

6.2.4. *Barysilite*

A specimen of the rare mineral barysilite $Pb_3Si_2O_7$ from the contact-metasomatic ore deposit at Franklin, U.S.A., has been investigated by Haack (1973). It revealed three fossil fission tracks on a scanning area of $0.08 \, cm^2$. The uranium content was determined as $0.1 \, \mu g/g$ together with a fission-track age of 0.03 Ma, which is four orders of magnitude lower than the expected Hercynian formation age of the deposit. Since no track annealing studies were reported on this mineral, the geological significance of this result cannot be assessed.

6.2.5. *Bastnäsite*

The mineral bastnäsite $CeCO_3F$ is one of the cerium- and rare earth-bearing fluorocarbonates which are commonly associated with carbonatites. Crystals, about $100 \, \mu m$ in size, from the Mountain Pass carbonatite body, California, were analyzed by Fleischer and Naeser (1972). A uranium content of about $20 \, \mu g/g$ was determined. From the fossil and induced fission-track densities, an age of 78 Ma was calculated, much too young when compared with the expected age of 1400 Ma. Accompanying annealing experiments indicated a low thermal stability similar to apatite, suggesting track annealing during the Laramian orogeny.

166   *Chapter 6*

6.2.6. *Beryl*

Several gem-quality crystals of beryl $Be_3Al_2Si_6O_{18}$ from various pegmatitic sources were investigated by Lorenz (1984). In most specimens, fossil fission tracks were observed, however with track densities far below 1000 tracks/cm^2. The counting of induced fission tracks, after thermal neutron irradiation, revealed uranium contents below 0.5 ng/g. In one beryl specimen of unknown origin, two small clusters of fossil tracks were observed, in which the density reach 2×10^6 tracks/cm^2. Apart from the presence of such uranium-rich inclusions, the bulk uranium content in beryl appears to be too low for useful fission-track dating, unless the investigated samples are of a very high age, in the order of 10^9 years.

Biotite
(*see* mica)

6.2.7. *Calcite*

Although calcite $CaCO_3$ is a common rock-forming mineral, it has been given only minor attention by fission-track geochronologists. Sippel and Glover (1964) observed fission tracks in crystals from various localities. Instead of polished faces, they used cleaved rhombohedral faces of Iceland spar. The etched fossil fission tracks were distinguished from etched dislocation by their different shape and annealing characteristics. Dislocations are thermally much more resistant and their etch pits are uniformly shaped. For a yellow calcite specimen from Chihuahua, Mexico, with 42 μg/g uranium and 5.4×10^3 fossil tracks/cm^2, a fission-track age of 2.5 Ma was calculated in accordance with the sample's presumably late Tertiary formation age. From annealing experiments, it was concluded that at ambient surface temperatures, the fission tracks are stable over several million years. This optimistic assessment of the potential for fission-track dating of calcite was somewhat demolished by Macdougall and Price's (1974) study on calcite crystals extracted from narrow cavities in bones found in hominid-bearing breccias at Makapansgat and Swartkrans, South Africa. The latter authors applied a more appropriate etchant, which produces characteristic fission-track etch pits with narrow channels. Although the uranium concentrations in the calcite specimen ranges up to 44 μg/g, corresponding to a density of 4×10^4 fossil tracks/cm^2 for the presumed formation age of about 2 Ma, no fossil tracks were observed. This discrepancy between expected and observed track density was solved by annealing experiments. Their data, extrapolated to 2 Ma, indicate that fission tracks in calcite will anneal completely at temperatures around 20°C. However, before concluding that the complete loss of tracks at ambient surface temperature is a general phenomenon of calcite and, thus, prohibits calcite fission-track dating, more systematic investigations on the subject are needed.

Chabazite
(*see* zeolite)

6.2.8. *Chlorite*

The phyllosilicate chlorite $(Mg,Fe,Al)_6(Al,Si)_4O_{10}(OH)_8$ is an important mineral in low-grade metamorphic rocks. Sharma *et al.* (1977) etched the cleaved faces in concentrated hydrofluoric acid, in order to reveal fossil fission tracks. The uranium content of investigated samples from India ranges from 2 to 4 ng/g with fossil-track densities of *ca.* 1.3×10^3 tracks/cm^2. Fission-track ages of *ca.* 570 Ma were calculated. Since the fossil fission tracks were 11% shorter than the induced ones, the authors applied an age correction derived from track-annealing experiments. Extrapolations of these experiments suggest that the chlorite fission-track ages may be geologically related to cooling around 150°C. This tentative conclusion is based on the above study only, and needs to be confirmed by additional laboratory and geological evidence.

6.2.9. *Epidote*

Epidote $Ca_2(Al,Fe)_3Si_3O_{12}(OH)$ is an important constituent of low- and medium-grade metamorphic rocks, as well as of contact-metamorphosed limestones. Also epidotization along fault planes is common. The wide geological occurrence and the relatively high uranium content of up to 200 μg/g make epidote principally very suitable for fission-track dating (Figure 6.4). The first fission-track age was determined in 1968 by Naeser *et al.*, but, a total of only 50 epidote fission-track ages were reported during the next decade (Bar *et al.*, 1974; Haack, 1976a; Harrison *et al.*, 1979; Naeser *et al.*, 1970; Reimer, 1972; Reimer and Wagner, 1971). Unfortunately, fission-track activities on epidote seem to have ceased during the past decade. The sample requirements are minimal, since the high uranium contents may allow the dating of single crystals. Epidote has been dated with the population as well as with the external-detector technique. The strong uranium-inhomogeneity, which is observed among grains but also within grains, often requires the use of external detectors for recording the induced fission tracks. Occasionally, inclusions and etched dislocations may hamper track identification (Naeser *et al.*, 1970; Reimer, 1972).

   At present, the geological interpretation of epidote fission-track ages is still uncertain. On the one hand, extrapolations from annealing laboratory experiments predict high geological track-retention temperatures around 500°C (Naeser *et al.*, 1970; Reimer, 1972) to 400°C (Haack, 1976e), on the other hand, fission-track and Rb-Sr-dating on co-existing minerals point to lower retention temperatures, probably around 250°C. For example, epidote fission-track ages (10 to 16 Ma) from the Central Alps are generally older than the apatite fission-track ages and consistently younger than the biotite-Rb-Sr ages (Reimer and Wagner, 1971). Similarly, Harrison *et al.* (1979) found, for the Coastal Plutonic Complex, British Columbia, concordant epidote and sphene fission-track ages, which are both lower than biotite-Rb-Sr ages and higher than the apatite and zircon fission-track ages. From his studies of the Damara Orogen, S.W. Africa, Haack (1976b) infers for epidote an effective retention temperature below 300°C, presumably around 260–

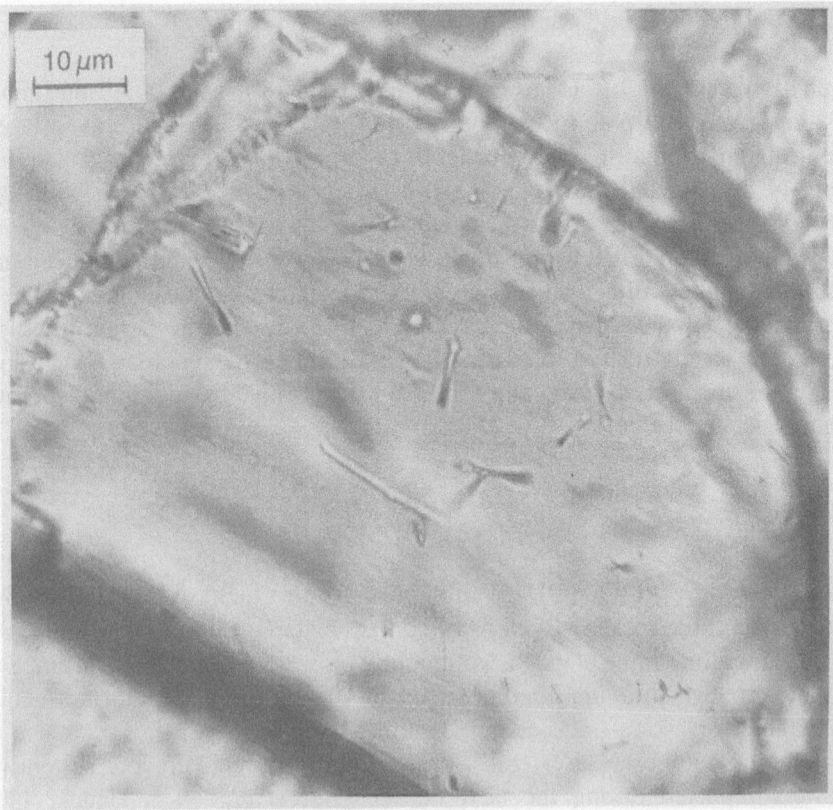

Figure 6.4.   Epidote crystal from the Central Alps etched for 80 min in 50 N NaOH at 150°C in order to reveal spontaneous fission tracks.

280°C. The discrepancy between experimental prediction and geological evidence of the effective retention temperature may be caused by variation of the chemical composition, by accumulated alpha-radiation damage in the crystal lattice, or even by low temperature growth of the epidote crystals. Apart from being used as a geothermo-chronometer, epidote fission-track ages have also been applied for dating Cretaceous and Tertiary movements, with accompanying epidotization, along faults and fissures in the Sinai Peninsula (Bar *et al.*, 1974). The length distribution of the fossil fission tracks helped to distinguish between genuine event ages and mixed ages due to track fading by some later event. These few prevailing applications of epidote fission-track dating are promising and certainly justify further investigation.

Feldspar
(*see* amazonite)

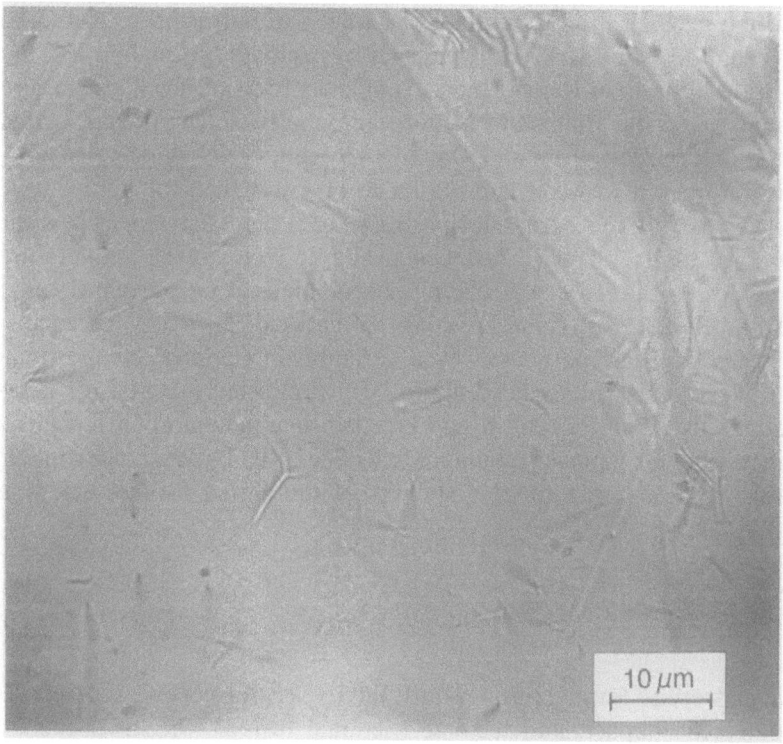

Figure 6.5. Garnet (grossular) crystal from the Hohe Waid, Odenwald etched for 5 h in 75 N NaOH at 180°C in order to reveal spontaneous fission-tracks.

6.2.10. *Garnet*

Garnets are typical constituents of metamorphic rocks. Their composition may vary widely. For fission-track studies, mostly garnets from the almandine $Fe_3Al_{12}(SiO_4)_3$, spessartine $Mn_3Al_2(SiO_4)_3$, grossular $Ca_3Al_{12}(SiO_4)_3$, and andradite $Ca_3Fe_2(SiO_4)_3$ group have been selected. Altogether, about 50 garnet fission-track ages have been reported (Naeser and Dodge, 1969; Banks and Stuckless, 1973; Haack, 1976a, 1977a; Nand Lal *et al.*, 1976a; Sharma and Nagpaul, 1978; Brandl, 1991). Haack and Gramse (1972), Reimer and Wagner (1971) and Reimer (1972) observed, in various garnets, fossil fission tracks but restrained themselves from age calculation owing to low track densities. The handling and etching of the sample is similar to epidote (Figure 6.5). The major obstacle for using garnets in fission-track dating is their low uranium content in the order of $\mu g/g$ and below. Haack and Gramse (1972) recommend andradite- and spessartine-rich varieties because of their higher uranium content. However, Nand Lal *et al.* (1976a) found for such garnets, uranium contents between 0.01 to 0.2 $\mu g/g$. For garnets from Palaeozoic and Precambrian basement rocks, Haack (1976a) and Nand Lal *et al.* (1976a) reported fossil track densities of 10^4 to 10^5 tracks/cm^2. Some of the garnets are not suitable for fission-track dating owing to the nonuniformity of uranium distribution.

As to the geological interpretation of garnet fission-track ages, Nand Lal *et al.* (1976a) claim to date the Proterozoic ages of pegmatite formation and metamorphism in the Rajasthan Range, India. Similar high fission-track ages are reported for garnets from the Bihar and Nellore mica belts, India (Nagpaul, 1982). For andradite-rich garnets from the Damara Orogen, S.W. Africa, Haack (1976b) finds fission-track ages which are similar to or lower than the boitite K/Ar ages. He concludes that the garnet fission-track ages date the time when the rocks cooled down to 260–280°C. This geologically derived closure temperature is significantly lower than the 400–450°C extrapolated from annealing experiments (Haack and Potts, 1972; Nand Lal *et al.*, 1976a) and 350 ± 50°C (Nagpaul, 1982). This discrepancy may be caused by a variation in chemical composition of the garnets. Nevertheless, garnet seems to be a promising material for deciphering the thermal history of Palaeozoic and Precambriam metamorphic terrains. Garnet is impracticable for younger mountain belts due to its low uranium content. More systematic studies of the thermal stability of the garnet fission-track system are needed.

6.2.11. *Glass*

Natural glasses are formed during short, intense geological events, such as volcanism, meteoritic impact, and tectonic frictional fusion, when rock-melts solidify rapidly without crystallization. Owing to their favourable uranium contents of, typically, a few $\mu g/g$, natural glasses are well-suited for fission-track dating (Price and Fleischer, 1964c). For large lumps of homogeneous glass, such as in the case of obsidian and tektite, the sample preparation is very simple and consists essentially of cutting and polishing. Small-sized glass fragments, as they occur in tephra layers, as well as foamy and inhomogeneous glasses, such as pumice and impact glasses, may first require crushing, separation, and concentration procedures. Most glasses are easily etched in hydrofluoric acid at room temperature. The etching time depends on the chemical composition and increases with the silica content of the glass. For some glasses, in particular basaltic deep-sea glasses, hydrofluoric acid is unsuitable, because unsolvable etch products can precipitate on the glass surface (Section 2.2.5). Therefore, other etchants have been developed (Appendix A). Track identification in glass is discussed in Section 2.4.

A major problem in applying fission-track dating to very young samples ($< 10^5$ years) is the low areal density of fossil tracks. This renders this dating technique very time-consuming because large areas have to be scanned under the microscope. Due to the limited number of fossil fission tracks, the resulting ages have a low precision. As a consequence, before starting the age determination of very young samples, one should first consider if the expected precision is sufficient in order to answer the geochronologic question. Perhaps, in the near future, new microcomputer-based image-processing and analyzing systems may considerably speed up the track counting and improve precision.

Another common and serious problem is track fading. There is ample evidence. from track diameter studies that in most natural glasses, fossil fission tracks have undergone some degree of fading. This track fading may be caused by the sample's exposure to the sun over long periods of time, heating by later volcanic events, artificial heating by early man, or simply by normal ambient temperatures. Most glasses have rather sensitive thermal annealing characteristics (see Appendix B). Generally, the track stability increases with the silica content of the glass. In order to evaluate the degree of track fading, it is compulsory, for age determination of glasses, to routinely measure the size of fossil and induced fission tracks. With the *size-correction* and *plateau-correction* techniques for glasses, two procedures are available which allow the correction of the lowered fission-track ages for track loss (Section 5.7).

Volcanic glasses are frequently used for fission-track dating. More than 400 successful age determinations have been reported. In volcanic glasses, the uranium content generally increases with the silica content. Basaltic glasses have uranium contents from about 0.1 $\mu g/g$ to a few $\mu g/g$ and silicic glasses from a few $\mu g/g$ up to about 20 $\mu g/g$. In size, the dated samples range from 60 μm glass shards up to single pieces of obsidian several cm across.

Due to their compact character and their large size, *obsidians* are the most favoured volcanic glasses (Fleischer *et al.*, 1965g). The large counting areas available allow the determination of track densities as low as a few tracks/cm^2 and, thus, although very time-consuming, the determination of obsidian ages as young as a few thousand years is possible (Miller and Wagner, 1981). Sometimes single fragments of about 1 cm^2 in area are accessible. By repeating the steps of grinding ($>20\ \mu m$ off), polishing and etching, large total counting areas become available. Although obsidians usually contain sufficient fossil-track densities for dating, they may present two particular difficulties. Apart from track fading, they often contain needle-like crystallites and small gas bubbles which, after etching, are easily confused with fission tracks (Figure 6.6). Both of these difficulties can be overcome by a one-step pre-annealing technique (Miller and Wagner, 1981; Westgate, 1989) whereby the fossil and induced fission tracks are annealed to a degree by which their size and density ratios reach a constant plateau value. Then, by their smaller size, the fission-track etch pits can be distinguished from other etched defects in the glass. In addition, the fission-track age, which one calculates from the plateau value, is already corrected for thermal track fading. Examples of obsidian dating have been presented by Arias *et al.* (1981), Bigazzi *et al.* (1971a), Kaneoka and Suzuki (1970), Komarov *et al.* (1972), Leach *et al.* (1981), Naeser *et al.* (1980), Suzuki (1970, 1973), Wagner and Weiner (1986). The ages cover various volcanologic and stratigraphic applications in the Quaternary and Tertiary periods. The oldest obsidian-like glasses which have been tackled by fission-track dating are Permian *pitchstones* (Storzer, 1970b). For prehistoric artifacts made of obsidian, the importance of fission-track dating is two-fold. The fission-track age may date either the age of the geological formation, which may be of interest for provenance studies, or the time of the archaeological working which involved track annealing by strong heating (Fleischer *et al.*, 1965g; Miller and Wagner, 1981; Watanabe and Suzuki, 1969).

Another volcanic material used for fission-track dating are *glass-shards* from

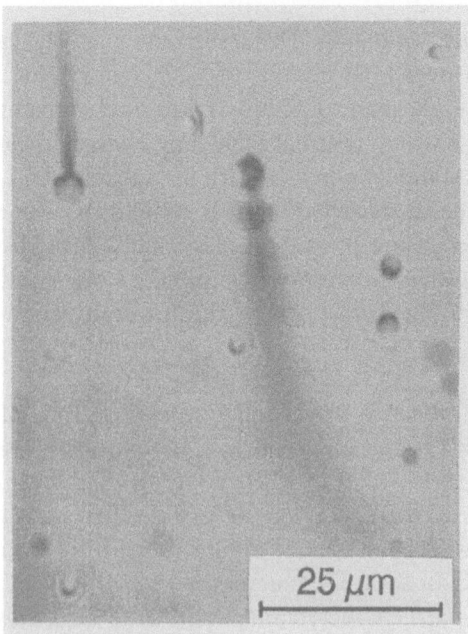

25 µm

Figure 6.6. Obsidian from Gutansar etched 4 min in 16% hydrofluoric acid at 23°C. Before etching the sample was pre-annealed for 63 h at 200°C. The etch pits on the partially annealed induced fission tracks are somewhat smaller than those developed on microlithic inclusions.

tephra-layers, tuffs and pumice-rich horizons (Figure 7.1). Size fractions of about 100 µm are sufficient for fission-track dating which is commonly done using the population technique. In this way, important stratigraphic marker horizons of Tertiary and Quaternary ages have been dated (Fleischer *et al.*, 1965f; Boellstorff and Steineck, 1975; Boellstorff and Te Punga, 1977; Hurford, 1974; Hurford and Hammerschmidt, 1985; Izett *et al.*, 1981; Izett and Naeser, 1976; Kohn, 1979; MacFadden *et al.*, 1979, 1985; Manley and Naeser, 1977; Naeser *et al.*, 1987; Nishimura, 1981; Seward, 1974, 1975, 1976; Suzuki and Yamanoi, 1970; Tahirkheli and Naeser, 1975; Westgate *et al.*, 1985). High vesicularity with many tiny bubbles, particularly in fine-grained pumiceous shards, disturb the identification and counting of fission tracks. In such cases, a point count procedure for the tracks has been recommended by Seward (1974). Systematically lower fission-track ages of glass shards compared to zircons from the same ash layers (Izett and Naeser, 1986) indicate natural track fading. Thus, track-size measurements and age corrections need to be taken into account (Seward *et al.*, 1980; Westgate *et al.*, 1987; Westgate, 1988 and 1989) when dating volcanic glass shards.

Another important material for fission-track dating is *basaltic deep-sea glass* for which more than 60 ages have already been reported. Deep-sea basaltic pillows often have a glassy skin, which formed when the lava was chilled on contact with cool ocean water. But glass is also found as shards in marine sediments. With increasing age, the deep-sea glasses undergo alteration which is accompanied by uranium uptake. Therefore, only fresh, unweathered glass remnants must be selected for fission-track dating. The uranium content of deep-sea glass is, unfortunately, rather low, mostly in the order of 0.1 $\mu g/g$. Nevertheless, basaltic extrusions as young as a few thousand years and up to 35 Ma have been successfully dated (Aumento, 1969; Fleischer *et al.*, 1968b; Luyendyk and Fisher, 1969; Macdougall, 1971b; Selo and Storzer, 1979). However, the microscopic scanning of several cm^2 requires time and patience. It came as a surprise that, even in the cool sea-bottom environment, significant track fading is a common phenomenon owing to the low thermal track stability of basaltic glasses. For instance, all the fission-track ages of various glasses from the Entrecasteaux area in the Pacific required correction (Selo and Storzer, 1981). In accordance with the concept of sea-floor spreading, the fission-track ages on samples from both sides of the mid-Atlantic ridge increase from several thousand years up to 20 Ma with distance from the ridge axis (Aumento, 1969; Fleischer *et al.*, 1968, 1971; Storzer and Selo, 1976) Spreading rates between 0.6 and 4 cm/a were calculated and temporal changes detected. With regard to the lack of other dating methods, especially for the last few million years, the fission-track method – although being tedious and time-consuming has a great potential for the chronology of the young basaltic deep-sea floor. Several fission-track ages on glass shards from deep-sea sediment cores were reported by Macdougall (1971a).

Tektites and other impact glasses were among the first materials that the fission-track dating method has been applied to (Fleischer and Price, 1963, 1964b; Fleischer *et al.*, 1965c; Wagner, 1966). Meanwhile, more than 250 specimens have been dated. In tektites, the uranium typically amounts to a few $\mu g/g$ and is generally homogeneously distributed. Owing to their homogeneous composition and to the absence of disturbing inclusions, tektites are probably the most suitable material for fission-track counting (Figure 6.7). On the other hand, impactite glasses were often less thoroughly melted and, consequently, contain inclusions and inhomogeneous uranium distribution which may both impede fission-track dating. Their uranium content varies with the rock source from which they were fused, but also lies mostly in the $\mu g/g$-level. The preparation and dating techniques for such glasses are similar to inhomogeneous volcanic glasses and may first require crushing of the sample with separation of suitable glass fragments. Due to low thermal track stability characteristics, tektites and impact glasses commonly show signs of track loss (Fleischer and Price, 1964b; Storzer and Wagner, 1969). Some track annealing may have already occurred at normal surface temperatures, but is certainly enhanced by long exposure to the sun or by accidental surface heating, as is the case for tektites. Therefore, track-size analysis and corresponding age correction – either with the track size or the plateau techniques – have to be routinely applied.

All major tektite groups have been dated and the fission-track ages agree well – after correcting for thermal track fading – with independent ages (Durrani and Khan, 1970; Fleischer *et al.*, 1965c, 1969a; Gentner *et al.*, 1967, 1969a,b; Storzer *et al.*, 1973; Storzer 1985; Storzer and Wagner, 1969, 1980; Wagner, 1966). Also, the sub-mm-sized *microtektites*, which were recovered from deep-sea sediment cores, were successfully dated (Durrani and Khan, 1971; Gentner *et al.*, 1970; Glass *et al.*, 1973). For some of the tektite-strewn fields, concordant fission-track

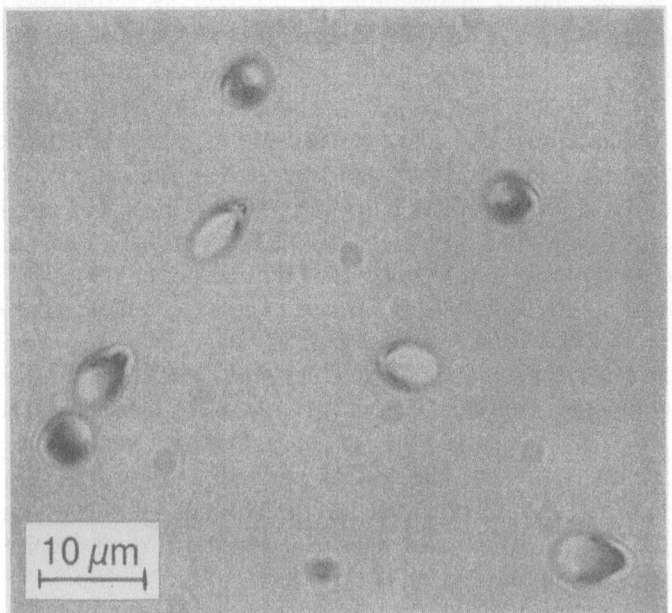

Figure 6.7. Moldavite from Bohemia etched 15 s in 48% hydrofluoric acid at 21°C. Depending on the intersection angle beween track and polished surface circular to elliptical etch cones develop. (From Wagner, 1976b.)

ages between tektites and the supposed source craters were found (Fleischer and Price, 1964b; Fleischer *et al.*, 1965c; Wagner, 1966; Gentner *et al.*, 1967, 1969a,b, Storzer and Gentner 1970). A large number of glasses from various meteorite craters have also been successfully dated and, thus, fission-track dating contributed significantly to a wider knowledge of the frequency of giant terrestrial impact events (Fleischer *et al.*, 1969d; Storzer and Wagner, 1977, 1979). Consequently, the fission-track method has turned out to be the most important dating tool for meteorite impact structures.

Frictionite glass (or *pseudotachylite*) occurs along thrust planes of huge mountain slides. It consists of more or less fused silicate material and is rich in bubbles – with pumice-like appearance – and in mineral inclusions due to incomplete melting. The origin as frictionite might not be generally accepted and, as in the case of the Köfels structure in the Tyrol, may also be due to fusion by meteorite impact. Notwithstanding that, fission-track dating can be applied to the glass phase of frictionite and, thus, contribute to the discussion of its origin. As already mentioned for impact glasses, inhomogeneous composition and extensive devitrification may impede fission-track dating. The 40 to 90 μm glass fraction from Köfels yielded a fission-track age of 8000 years (Storzer *et al.*, 1971). This age agrees with the time of large glacier melting in the Alps and, thus, lends support for the idea of Köfels' origin as a mountain slide. Hitherto, other attempts of dating frictionite glass with fission tracks are not known.

6.2.12. *Glauconite*

Glauconite $K(Fe,Mg,Al)_2(Si_4O_{10})(OH)_2$ is an authigenic mineral of marine sedimentary rocks that was first tested for fission-track dating by Srivastava *et al.* (1983). Fossil-track densities of 0.5 and 1.1×10^4 tracks/cm^2, uranium contents of 0.12 and 0.03 μg/g and fission-track ages of 87 and 680 Ma, respectively, were determined. These ages apparently agree with the expected stratigraphic ages of the sandstone. Although the above results represent an interesting attempt at dating sedimentary rocks, further experimental and geological evidence is needed before the potential of the fission-track method for dating glauconite can be assessed with confidence.

Heulandite
(*see* zeolite)

6.2.13. *Hornblende*

Hornblende $NaCa_2(Fe,Mg,Al)_5(Si,Al)_8O_{22}(OH)_2$ can vary greatly in composition and is a widespread rock-forming mineral. Only a few hornblende specimens have been investigated with regard to fission-track dating (Fleischer and Price, 1964d; Welin *et al.*, 1972). The uranium content of the two specimens studied by Fleischer and Price (1964d) turned out to be about 0.4 μg/g. The tracks were revealed on cleaved faces with warm hydrofluoric acid. The fission-track ages essentially agree with the independently known Precambrian ages. A high thermal stability of tracks in hornblende is indicated by the accompanying annealing experiments (Fleischer and Price, 1964d). On the other hand, Nagpaul (1982) mentions an experimentally derived, closure temperature of only $145 \pm 25°C$. Welin *et al.* (1972) found very different fission-track ages for two halves of a large euhedral crystal. They attributed this difference to the combined action of thermal track fading and a recent uranium loss, which are both difficult to evaluate. Because of its chemical variation, one can expect systematic changes in characteristics such as etching behaviour and track annealing. Nevertheless, it certainly seems worthwhile to continue fission-track work on this mineral.

6.2.14. *Hübnerite*

Hübnerite $(Fe,Mn)WO_4$ specimens from various occurrences were analyzed by Haack (1973) who reports numerous, but inhomogeneously distributed fossil tracks.

6.2.15. *Kyanite*

Kyanite Al_2SiO_5 – also known as disthene – occurs characteristically in medium-grade shists and gneisses. A fission-track study on six specimens from the Himalayas was reported by Saini *et al.* (1983). Uranium contents of between 0.11 and 0.16 $\mu g/g$ were found. Fission-track ages of between 9 to 17 Ma were determined for five samples from NW Himachal Himalaya and 320 Ma for one sample from Assam. From accompanying annealing experiments, a track-retention temperature of 80°C was estimated, suggesting that the reported fission-track ages are post-metamorphic cooling ages.

Lepidolite
(*see* mica)

6.2.16. *Mica*

The mica group consists of phyllosilicates with perfect basal cleavage. The simplified compositions of these species, which have been used for fission-track studies, are *muscovite* $KAl_2(AlSi_3O_{10})(OH)_2$, *phlogopite* $KMg_3(AlSi_3O_{10})(OH)_2$, *biotite* $K(Mg,Fe)_3(AlSi_3O_{10})(OH)_2$, and *lepidolite* $KLi_2Al(Si_4O_{10})(OH)_2$. These minerals were the first in which fossil fission tracks were observed and for which fission-track ages were determined (Price and Walker, 1962d, 1963a). The uranium content can reach up to 10 $\mu g/g$ in muscovite and 30 $\mu g/g$ in biotite, but is usually much lower. Large mica flakes or sheets, which have an area of at least several mm^2, are chosen for fission-track dating. No grinding and polishing is necessary for sample preparation, since cleaved surfaces can be used. In order to minimize the danger caused by secondary uranium migration along the cleavage planes, the mica sheets should be cleaved along their tightest planes. When applying the re-etch technique, the mica sheets can be only partially split open – like a book – for revealing the spontaneous fission tracks and closed again for neutron irradiation; this procedure enables the registration of the induced fission tracks in 4π geometry. The fission tracks are etched in hydrofluoric acid at room temperature for various times ranging from seconds to a few hours. Depending on the etching time, the tracks are revealed as needle-like (Figure 6.8) or rhombic etch-pits (Figure 6.9). Fission-track dating has been applied to numerous samples from different rock types, such as plutonic rocks, pegmatites, metamorphic shists, and volcanic rocks. Altogether, about 300 fission-track ages have been reported in the literature. Mostly muscovite (Abdullajev *et al.*, 1966; Barman and Nandi, 1978; Berzina *et al.*, 1966; Bigazzi, 1967; Gupta *et al.*, 1971a; Kashukejev *et al.*, 1970; Maurette *et al.*, 1964; Mehta and Nagpaul, 1971; D. K. Miller *et al.*, 1968; D. S. Miller and Jäger, 1968; Nagpaul *et al.*, 1974a; Shukoljukov *et al.*, 1965; Shukoljukov and Komarov, 1966) as well as biotite and phlogopite (Abdullajev *et al.*, 1966; Barman and Nandi, 1978; Bigazzi *et al.*, 1971b; Danis *et al.*, 1969; Gupta *et al.*, 1971b; Hashemi-Nezhad and Durrani, 1986; Maurette *et al.*, 1964; D. K. Miller *et al.*, 1968; D. S. Miller, 1968; Nagpaul *et al.*, 1974a; Shima *et al.*, 1969; Shukolju-

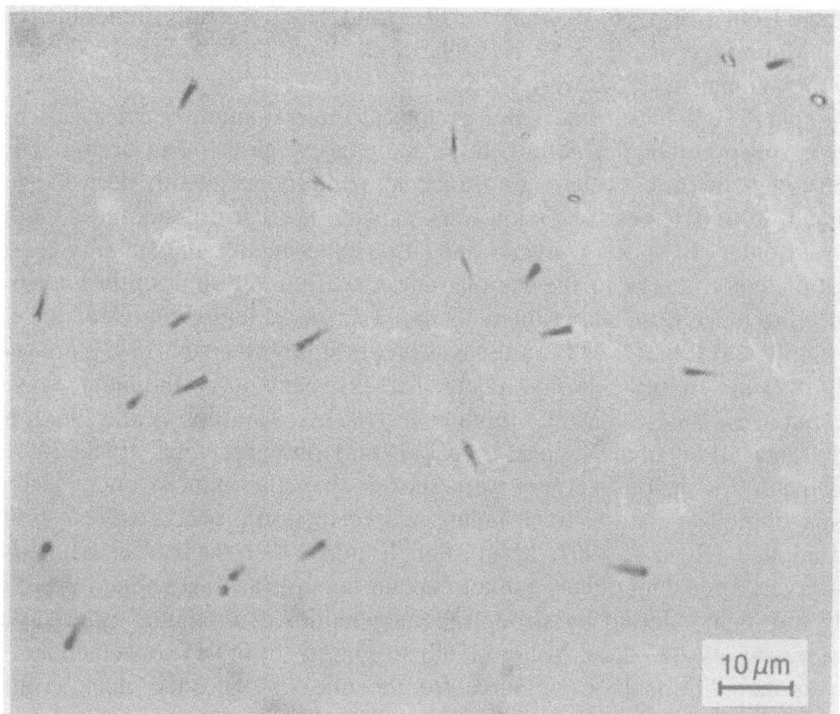

Figure 6.8. Biotite from Bornholm etched for 5 s in 48% hydrofluoric acid at 21°C in order to reveal fission tracks. (From Wagner, 1969b.)

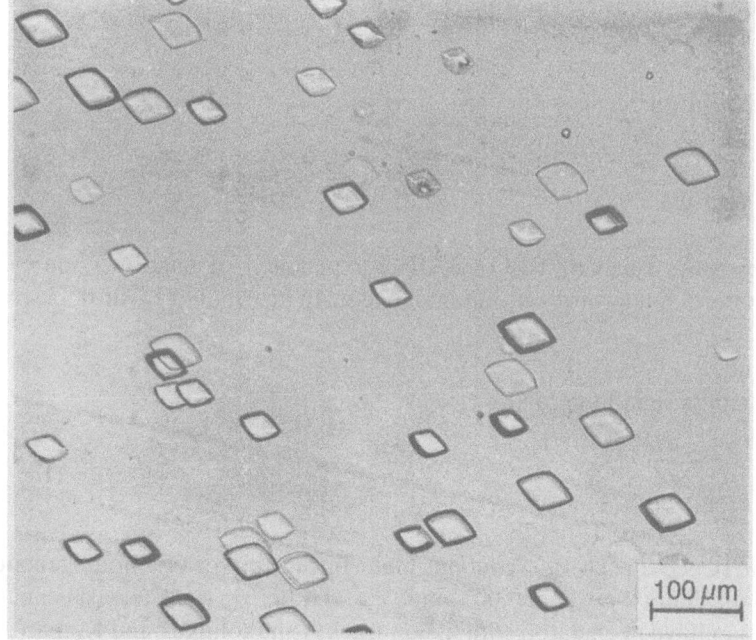

Figure 6.9. Spontaneous fission tracks in muscovite after etching in 40% HF for 15 h at 25°C.

kov and Komarov, 1966; Welin *et al.*, 1972) and, less frequently, lepidolite (Haack, 1973; Maurette *et al.*, 1964; D. K. Miller *et al.*, 1968; D. S. Miller, 1968; Shukoljukov and Komarov, 1966) were used.

In spite of all these dates, the geological interpretation of mica fission-track ages is still problematic. Although, in many cases, good – and occasionally 'too good' – agreement between fission-track ages and independently derived ages was claimed, in most cases the fission-track ages do not correspond and are younger or even older. In spite of several annealing experiments, the stability characteristics of fission tracks in the various micas are not yet sufficiently known. It is difficult to judge if the experimentally derived closure temperatures of $100 \pm 20°C$ for biotite and $170 \pm 25°C$ for muscovite, reported by Nagpaul (1982), are geologically realistic. In any case, loss of fossil tracks seems to be the major reason for lowered ages. Bigazzi (1967), Gupta *et al.* (1971a), Maurette *et al.* (1964), Mehta and Rama (1969) and Nagpaul *et al.* (1974a) observed fossil tracks in several muscovites which, on average, were shorter than the induced ones. This effect was attributed to natural track fading and correction procedures were proposed and applied (Bigazzi, 1967; Mehta and Rama, 1969; Nagpaul *et al.*, 1974a,b). Howerver, annealing effects cannot explain fission-track ages which are geologically too high. Alerted by strong inhomogeneities of uranium, especially along fractures or crystal edges, Miller (1968) suggested secondary movements of uranium as an additional error source for the mica fission-track clock. Thus, the assumption of the *closed system* is violated, resulting in anomalously high or low fission-track ages. Similar inhomogeneities have been repeatedly observed (Mehta and Nagpaul, 1971). As a consequence, fission-track dating of micas was practically stopped during the last decade but low uranium mica is still intensively used in the dating method as an external detector.

Microcline
(*see* amazzonite)

6.2.17. *Microlite*

For a microlite $(Na,Ca)_2Ta_2O_6(O,OH,F)$ specimen of unknown source, Haack (1973) reports numerous, but inhomogeneously distributed fossil tracks.

Mimetisite
(*see* secondary lead minerals)

6.2.18. *Monazite*

Monazite $(Ce,La,Y,Th)PO_4$ contains high concentrations of inhomogeneously distributed uranium between 0.005 and 5%. Hitherto, only two specimens from Kazakhstan have been dated with fission tracks (Shukoljukov and Komarov, 1970). The determined fission-track ages of *ca.* 25 Ma turned out to be much lower than the independently known age of 200 Ma. Shukoljukov and Komarov (1970) attributed this discrepancy to the experimentally investigated, low thermal stability of the tracks in monazite. The sensitive annealing characteristic of monazite is not surprising because the *phosphate* apatite shows a similar behaviour. However, the effects of metamictization and fast neutron-induced fission of thorium – another common constituent in monazite – also need to be considered before one can assess the utility of this mineral for fission-track dating. Finally, the high REE content can cause significant neutron absorption during irradiation.

Muscovite
(*see* mica)

Obsidian
(*see* Glass)

6.2.19. *Orpiment*

Orpiment (auripigment) As_2S_3 is the only sulphide mineral in which fossil fission tracks have been observed (Jakupi *et al.*, 1982). Transparent crystals of 100 to 200 μm size were separated from the ores of the Alshar Mine, Yugoslavia, and investigated for fossil fission tracks which were difficult to identify due to the existence of disturbing dislocation etch pits. A fossil-track density of 70 tracks/cm^2 was determined. After irradiation with thermal neutrons and counting the induced fission tracks, a uranium content of 0.04 μg/g and a fission-track age of 5 Ma were calculated. Jakupi *et al.* (1990) reported, for orpiment from the Alshar region, a fission-track age of 13 Ma. They also consider this age to be too low owing to low thermal track stability (Todorovic *et al.*, 1980). Due to the lack of independent age information on the Alshar deposit, it is at present difficult to form an opinion on the significance of these results. Certainly, fossil tracks in orpiment and related sulphide minerals deserve further attention.

Phlogopite
(*see* mica)

Pyromorphite
(*see* secondary lead minerals)

6.2.20. *Quartz*

Quartz SiO_2 certainly has a great appeal for fission-track dating because of its widespread presence as a rock-forming mineral, its resistance against weathering, and its high thermal track retention (Fleischer *et al.*, 1968a). Unfortunately, its uranium content is rather low ($\ll 0.1$ μg/g) but some quartz grains bear uranium-rich areas and inclusions in which the uranium content may reach several hundreds of μg/g (Sutton and Zimmerman, 1978). For quartzes from the Oklo uranium deposit, Gabon, up to 0.2% uranium in clusters was reported (Durrani *et al.*, 1975). There have been several attempts to date quartz with the fission-track method. Due to changes in etching efficiency with crystallographic orientation (Khan and Ahmad, 1976), crystal faces perpendicular to the *c*-axis were recommended for track revelation (Weiland *et al.*, 1980). Some quartzes from uranium deposits contained fossil-track densities of 10^9 tracks/cm^2, which necessitated scanning or transmission electron instead of optical microscopy for track counting. In the case of the Midnite uranium mine, Washington, such high track densities were found in quartz grains adjacent to uraninite, that is, the uranium-free quartz acts as a detector for the fission tracks originating from the uraninite and a fission-track age of 51 Ma was determined (Weiland *et al.*, 1980). A similar investigation was reported by Cunningham *et al.* (1982). In the case of the Oklo uranium deposit, the tracks were caused by the high uranium inclusions within the quartz grains and a fission-track age of 1.73 Ga was calculated – after correcting for track fading due to a size reduction of the fossil tracks. In another investigation (Vincent *et al.*, 1984) of volcanic quartz grains with uranium-rich vitreous inclusions, the fossil tracks were etched and counted within the glass itself as well as in the surrounding quartz host mineral. Owing to the different thermal retention properties of the different detectors, discordant fission-track ages were found, namely for a Precambrian Moroccan sample 108 Ma for the glass and 246 Ma for the quartz detector. These few available fission-track results justify further pursuit of the fission-track clock in quartz.

6.2.21. *Scheelite*

For scheelite CaWO$_4$ specimens from various occurrences, Haack (1973) reports numerous, but inhomogeneously distributed fossil tracks.

6.2.22. *Secondary lead minerals*

For secondary lead minerals such as mimetite Pb$_5$(AsO$_4$)$_3$Cl, vanadinite Pb$_5$(VO$_4$)$_3$Cl, and pyromorphite Pb$_5$(PO$_4$)$_3$Cl from various occurrences, Haack (1973) reports numerous, but inhomogeneously distributed fossil tracks. These minerals may have a potential for fission-track application due to their structural similarity to apatite.

6.2.23. *Sphene*

Sphene (= titanite) CaTiSiO$_5$ occurs widely as an accessory mineral in plutonic and metamorphic rocks but is less common in pegmatites and volcanic rocks. Its chemical composition is somewhat variable. Generally, the uranium content in sphene is rather high, commonly between 100 and 1000 μg/g but sphenes of Late

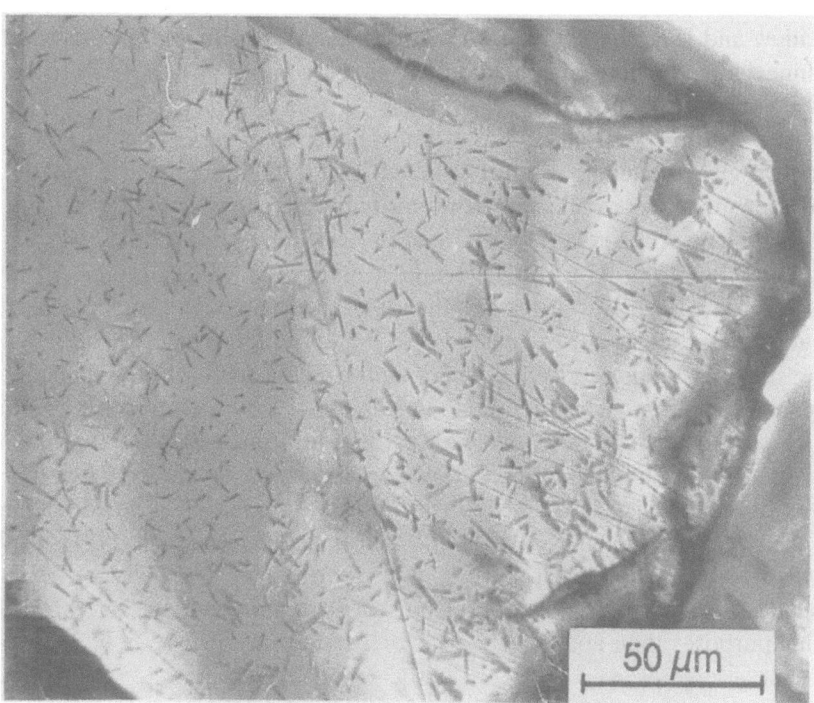

Figure 6.10. Sphene from a syenite intrusion in Central Burundi etched for 15 min in an acidic mixture (HF—HCl—HNO₃—H₂O) at 25°C. Note that the crystal is a twin as revealed by the intersection line of the twin plane (centre). The two faces separated by the twin plane exhibit slightly differently etched tracks and etched polishing scratches can only clearly be seen on the face at the right.

Precambrian age have also been found with uranium contents low enough to be successfully dated (Van den haute, 1986a).

The uranium abundance enables fission-track dating of single grains. Regarding preparation, sphene is usually separated together with other heavy minerals, such as apatite and zircon, from about a 1 or 2 kg rock sample, but is commonly magnetic, while apatite and zircon are not. The sphene concentrates are dated with either the external detector or with the population technique. In the cases of uranium inhomogeneity or detrital origin, it is advisable to use a single-grain determination technique with external detectors (Figure 6.10). A difficulty which arises for sphene – as it probably does for other minerals sensitive to metamictiz-ation – is the change of etching efficiency with increasing radiation damage (Glea-dow, 1981). To overcome this effect, one should only select grains with a certain degree of radiation damage, for instance, only grains with fossil-track densities of between 10^5 and 10^7 tracks/cm². The experimentally derived thermal annealing characteristics indicate rather high geological retention temperatures of between 450 and 300°C (Fleischer *et al.*, 1969b; Naeser and Faul, 1969; Watt and Durrani, 1975). However, these laboratory-derived data are not supported by geological evidence. Sphene fission-track ages of slowly cooling rocks are usually lower than Rb-Sr and K-Ar biotite ages but higher than apatite and zircon fission-track ages, which confine the effective track retention temperature of sphene to about 250°C

(Gleadow and Lovering, 1978b; Harrison *et al.*, 1979). Supplementary studies on the annealing behaviour of sphene under *in-situ* conditions in deep bore holes are needed.

Sphene fission-track dating has been applied to a variety of rock types and geological environments. A total of about 200 dates have been reported. Most dated specimens were collected from plutonic and metamorphic basement rocks (Fitzgerald and Gleadow, 1988; Gleadow and Lovering, 1978b; Green, 1986; Harrison *et al.*, 1979; Hurford, 1977a,b; Marvin *et al.*, 1980; Miller *et al.*, 1978; Naeser *et al.*, 1968; Seward and Rhoades, 1986; Stuckless and Naeser, 1972; Wagner and Storzer, 1975; Zeck *et al.*, 1988; Zeitler *et al.*, 1982a,b; Van den haute, 1986a). In these studies, the sphene fission-track ages, especially when compared to apatite and zircon fission-track ages, as well as to other radiometric ages on co-existing minerals, reveal a wealth of information on the thermal history of rocks from several hundred °C down to surface temperatures. Further applications of sphene fission-track dating comprise carbonatite intrusions (Fleischer and Naeser, 1972; Wagner, 1976a) and volcanic rocks (Naeser, 1971; Lindsey *et al.*, 1975). Owing to their relatively high thermal track stabilty, sphenes from volcanic rocks and tuffaceous marker-horizons are more suitable for chrono-stratigraphic purposes than apatites.

6.2.24. *Stibiotantalite*

Stibiotantalite $Sb(Ta,Nb)O_4$ is a transparent ore mineral associated with pegmatitic niobium and tantalum ores. A 2 cm large, honey-yellow crystal from Alto Ligonha, Mozambique, was investigated for fossil fission tracks by Haack (1970). The tracks were confined to brown-yellow shaded, irregular shaped narrow zones with well-defined boundaries. In these zones, the uranium content amounts to about 10 $\mu g/g$. The determined fission-track age of 232 Ma turned out to be much lower than other radiometric ages from the same pegmatite district. This discrepancy may be due to thermal annealing of fossil tracks as inferred from accompanying annealing experiments (Haack, 1970).

Stilbite
(*see* zeolite)

6.2.25. *Tanzanite*

Eight crystals of tanzanite, a gem variety of zoisite $Ca_2Al_3Si_3O_{12}(OH)$, from pegmatites at Ally Mine, Tanzania, were etched in NaOH at 140°C for various durations in order to reveal fossil fission tracks (Naeser and Saul, 1974). High track densities of up to 4.4×10^7 tracks/cm^2 were found. After neutron irradiation, uranium contents of between 0.15 and 15 $\mu g/g$ were determined from the induced track densities. The mean fission-track age of 585 Ma may be related to the period of pegmatite emplacement. The track-annealing characteristics of tanzanite turned out to be very similar to those of epidote.

Tektite
(*see* Glass)

Vanadinite
(*see* secondary lead minerals)

6.2.26. *Vermiculite*

Vermiculite $KAl_2(AlSi_3O_{10})(OH)_2$ is a hydrothermal alteration product of biotite and phlogopite. Fossil fission-track densities of about 2×10^4 tracks/cm^2 in samples from Kasipatnam, India, were observed by Sharma *et al.* (1979). From the induced fission-track densities, a 0.1 μg/g uranium concentration was determined. The calculated fission-track age of 544 Ma takes into account a correction due to a 14% length reduction of the fossil tracks. The extrapolation of experimental annealing data suggests an effective track-retention temperature of *ca.* 125°C. As is always the case with such isolated studies, the potential of vermiculite for fission-track application is at the moment difficult to assess.

6.2.27. *Vesuvianite*

Vesuvianite (= idocrase) $Ca_{10}Mg_2Al_4(Si_2O_7)_2(SiO_4)_5(OH)_4$ occurs mainly in contact-metamorphosed limestones. Applying the population technique, about 20 vesuvianite specimens from skarns at various localities were dated (Haack, 1976a,b, 1977a; Brandl, 1991). The uranium distribution in these samples turned out to be mostly zoned but, on average, the uranium content reached a few μg/g. The tracks are clear and easy to identify (Figure 6.11). Fossil-track densities in the order of 10^5 to 10^6 tracks/cm^2 were observed. According to Haack's studies, the vesuvianite fission-track ages either agree with other radiometric ages or, in the case of the slowly cooled down Damara orogen, S.W. Africa, disagree in the following age sequence: biotite K-Ar > garnet fission-track > vesuvianite fission-track > apatite fission-track age. Brandl (1991) determined ages of $\approx$250 Ma for several samples from the crystalline basement of the Bavarian Forest and related them to post-Variscan hydrothermal activities. Based on annealing experiments, effective geological track retention temperatures of 160°C and *ca.* 270°C were estimated by Haack (1976c) and Brandl (1991), respectively, which are essentially in accordance with the observed age pattern, but a better knowledge of the effective track-retention temperature is certainly needed. Nevertheless, in spite of its rare occurrence, vesuvianite seems to be a useful mineral for reconstructing the thermal history of rocks.

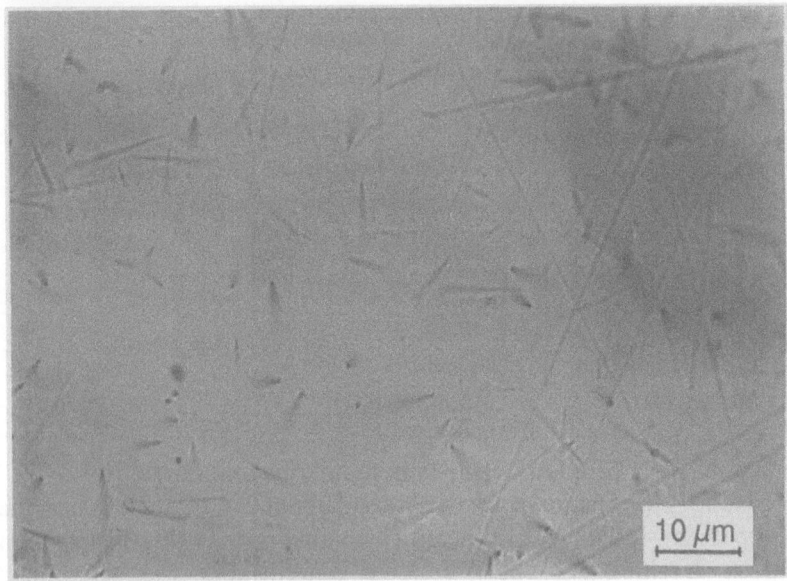

Figure 6.11. Vesuvianite crystal from Kropfmühl, Bayerischer Wald, etched for 3.5 h in 75 N NaOH at 180°C in order to reveal induced fission tracks.

6.2.28. Zeolite

As regards the zeolite group, fission-track dating has been reported by Koul *et al.* (1983) for chabazite $(CaAl_2Si_4O_{12}\cdot 6H_2O)$, heulandite $(CaAl_2Si_7O_{18}\cdot 6H_2O)$ and stilbite $(CaAl_2Si_7O_{18}\cdot 7H_2O)$. These minerals are commonly associated with basaltic rocks. Often, they are alteration products of feldspars. Koul *et al.* (1983) worked on various zeolites sampled from basalts on the Faroe Islands. The dating results were interpreted in terms of the age of the volcanic activity. The insufficient experimental data presented do not yet allow the assessment of the reliability of those results and the suitabilty of zeolites for fission-track dating. Annealing experiments (Koul *et al.*, 1986) on chabazite suggest a relatively low thermal track stability.

6.2.29. Zircon

Zircon $ZrSiO_4$ is a common accessory mineral in igneous plutonic and volcanic rocks. Because of its resistance to mechanical and chemical disintegration, it is also a common heavy mineral in sandstone. Zircon generally has a high uranium content of between 300 and 6000 $\mu g/g$ and is, therefore, well-suited for single-grain fission-track dating. Various etchants have been proposed and applied (Appendix A), but mostly a eutectic melt of KOH-NaOH at 215°C is used to reveal the fossil fission tracks (Figure 6.12). The etching duration can vary from crystal to crystal of the same sample and is essentially controlled by the degree of radiation damage. The higher the spontaneous fission-track density, the shorter the etching

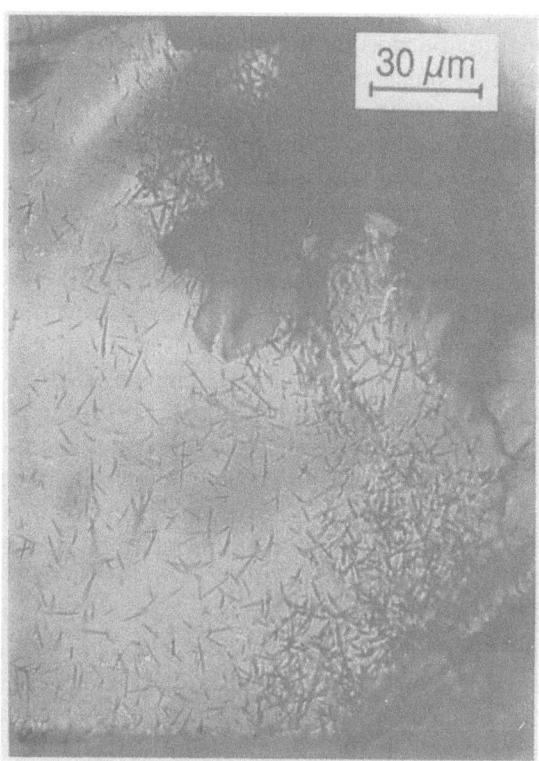

Figure 6.12. Spontaneous fission tracks in zircon (*ca.* 0.2 mm across) from the Oberpfalz (Bohemian Massif), etched in an eutectic mixture of NaOH, KOH and LiOH (6:14:1) for 3.5 h at 200°C. The track density distribution reveals a strong heterogeneity of uranium in this crystal.

time. The rigorous etching treatment requires special embedding techniques in Teflon sheets (Section 3.7.1). Due to the strong uranium inhomogeneities among and within the grains, the external track-detector technique is recommended. In some exceptional cases of homogeneous uranium distribution, the population technique has been used. The annealing characteristics of tracks in zircon were investigated in several laboratory experiments (Fleischer *et al.*, 1964c, 1965b; Krishnaswami *et al.*, 1974; Nishida and Takashima, 1975; Tagami *et al.*, 1990). The extrapolation to geological times indicates track-retention temperatures of 300 to 400°C. But, analogous to epidote and sphene, the retention temperatures which are derived by fission-track dating on zircons from slowly cooling rocks and from deep bore-hole rocks, point to much lower effective retention temperatures of around 200°C (Harrison *et al.*, 1979; Hurford, 1986; Naeser, 1979; Zaun and Wagner, 1985). This picture might be even more complicated by the radiation damage phenomenon. It is conceivable that the degree of radiation damage not only affects the etching efficiency but also the track-retention properties of zircon. Due to insufficient knowledge of track-length reduction under the influence of annealing, track-length analysis in zircon has – apart from sporadic studies (Ito *et al.*, 1989) – not yet been commonly used in order to decipher the thermal past of rocks.

Zircon is, after apatite, the most frequently used mineral for fission-track dating. The first fission-track ages on zircon had been reported by Fleischer *et al.* (1964c), Carbonnel and Poupeau (1969), Naeser and McKee (1970) and Wagner and Storzer (1971). So far, altogether about 850 fission-track ages on zircons from different rock types and various geological settings have been published. Owing to its good track-retention properties at ambient surface temperatures and its common occurrence, zircon is the favoured fission-track mineral for the chronology of volcanism and the chrono-stratigraphy of tuffaceous marker horizons. Some of the major contributions in this respect are those by Bryant *et al.* (1980), Comer *et al.* (1980), Haggerty *et al.* (1983), Hurford and Watkins (1987), Kasuya (1987), Komarov *et al.* (1973), Lindsey *et al.* (1980), Lipman *et al.* (1986), Ross *et al.* (1982), Steven *et al.* (1979), Suzuki and Yamanoi (1970), and Tamanyu (1975). Another large field of zircon fission-track dating are plutonic rocks, especially in connection with apatite fission-track and other radiometric dating on co-existing minerals. By such combined data sets, the thermal evolution of igneous rocks can be traced backward. Pertinent examples were reported by Benjamin *et al.* (1987), Bigazzi *et al.* (1972), Bookstrom *et al.* (1987), Flisch (1986), Hurford (1977a,b, 1986), Green (1986), Gleadow and Lovering (1978b), Ito *et al.* (1989), Naeser *et al.* (1971), Zeitler *et al.* (1982a,b). In some cases, the ages were connected to the time of mineralization (Lipman *et al.*, 1976; Ludwig *et al.*, 1981;, Naeser *et al.*, 1979a). Detrital zircons in sediments have also been dated. Examples, in which the detrital zircons bear still the age of their source regions, were presented by Gleadow and Duddy (1980), Hurford *et al.* (1984) and Hurford and Carter (1991). Another interesting demonstration of the potential of zircon fission-track dating was given by Coates and Naeser (1984). The zircon fission-track clock in sandstone had been thermally reset by the burning of an adjacent clinker. Thus, the zircon fission-track ages revealed the age and rate of the natural burning of the coal bed. All these various examples show that fission-track dating of zircon is a versatile geochronological tool.

Zoisite
(*see* tanzanite)

CHAPTER 7

Application

7.1. Tephrochronology

Tephra deposits consist of pyroclastic materials that are ejected during a volcanic explosion and transported through the air. Due to rapid deposition, tephra essentially form synchronous layers in continental as well as in marine sediments. Ash layers may extend over 1000 km and more. If they have distinct geological and physico-chemical characteristics, tephra beds can serve as important marker horizons for stratigraphic correlation, particularly of Quaternary and Tertiary sections (Westgate and Gorton, 1981). The usefulness is greatly enhanced if the tephra are absolutely dated with geochronological methods. In this way, a tephrochronology may be established that serves as a chronostratigraphic framework for sedimentary sequences, even in areas remote from the volcanic centres.

Once a tephra layer is deposited, a variety of secondary processes, such as redeposition by water or wind, creeping, slumping, cryoturbation, and bioturbation, commonly mix contaminants into the tephra materials. In order to recognize such contaminating constituents, grain-discrete techniques are preferable to bulk analyses. Obviously, this holds true for the physico-chemical as well as for the geochronological studies of tephra. The problem of mixing is particularly serious for thin distal tephra beds in contrast to thicker proximal tephra occurrences.

Fission-track dating seems to be ideally qualified for tephrochronology. Firstly, tephra commonly contain glass shards and minerals which are suitable for track revelation. Secondly, these materials have sufficient uranium content for applying the technique, even to young Quaternary events. Thirdly, fission-track dating is grain-discrete, since each individual grain is routinely scanned under the microscope. Indeed, fission-track dating – apart from K-Ar dating – has recently become the most frequently used dating technique in tephrochronological studies (Seward 1976; Westgate and Briggs, 1980; Naeser and Naeser, 1984). Fission-track dating has the advantage over the K-Ar method that glass, a common constituent of tephra, can be dated. Unfavourable properties, such as excess argon, argon loss, and potassium exchange, discouraged the use of glass shards for K-Ar age determination. Due to their common occurrence and their easy experimental handling, glass shards (of usually between 100 and 200 μm) are the most preferred material in tephrochronological fission-track applications (Figure 7.1). Specific problems,

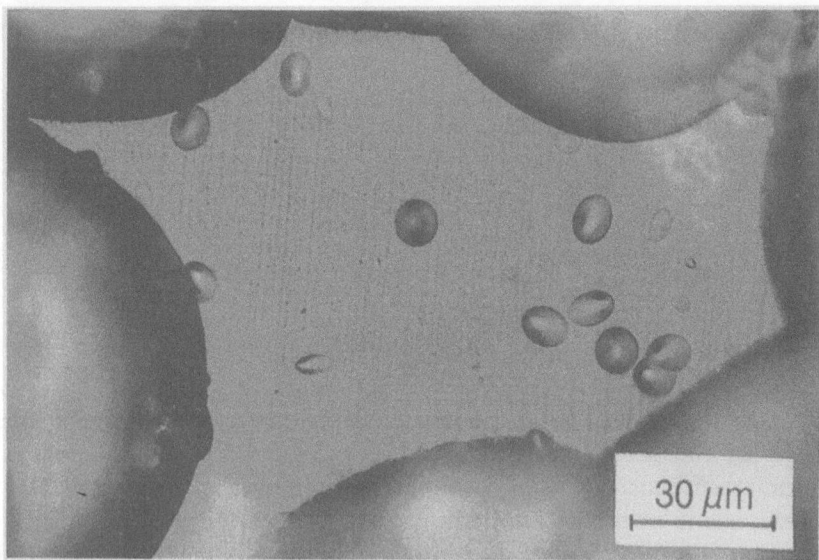

Figure 7.1. Hydrated, silicic glass shards from Banks Island tephra, western Arctic, with induced, pre-annealed (3 months at 100°C) fission tracks. (Courtesy by J. Westgate.)

such as very low areal densities of fossil tracks and the presence of microlites that can be encountered when dating volcanic glasses, are discussed in Section 6.2.11. If substantial fading of the fossil tracks is detected from their reduced size, compared to the induced tracks, the fission-track age determinations are likely to yield results which are too low. Accordingly, appropriate correction procedures have been proposed (Section 5.7), namely the track-size technique (Storzer and Wagner, 1969) and the plateau technique (Storzer and Poupeau, 1973). The general validity of these correction techniques for volcanic glasses has been demonstrated in numerous case studies, and most recently by Westgate (1989) for hydrated glass shards.

Besides glass, zircon is the other tephra constituent frequently used for fission-track dating. Zircon has the advantages over glass of being less susceptible to track annealing and of having a higher uranium content, facilitating grain-by-grain age determination. On the other hand, zircon does not occur in all tephra, the silicic tephra generally being more productive than basic ones. In distal tephra layers, the grain size of zircon tends to be too small ($<75 \mu m$) for fission-track dating.

An early example of fission-track tephrochronology is the study by Seward (1974) on the Wanganui Basin, North Island of New Zealand. The Pleistocene marine sediments of this basin contain many tephra layers and pumice-rich horizons with good stratigraphic and palaeontological control. The tephra are in a distal position with respect to their source in the central volcanic district more than 100 km to the north. Ten tephra horizons from the 2450 m thick Rangitikei section were dated. They contain glass shards of rhyolitic composition without any sign of devitrification. Figure 7.2 shows the results. The precision of the dates lies between 10 and 20%. The fission-track ages clearly correlate with stratigraphic position. The ages are also consistent with palaeomagnetic studies which indicate that the Brunhes-Matuyama boundary (730 ka) is situated between the Potaka and the Rewa pumices. Seward (1976) supplemented this tephrochronology by additional fission-track ages and, thus, established a solid base for chronostrati-graphic correlations within the Wanganui Basin. Further fission-track studies on glass shards from tephra beds in New Zealand were reported by Seward (1975) and Kohn (1979), including a Pleistocene age of 20.3 ± 7.1 ka.

In his study of various North American distal tephra occurrences, Westgate (1989) stresses the potential and reliability of fission-track dating of hydrated glass shards which are often the only datable material in such deposits. Many distal tephra beds appear difficult to date owing to their thin and discontinuous character, fine-grain size, low abundance of crystals, and common presence of detrital grains. The dated glass shards of rhyolitic-to-dacitic composition, had water contents of 5–6% and, in the case of the Banks Island tephra, even up to 9%. Their high water content was mainly caused by secondary hydration. Less than 25 mg of separated, fine sand-sized glass shards were sufficient for dating. All samples exhibited signs of partial fading of the spontaneous fission tracks as indicated by their distinctly smaller track size (d_s) compared to the induced tracks (d_i, Table 7.1) and, thus, the raw fission-track ages must be regarded as minimum estimates for the volcanic eruption. Therefore, an isothermal plateau-correction procedure was applied to all samples. The corrected fission-track ages that are characterized by similar sizes for spontaneous and induced tracks (d_s/d_i^* after annealing) agree well with independent age estimates on co-existing minerals, as determined by K-Ar, $^{40}Ar/^{39}Ar$, and zircon fission-track dating. This application demonstrates the great potential of fission tracks for dating Pleistocene sedimentary sequences, both continental and marine, even in areas distal to volcanic centres. Glass shards from volcanic ash layers in deep-sea sediments have been successfully dated with fission tracks by Macdougall (1971a).

In Table 7.1, the zircon fission-track ages do not require correction due to full track stability at ambient temperatures. However, this apparent advantage of zircon over glass cannot often be exploited due to the absence of sufficiently large zircon grains in widespread distal tephra beds. Nevertheless, important contributions to the North American tephrochronology have been achieved by zircon fission-track studies, such as the recognition that the well-known Pearlette tephra in North America consists of at least three distinct beds with different ages – namely 1.9, 1.2, and 0.6 Ma (Naeser *et al.*, 1973; Boellstorff, 1976) rather than of the formerly believed single unit. Of particular interest are zircons for proximal tephra beds. Whenever sufficiently sized ($>75 \mu m$) zircon grains occur in tephra beds, they should be included in the dating program. In tephrochronology, it is generally advisable to apply as many different dating techniques as possible in order to establish a firm chronostratigraphic framework. Other extensive zircon fission-track dating programs of tephra beds have been carried out in Japan (Koshimizu and Kim, 1987; Tamanyu, 1975).

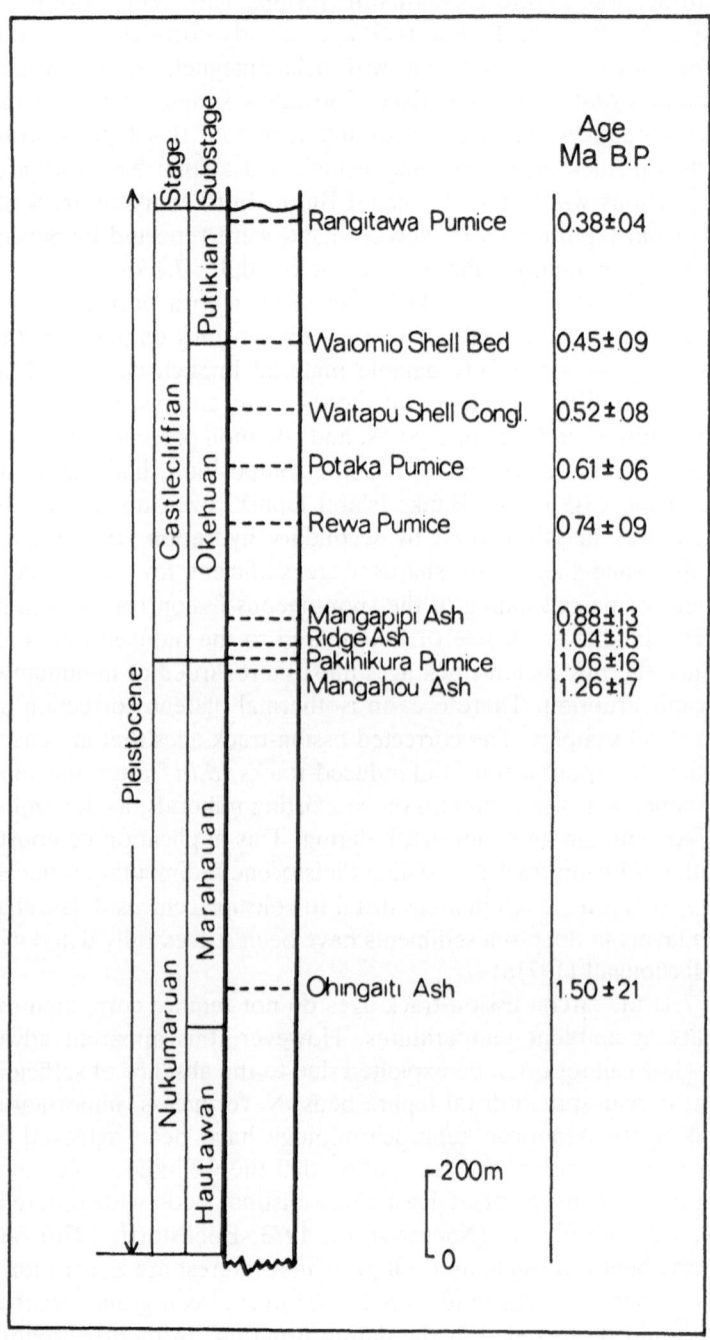

Figure 7.2.   Dated tephra and pumice-rich horizons in Pleistocene marine sediments of the Wanganui Basin, New Zealand. The fission-track ages of the glass shards from these tephra beds decrease with decreasing stratigraphic **position.** (After Seward, 1974.)

Table 7.1. Fission-track ages on hydrated glass shards and zircons from some North-American tephra beds (after Westgate, 1989)

Tephra sample	Glass (raw)		Glass (corrected)		Zircon age (Ma)
	d_s/d_i	age (Ma)	d_s/d_i^*	age (Ma)	
Banks Isld. tephra, N.W.T.	0.66	38.50	1.00	65.67 ± 1.68	
Borchers Ash, Kansas	0.78	1.16	1.01	2.04 ± 0.16	1.9 ± 0.1
Lake Tapps. tephra, Washington	0.76	0.65	0.95	1.06 ± 0.11	0.9 ± 0.3
Fort Selkirk tephra, Yukon	0.81	1.1	1.09	1.48 ± 0.19	
Bishop tuff, California	–	0.54	1.14	0.72 ± 0.05	0.74 ± 0.03
Old Crow tephra, Alaska	0.80	0.13	1.00	0.15 ± 0.15	

7.2. Post-Orogenic Uplift of Mountain Belts

The measurement of uplift rates in the Central Alps by means of apatite fission-track dating (*fission-track tectonics* by Wagner and Reimer, 1972) demonstrated for the first time the tectonic potential of the fission-track method. With this new application, the method ceased to be purely an 'age-determination' technique and developed into a unique thermo-tectonic tool. In the meantime, the technique has been further refined – in particular by the introduction of track length analysis – and has been applied to many of the world's young orogenic belts. At present, fission-track analysis is an established technique for reconstructing the uplift-evolution of active mountain belts. It is also unique, since it provides depth-time paths for rocks during their upward movement through the upper kilometers of the Earth's crust. Such data are of fundamental importance for understanding mountain building processes.

7.2.1. *General*

The basic idea which underlies the use of fission tracks for uplift studies is simple. During the orogeny, the rock temperatures become sufficiently high in order to completely reset the fission-track system. When the orogenic belt undergoes uplift and denudation, the rocks cool gradually on their way upward towards the surface whereby, at a certain depth, the rock temperature drops below the track-retention temperature of the fission-track recording mineral. At that moment the fission-track clock starts to run. Finally, the rock reaches the surface. Its fission-track age gives the time elapsed since the mineral emerged from the depth of the track-retaining isotherm. In this concept, the fission-track age is clearly a *cooling age* to a certain *retention temperature* (Sections 4.3.3, 5.5.2). The average *cooling rate* for the respective period can be calculated from the cooling age and the retention temperature (Section 5.2.2). In order to assess the depth corresponding to this retention temperature as well as the average *uplift rate*, one has to know the geothermal gradient. Then the uplift rate is calculated according to Equation (5.2).

The information on the uplift history is much improved if several fission-track systems, as well as other radiometric systems with different retention temperatures, are available for the same rock sample.

An independent approach, which does not require the knowledge of the geothermal gradient and the retention temperature, exploits the increase of the cooling age with topographic elevation. The higher samples within the same rock column move earlier through the isotherm of the retention temperture than the lower ones. The cooling rate is calculated directly from the slope of age-elevation profiles according to

uplift rate = elevation difference/age difference.   (7.1)

Both outlined approaches require that the isotherm of the retention temperature resides at a constant depth with respect to sea-level. This is by no means a simple prerequisite, since the uplift movement – in particular for very fast uplift rates >0.5 km/Ma (Parrish, 1983, 1985) – tends to drag the isotherms along (Section 5.5.2, Figure 5.10). Another complication is introduced by the existence of a broad *partial annealing zone* (PAZ) instead of a sharp track-retention temperature. The gradual annealing in this zone causes slopes which must not be mistaken for uplift rates (Section 5.6).

Both approaches have been applied to decipher the uplift history of active mountain belts. Most commonly, apatite (corresponds to ≈100°C track retention temperature) and, less frequently, zircon (≈210°C), sphene (≈250°C), garnet (≈270°C), vesuvianite, and kyanite have been used. Apart from the Alps, for which some results are presented below, uplift studies were reported for the Tatra Mountains (Burchart, 1972), the Pyrenees (Yelland, 1990), the Himalayas (Nagpaul, 1982; Saini *et al.*, 1983; Sharma *et al.*, 1980; Zeitler *et al.*, 1982a,b, 1985), the Tibetean plateau (Lewis, 1990), the Rocky Mountains (Bryant and Naeser, 1980; Bryant *et al.*, 1981; Dokka, 1984; Harrison *et al.*, 1979; Naeser, 1979; Naeser *et al.*, 1983; Parrish, 1983), the Andes (Benjamin *et al.*, 1987; Crough, 1983; Kohn *et al.*, 1984), Japan (Agar *et al.*, 1989) and the New Zealand Alps (White and Green, 1986; Kamp *et al.*, 1989).

7.2.2. *Central Alps*

The Alps are a complexly built orogen formed by the collision of the African and Eurasian tectonic plates during the Cretaceous and the Tertiary. In the course of crustal shortening, the plate boundaries with the intervening sediments of the Tethyan geosyncline suffered polyphase deformation and metamorphism. In order to unravel the age and style of uplifting as the final orogenic stage, fission-track dating has been carried out since the early 1970s in the Alps, mainly in the Central Swiss Alps (Schaer *et al.*, 1975; Wagner and Reimer, 1972; Wagner *et al.*, 1977, 1979). In the southern and western Alps, this work has been continued more recently by Carpena (1985), Giger (1991), Giger and Hurford (1989), Hurford (1986b), Hurford and Hunziker (1985), (1989) and Hurford *et al.*, 1989, 1991). Also, from the eastern Alps, extensive apatite fission-track uplift investigations have been reported by Flisch (1986), Grundmann and Morteani (1985), Hejl and Grundmann (1989), Hejl and Wagner (1991), and Staufenberg (1987). With several

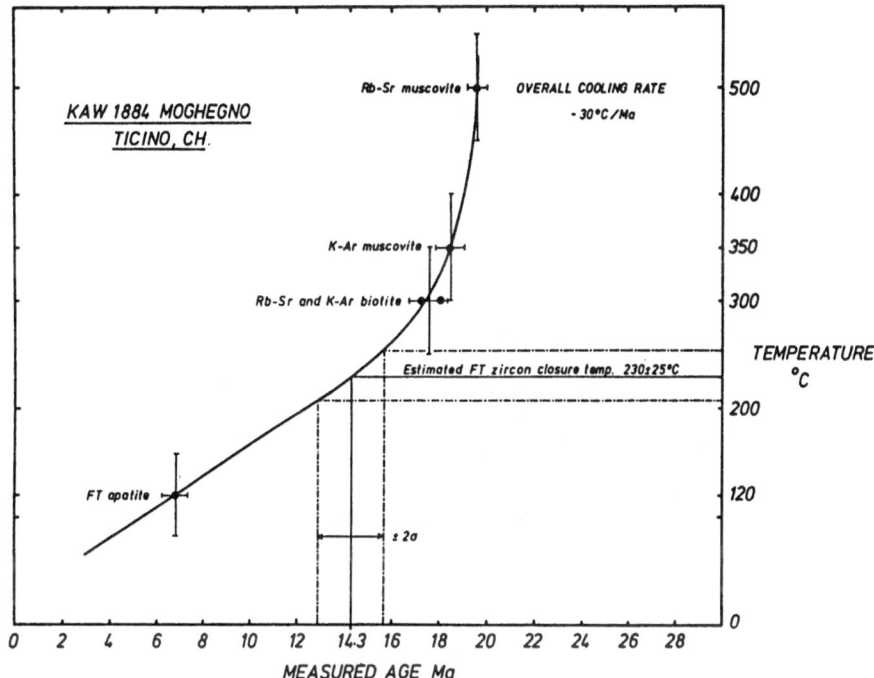

Figure 7.3. Different radiometric mineral ages versus respective retention temperatures for a gneiss from Maggia Valley, Switzerland. The data points trace the temperature-time path during the rock's cooling history. The retention temperature (≈ 230°C) of the zircon fission-track system is estimated from the intercept of measured age with the curve. (From Hurford, 1986b.)

hundred fission-track age determinations, the Alpine chain is by far the most densely studied orogenic belt in this respect.

The Swiss Central Alps may be considered as the classical area in which the concepts of unravelling the uplift history by apatite fission-track analysis were developed (Wagner and Reimer, 1972; Wagner *et al.*, 1977). Apatite fission-track ages of the Central Alps between 2.2 and 17.4 Ma were found. From these ages, several regularities were noticed:

(1) The apatite fission-track ages are always lower than the Rb-Sr and K-Ar mica-ages of the same rock samples. The age sequence muscovite Rb-Sr > muscovite K-Ar > biotite Rb-Sr ≈ biotite K-Ar > apatite fission-track is consistent with the sequence of the respective retention temperatures (Figure 7.3).

(2) A clear correlation exists between apatite fission-track ages and topographic elevation: the ages increase with elevation (Figure 5.14).

(3) The apatite fission-track ages show a regional distribution pattern. This pattern becomes more distinct when the ages are reduced to the same level of elevation (Figure 7.4).

(4) The regional distribution of the apatite fission-track ages does not conform to the general tectonic structures and metamorphic isogrades of the area.

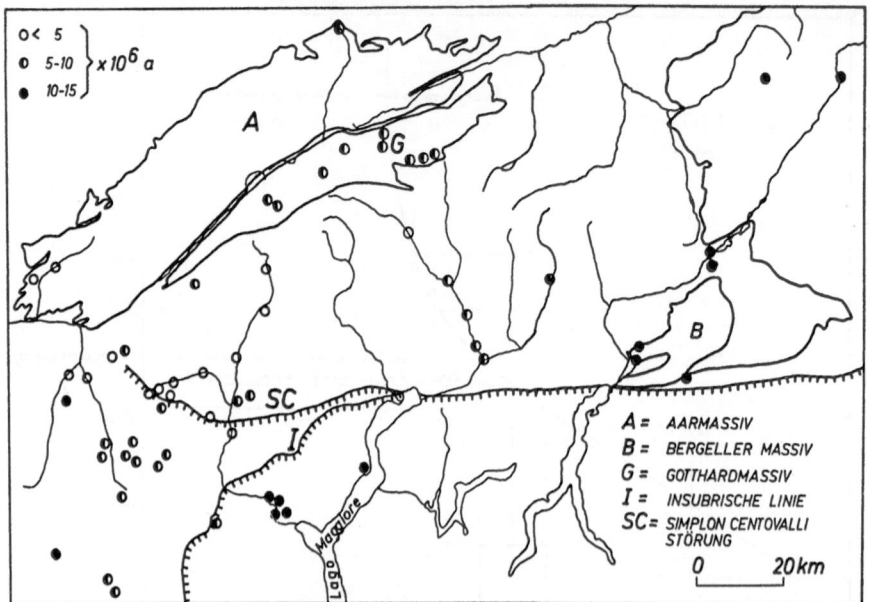

Figure 7.4. Regional distribution of apatite fission-track ages in the Swiss Central Alps. They date the cooling to 120°C and reflect the young uplift pattern of the Central Alps. (From Wagner, 1972a.)

All these observations are compatible with the idea of apatite fission-track ages being uplift- and erosion-controlled cooling ages subsequent to Alpine metamorphism. This interpretation has been further supported by confined tracklength measurements: for apatites from the Maggia valley, Hurford (1986b) observed unimodal length distributions with mean lengths of between 11.7 and 13.7 μm. As to the explicit value of the effective track retention temperature, note that due to the fast cooling rates of the Swiss Alps, 120°C is a more realistic value than the conventional 100°C (Section 5.2).

Provided young (more precisely, post-120° cooling) vertical tectonic displacement between the sampling locations is absent, samples from nearby locations of different elevations can be combined and the palaeo-uplift rate at the time given by the ages can be directly read from the slope of the *age-versus-elevation* curve. By doing this, uplift curves for some areas of the Central Alps have been constructed (Figure 5.13). The respective uplift rates are listed in Table 7.2. Temporal changes of the uplift rate should be accordingly recorded by slope changes of the curve. Note that sudden changes of uplift do not appear as breaks but as bends in the slope (Section 5.6.4).

An alternative way of illustrating the information contained in the age-versus-elevation relationship are sections such as the one across the Gotthard massif (Figure 7.5). After recording the apatite fission-track age for each sample in the section, isochrons are interpolated. The isochrons mark the present position of the uplifted palaeo-120°C isotherm which was originally horizontal and ≈3 km deep. They trace the amount of differential uplift since 10 to 6 Ma ago. In the case of the Gotthard section, the curved isochrons vividly illustrate the updoming of the massif towards the south, which can still be registered by precise levelling.

Table 7.2. Uplift rates (mm/a) of some areas in the Swiss Central Alps given for consecutive cooling intervals as derived from apatite fission track as well as mica Rb-Sr and K-Ar dating (Wagner *et al.*, 1977; Hurford, 1986b)

Area	300°C to 120°C	Around 120°C	120°C to surface
Bergell	0.7	0.4	0.2
Ticino	0.6	0.4	0.4
Gotthard	0.75	0.5	0.5
Monte Rosa	0.3	0.4	0.7
Simplon	0.7	0.9	1.1
Maggia	0.5	0.5	0.5

According to the tectonic model (Section 5.5.2), the *horizontal variation* of the apatite fission-track ages reveals the regional uplift pattern. An average cooling rate can be individually calculated for each sample and – if the geothermal gradient is known – these cooling rates can be transformed into uplift rates (Equation (5.2)). In mountainous areas such as the Alps, the problem arises as to which degree the geothermal gradient and, thus, the depths of the critical isotherms is modulated by the surface morphology (Parrish, 1983, 1985). Although the possibility of bulged isotherms below the Central Alps cannot be completely ruled out, it was neglected for the calculations of the uplift rates in Table 7.2. The youngest apatite fission-track ages (2 to 3 Ma) of the Alps were found in the Simplon area (Figure 7.4). They point at a strong young uplift (1.1 mm/a, assuming 30°C/km) which is in agreement with precise levelling results.

Additional information on the cooling and uplift history is derived from mica Rb-Sr, and mica K-Ar ages on the same rocks. In particular, the biotite Rb-Sr

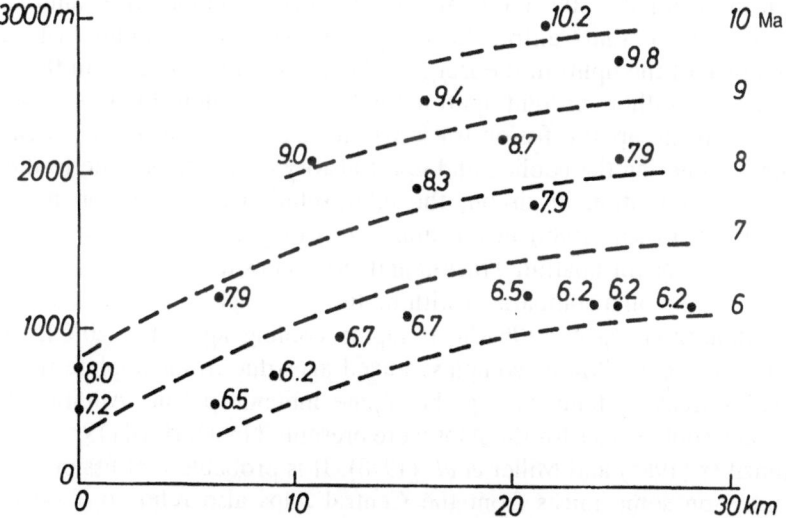

Figure 7.5. Apatite fission-track ages along a topographic section (elevation versus horizontal distance) across the Gotthard Massif, Switzerland. The isochrons represent the uplifted former 120°C isotherms and illustrate the massif's young updoming at 20–30 km distance. (After Schaer *et al.*, 1975.)

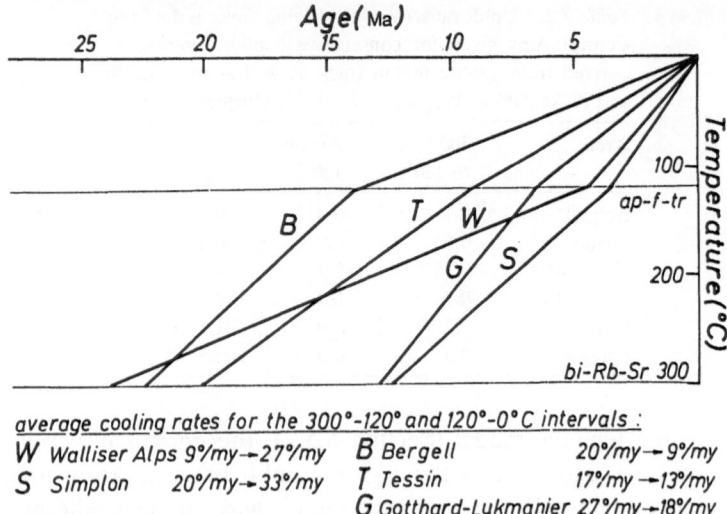

Figure 7.6. Biotite Rb-Sr and apatite fission-track ages versus their respective retention temperatures for some areas in the Swiss Central Alps. The curves trace the *T–t*-paths during cooling. (After Wagner *et al.*, 1977.)

and apatite fission-track ages have been used for reconstructing, for the Central Alps, the *T–t*-cooling paths from 300°C to 120°C and 120°C to surface temperature (Figure 7.6). Assuming a geothermal gradient of 30°C/km, the respective uplift rates can be calculated (Table 7.2). These uplift rates are in good agreement with those independently derived from the slope of the age-elevation profile. Since the latter uplift rate corresponds to the 120°C cooling age, it falls chronologically in between those for the 300–120°C and 120°C–surface temperature intervals. The various regions of the Central Alps experienced different uplift histories: the slowing down of the uplift in the Bergell, the acceleration of uplift in the Simplon region, and the rather constant uplift in the Gotthard region (Figures 7.3 and 7.5).

Apart from the apatite fission-track system, zircon has also been occasionally used for deciphering the cooling and uplift history of the Alps. Due to the higher thermal track-retention in zircon, the interpretation of the zircon fission-track ages found by Flisch (1986) in the upper austroalpine Silvretta nappe is not as straightforward as for apatite. Track-length criteria as for apatite, do not yet exist for zircon. Based on a comparison with biotite K-Ar ages, Flisch (1986) interprets zircon fission-track ages <106 Ma as alpine cooling ages (to 220°C), but ages between 131 and 184 Ma as complex, mixed ages due to incomplete resetting of their fission-track systems during the alpine metamorphism. Additional zircon fission-track cooling ages for the Alps were presented by Hurford (1986b), Hurford and Hunziker (1985) and Miller *et al.* (1978). It is probable that fission-track ages determined on some micas from the Central Alps also refer to a stage in the Alpine cooling history (Miller and Jäger, 1968), although more work needs to be done.

7.3. Epeirogenic Uplift of Basements

In addition to the fast, post-orogenic uplift, a slow uplift associated with epeirogenic crustal movements can also be detected by fission-track studies. Epeirogenic movements may involve periods of slow uplift interrupted by long periods of tectonic quiescence or even subsidence. They are of decisive significance for the denudational, sedimentary and geomorphological evolution of continents.

Epeirogenic crustal movements are difficult to detect, especially in crystalline basements, by stratigraphic and facies analyses in adjacent sedimentary basins. Essentially, by exploiting the apatite fission-track system as a sensitive thermochronometer to $\approx 100°C$ – and, to a lesser degree, the zircon and sphene fission-track systems – the timing, rate, and amount of epeirogenic uplift movements can be traced. In contrast to fast post-orogenic uplift, the concept of the *partial annealing zone* (Section 5.2) of the apatite fission-track system plays a dominant role when deciphering slow uplift movements.

Of particular interest in this approach is the variation of the apatite fission-track age and the track-length distribution along vertical rock columns. As in the case of fast uplift, the increase of age with elevation may be directly interpreted under conditions of denudational equilibrium and constant geothermal gradient as an apparent uplift rate. However, a slope in the age-versus-elevation curve is less likely to represent a true uplift rate, especially if the apparent rates are very low <20 m/Ma (Gleadow and Fitzgerald, 1987; Van den haute, 1986b). It can be easily demonstrated (Section 5.6.4) that, due to the existence of a *partial annealing zone* (PAZ), such slopes develop, even under the condition of tectonic stability. Consequently, one must clearly distinguish between slopes which represent fossil uplift rates and those which represent uplifted fossil partial annealing zones. The *distribution of the track lengths* can be used as a reliable criterion for recognizing the true nature of the slopes in the age-versus-elevation curves because the apatite fission-track ages within the slopes are either cooling ages or complex ages, respectively. The advantage of combining ages from vertical sampling profiles with track-length measurements is illustrated in the following case of the Transantarctic Mountains.

7.3.1. *Transantarctic Mountains*

In several fission-track studies, the uplift history of South Victoria Land, which forms a section of the Transantarctic Mountains, has been investigated (Fitzgerald *et al.*, 1986; Fitzgerald and Gleadow, 1990; Gleadow and Fitzgerald, 1987; Gleadow *et al.*, 1984; Wagner *et al.*, 1989b). The Transantarctic Mountains extend some 4000 km across Antarctica from northern Victoria Land to the Wedell Sea. They mark the boundary between the Precambrian East Antarctic craton and the Mesozoic/Cenozoic West Antarctica. South Victoria Land geologically consists of Precambrian to early Palaeozoic metamorphic and granitic basement that was eroded during the Silurian and early Devonian periods. On this erosional surface, a thick sedimentary cover was deposited from the Devonian to the Triassic. The

Mid-Jurassic was characterized by extensive magmatic activity with intrusions of sills and dikes as well as with basaltic extrusions. Subsequently, uplift movement set in which produced mountain peaks that reach elevations of more than 4000 m.

The sampling strategy for fission-track dating included vertical profiles within the crystalline basement rocks, with elevation ranges of more than 1000 m generally sampled at 100 m intervals, and regional profiles intersecting tectonic structures. In all vertical sampling profiles, the apatite fission-track ages increase with elevation. As a characteristic feature of this region, a pronounced *break-in-slope* in the apatite fission-track age profiles was observed (Figure 5.15). The upper (older) section of the profile has a shallow age gradient of 15 m/Ma, while the lower (younger) section has a much steeper gradient of 100 to 200 m/Ma. This break also corresponds to a significant change in the confined track-length distributions. Above the break, they are relatively broad, even bimodal, with mean lengths of about 13 μm and, below the break, they have a mean length in excess of 14 μm and narrow distributions with very few short tracks.

From these fission-track age and confined track length results, Gleadow and Fitzgerald (1987) inferred a two-stage tectonic history for South Victoria Land, such as that illustrated in Figure 5.20. The first stage was a period of relative tectonic and thermal stability during the Cretaceous when a *complex age-elevation profile* (curve b in Figure 5.16) was established. The uplift responsible for the present mountain range began approximately 50 Ma ago and has continued at an average rate of about 100 m/Ma, creating an *uplift age-vs.-elevation profile* below the break. The observed break-in-slope is interpreted as the base of a PAZ prior to the onset of rapid uplift. Prior to this uplift, samples above the break lay within the PAZ for apatite, whereas the deeper samples had a zero apparent age. A total uplift of about 5 km since the early Tertiary, was inferred in this area.

Also, the projected lengths of spontaneous fission tracks were measured in apatites from the vertical sections at Mt. Doorly and Mt. Jason in South Victoria Land (Wagner *et al.*, 1989b). The t_f-ages of effective cooling to $\approx$60°C that are derived from the projected track lengths (Section 5.4) and measured t_m-ages are plotted as a function of elevation for Mt. Doorly and Mt. Jason in Figure 5.15. The break-in-slope of the t_m-age profile observed at Mt. Doory is not exhibited by the t_f-ages. Below the break, both age profiles are steep and parallel for the t_m ages. The t_f-profile extends with the same steep slope for the samples whose t_m-ages are above the break. These observations are completely consistent with the model predictions for an uplifted PAZ, where the break-in-slope of the t_m-ages marks the base of that PAZ with *ca.* 100°C cooling ages below, and mixed ages above the break. The t_f-ages are all effective cooling ages to $\approx$60°C. The uplift rate is revealed by the t_f-age profile and by the lower, steep part of the t_m-age profile, but as the age remains almost constant, this rate cannot be assessed precisely. The age profiles for Mt. Jason look different: they show no break-in-slope and their slopes are not parallel, the profile being steeper for the t_f- than for t_m-ages. At Mt. Jason, a break-in-slope should be found for the t_m-ages at lower elevations than presently sampled. All observed t_m-ages are mixed ages and the t_f-ages are again cooling ages to *ca.* 60°C. The uplift rate is only revealed by the slope of the t_f-age profile. In terms of apatite fission-track annealing, the Mt.

Jason section displays a shallower erosional level than Mt. Doorly. This agrees with geological evidence because, structurally, Mt. Jason has been uplifted less than Mt. Doorly (Gleadow and Fitzgerald, 1987). A similar study has been performed in the Antarctic Heimefrontfjella Mountains (Jacobs, 1991).

7.3.2. *Central European Basement*

With respect to the number of publications, the uplift history of crystalline basements seems to be a favourite geological subject of fission-track application. Regional uplift studies have been reported in *North America* for the Appalachian Mountains (Johnson, 1985; Lakatos and Miller, 1983; Miller and Duddy, 1989; Miller and Lakatos, 1983), the intracontinental Precambrian basement (Crowley *et al.*, 1985, 1986; Crowley and Kuhlman, 1988; Stuckless and Naeser, 1972); in *Africa* for the Damara Orogen (Haack, 1976, 1983), the south-western continental margin (Brown *et al.*, 1990), the Precambrian basement adjacent to the East African rift (Van den haute, 1984, 1986b; Wagner *et al.*, 1992), and to the Red Sea rift (Bohannon *et al.*, 1989; Kohn and Eyal, 1981; Omar *et al.*, 1989); in *Australia* for the western Victoria basement (Gleadow and Lovering, 1978b), the southeastern continental margin (Moore *et al.*, 1986; Morley *et al.*, 1980); and in *Europe* from the Precambrian Fennoscandian basement (Andriessen, 1990; Lehtovaara, 1976; Van den haute, 1977; Zeck *et al.*, 1988), the British Palaeozoic basement (Green, 1986, 1988, 1989), the Brabant Massif (Van den haute and Vercoutere, 1989) and the Variscan basement in Central Europe which is presented as an example below.

Following the early days of apatite fission-track dating in the Odenwald area (Wagner, 1968), the crystalline basement of *Central Europe* had only drawn little attention from fission-track workers. This picture has changed greatly during the past few years. Presently, more than a hundred apatite fission-track ages exist for crystalline outcrops in the Odenwald, the Oberpfalz, the Schwarzwald, and the Vogesen areas (Figure 7.7). The revived activity in the geological investigation of the crystalline basement has been triggered by the *German Continental Deep Crustal Program (Kontinentale Tiefbohrung KTB)* with a target depth of 12 km and a rock temperature of around 300°C to be reached in the mid 1990s (Emmermann, 1986). Fission-track studies, among other geological and geophysical surveys, have been carried out in the Schwarzwald and the Oberpfalz as a basis for deciding on the final KTB-location. The apatite fission-track studies detected a ≈30 Ma old geothermal activity underneath the proposed Schwarzwald location and, thus, contributed to the decision against the Schwarzwald: KTB is situated near Windisch-Eschenbach, Oberpfalz, on the north-western margin of the Bohemian Massif (Figure 7.7).

The mountainous areas of Central Europe north of the Alps are built up dominantly by an exposed and eroded crystalline basement. After its last strong heating during the Variscan orogeny and uplift/denudation in the late Palaeozoic, the basement subsided and was covered by 500–1500 m thick sediments during most of the Mesozoic. Since the upper Cretaceous, the sense of vertical motion

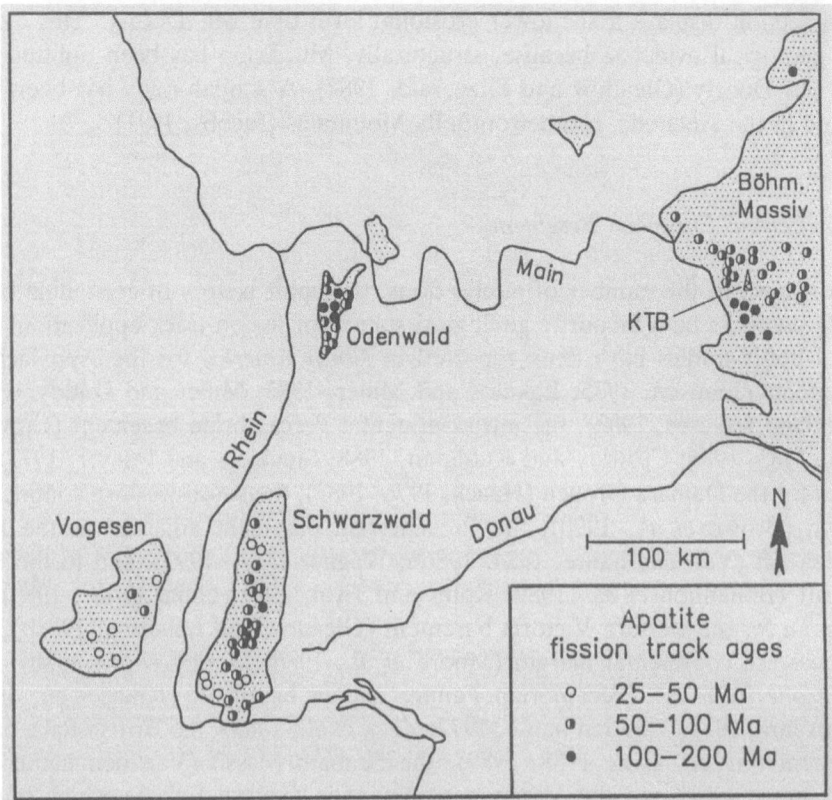

Figure 7.7. Apatite fission-track ages on the Variscan crystalline basement in Central Europe. 'KTB' marks the location of the continental deep drilling. (From Wagner, 1990.)

reversed, resulting in uplift and denudation. Large areas of denudated basement are now exposed in the Vogesen, Schwarzwald, and Odenwald on both shoulders of the Rheingraben rift and in the Bohemian Massif. The basement dominantly consists of gneisses and granites. The regional distribution of the apatite fission-track ages (Wagner, 1968, 1969a; Wagner and Storzer, 1975; Wagner *et al.*, 1989c; Hejl and Wagner, 1990) is shown in Figure 7.7. In the basement areas neighbouring the Rheingraben, the ages range between 29 and 107 Ma with some large changes over a short horizontal distance of a few km, especially in the Schwarzwald. In the Oberpfalz, the ages range between 50 and 200 Ma with a more smooth regional pattern.

Confined and projected track lengths were measured for the majority of the apatites. Nearly all apatite fission-track ages ($t_m < \approx 100$ Ma) are – according to the confined track lengths – *cooling ages*.

Therefore, during Mesozoic subsidence, almost all investigated basement apatites suffered sufficient heating (>130°C) and completely lost their older track record. Subsequently, the basement underwent cooling due to Cretaceous and Cainozoic uplift and denudation. The cooling ages to ≈100°C and ≈60°C are recorded in the fission-track data (t_m and t_f, respectively). Average uplift rates

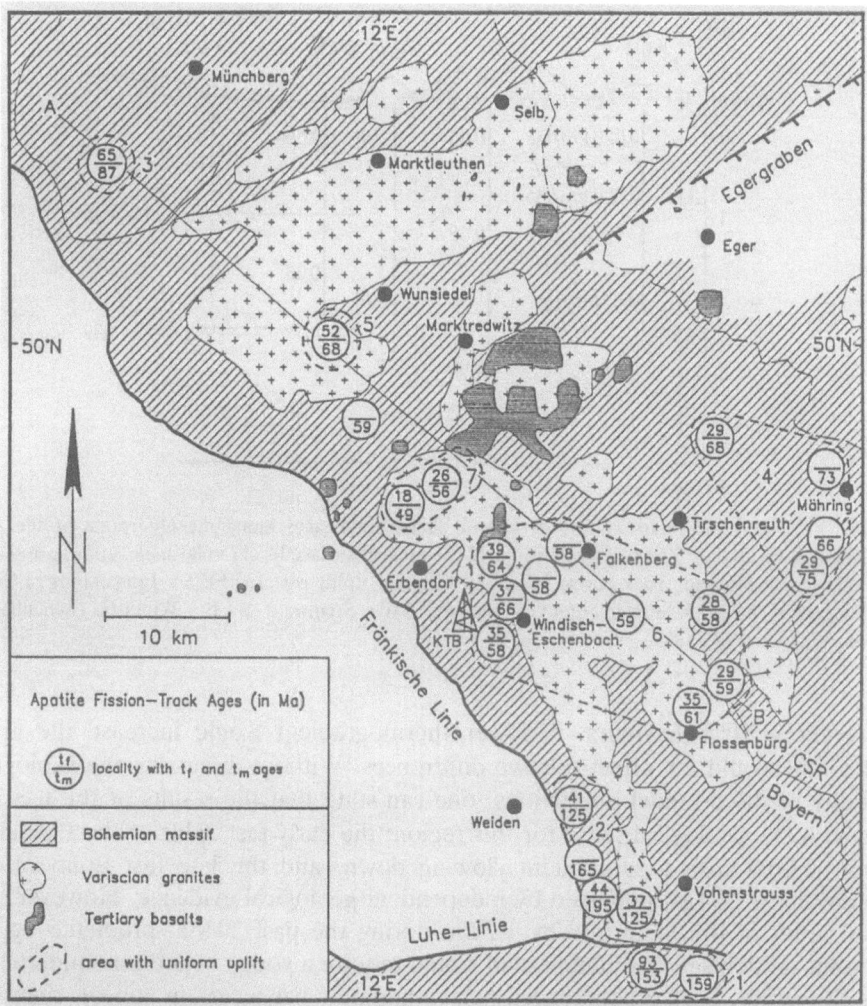

Figure 7.8. The Oberpfalz area on the western margin of the Bohemian Massif with apatite fission-track ages. 'KTB' marks the location of the continental deep drilling. Areas with uniform age patterns are encircled. The profile A-B refers to Figure 7.9. (From Wagner, 1990.)

range between 0.02 and 0.08 km/Ma. For the Schwarzwald, the data reveal pronounced block faulting and an increased palaeo-thermogradient during the last 50 Ma. Both phenomena are probably related to the formation of the Rheingraben during the Tertiary.

The t_m and t_f-cooling ages of the Oberpfalz are presented in Figure 7.8. This figure shows that there are distinct regions with similar age patterns which can be discussed, along a NW–SE profile shown in Figure 7.9, in terms of the tectonic model (Section 5.5.2) assuming an equal thermogradient of 30°C/km. Due to the known T–t-paths, the palaeo-positions of the present surface (taken as 500 m above sea level) can be calculated. Figure 7.9. shows such palaeo-positions 30, 60, and 90 Ma ago within the Earth's crust, from where they reached the surface by uplift and denudation. The fission-track data reveal several tectonic blocks with

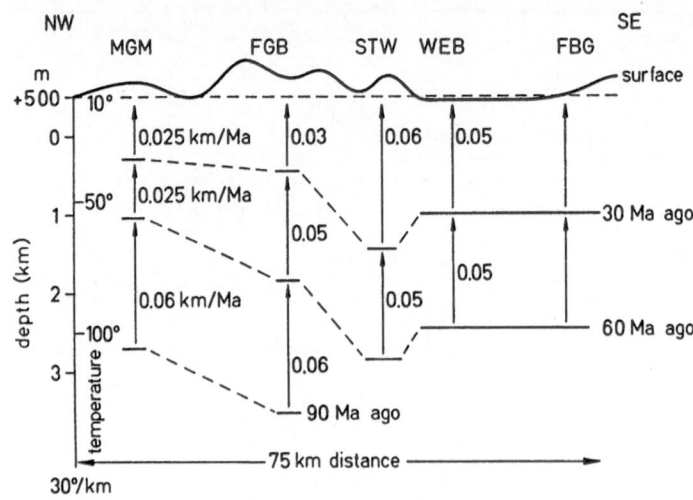

Figure 7.9. Tectonic interpretation of apatite fission-track ages along the Oberpfalz profile A-B indicated in Figure 7.8. The palaeo-positions of the present surface level (≈500 m elevation) are shown 30, 60, and 90 Ma ago. They reveal blocks of different uplift pattern (FBG = Flossenbürg, FGB = Fichtelgebirge, MGM = Münchberger Gneismasse, STW = Steinwald, WEB = Windisch-Eschenbach). (From Wagner *et al.*, 1989c.)

different uplift behaviours. A lower thermogradient would increase the uplift rates, but would not affect relative differences. Without discussing the geological validity of these model predictions, one can state that the results of the tectonic model make geological sense for this region: the early fast uplift of the Fichtelgebirge (FGB) with a subsequent slowing down, and the late fast uplift of the Steinwald (STW) is supported by independent geological evidence. However, the total uplift of the Steinwald by 1.2 km during the past 20 Ma, predicted by the tectonic model, seems too high, and would require a young heat dome underneath the Steinwald. Such a geothermal palaeo-anomaly has a certain appeal, since the Steinwald is situated along the SW extension of the Oligocene/Miocene Egergraben rift with its accompanying basaltic volcanism (Figure 7.8).

The information discussed here was obtained from the regional distribution of the apatite fission-track ages, the track-length measurements, and the interpretation of the ages as ≈100°C (t_m) or 60°C (t_f) cooling ages. Additional information on the uplift history is gained from the age-versus-depth variation along the KTB-drill core (Wagner *et al.*, 1989a). Although the fission-track analysis – including apatite as well as zircon and sphene – is still in progress, the apatite fission-track age profile already reveals an interesting uplift history (Figure 7.10). According to the criteria of confined track lengths (Section 5.4), all t_m-ages are cooling ages. In the upper 1000 m, the t_m-ages range between 60 and 70 Ma and decrease slightly with depth. They indicate a phase of relatively strong uplift (0.12 km/Ma) during the upper Cretaceous. This conclusion is supported by geological evidence. During that period, thick fans of coarse detritus have were shed from the eroding massif across the *Fränkische Linie*, a major tectonic line separating the basement from the sedimentary basin adjacent to the west. Between 1180 and 1400 m depth,

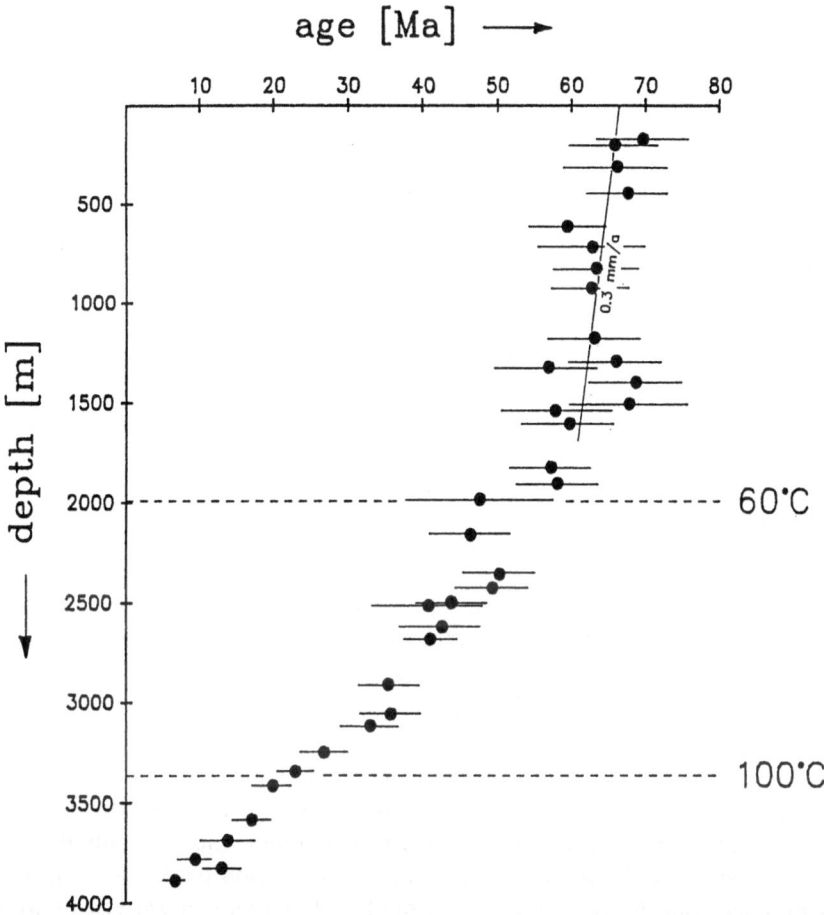

Figure 7.10. Variation of apatite fission-track ages with depth of the 4000 m deep KTB pilot drill-hole. Track-length measurements indicate that the ages at shallow depths (<2000 m) are cooling ages revealing an uplift rate of 300 m/Ma around 65 Ma ago. The age decrease between 2000 and 4000 m is caused by partial track annealing in the present PAZ. At a depth of 3387 m (bore-hole temperature = 118°C) the apatite fission-track age is reduced to approx. 6 Ma.

t_m-ages around 70 Ma are observed. This increase of age with depth is not yet fully understood, but may be caused by a young reverse fault with a vertical throw of several hundred meters. This information supplements that already gained from the regional age pattern in order to reconstruct the post-Variscan uplift history of the western Bohemian Massif.

7.4. Age and Amount of Displacement along Faults

Fission-track dating is occasionally used to study the sense and amount, as well as the timing, of vertical tectonic displacements along faults. This application is of particular importance for crystalline basements in which faults are usually difficult to detect from field evidence. It has been shown (Section 5.6) that a systematic increase of the fission-track age – in particular, for apatite – with

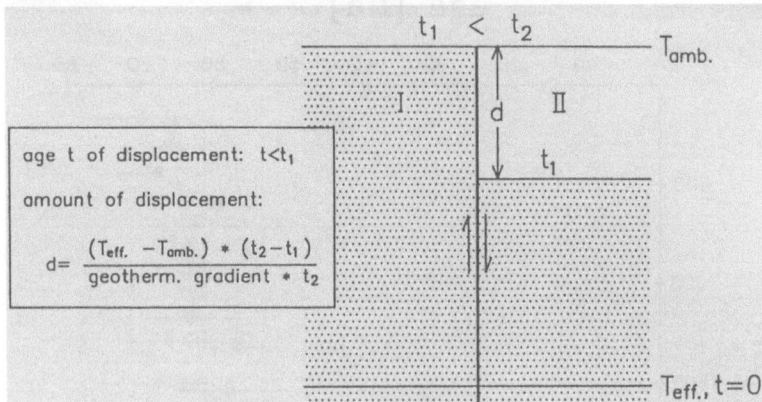

Figure 7.11. The age t and the amount d of vertical displacement on a fault as derived from cooling ages t_1 and t_2, which are different across the fault. A constant, known geothermal gradient for both tectonic blocks and a constant uplift rate for block II are assumed. T_{amb} is the ambient surface temperature and T_{eff} is the effective retention temperature.

elevation is often observed. Within an area of the same thermo-tectonic history, one expects identical vertical and no horizontal age variation. If an area with such a regular age pattern is subsequently affected by faulting, this pattern will be disturbed by the displacement of the rocks. In other words, on both sides of a young fault with vertical displacement, one expects different ages which may allow the determination of the sense and amount, as well as the timing of the displacement. In addition, frictional heating may be associated with the movement along the fault, causing partial or complete track fading and, thus, potentially enabling direct dating. Also, the dating of the vein mineralization has been attempted.

In practice, one is frequently confronted with a regional pattern in which an area with similar fission-track ages is separated from another one with different ages. Such a regional pattern for apatite fission-track ages from the Odenwald basement had been originally explained by Wagner (1969a) in terms of differential uplift. Although such an age pattern may be caused either tectonically or thermally (Figure 5.11), the tectonic interpretation is particularly convincing if the ages are offset across a visible fault line.

The principle of assessing the amount and the time of a displacement from the age pattern on a horizontal surface according to the tectonic model is illustrated in Figure 7.11. Across the fault, the ages t_1 (block I) and t_2 (block II) are different ($t_1 < t_2$). The ages are cooling ages to the effective retention temperature T_{eff}. The geothermal gradient is temporally and regionally constant. Then, tectonic block I must have experienced, on average, a faster uplift than block II. Depending on the dip of the fault plane, one deals with a normal (dipping towards block II) or a reverse fault (dipping towards block I). Assuming a constant uplift rate for block II, the amount of vertical displacement can be calculated from the age difference $t_2 - t_1$, the difference between the effective retention temperature T_{eff} and the ambient surface temperature T_{amb}, and the geothermal gradient according to the equation given in Figure 7.11. The fault must obviously be younger than the youngest age found on either side, i.e., $t < t_1$. A practical example of applying

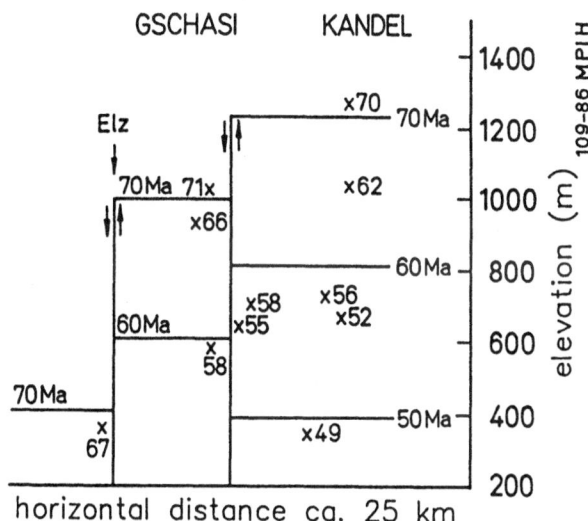

Figure 7.12. Apatite fission-track ages from various sampling locations of different elevations along a profile in the Central Schwarzwald, Germany. The offset isochrones reveal the direction and amount of vertical displacements between the tectonic blocks. (From Wagner *et al.*, 1989c.)

this model is illustrated for the apatite fission-track age variation in the Oberpfalz crystalline basement along the profile A–B in Figure 7.8. Palaeo-positions of the present surface level along this profile are shown in Figure 7.9, revealing several tectonic blocks with differential uplift behaviour separated by young faults. Another example has been reported by Zimmermann (1980) in a study across the Fall Line in the eastern United States, where apatite fission-track ages turned out to be different on either side of the fault line.

The information on fault movements becomes more precise if, in addition to the horizontal age variation, the vertical age variation is also available. The two levels of the disrupted fission-track isochron across the fault show the sense and the amount of the vertical displacement since the time indicated by this isochron. The youngest broken isochron puts an upper age limit, whereas the oldest unbroken one puts a lower age limit to the last movement on the fault. An example is the Gschasi-Kandel region in the Schwarzwald (Wagner *et al.*, 1989c). The apatite fission-track ages from various sampling locations are plotted within the profile (Figure 7.12). As revealed by the displaced isochrones, the Kandel block was uplifted by 200 m relative to the Gschasi block which, in turn, was uplifted by 600 m relative to the next block separated by the Elz fault. This picture is in agreement with geological observations. The movements must have occurred less than 60 Ma ago and were probably associated with the down-faulting of the nearby Rheingraben. Another example of this kind was reported by Fitzgerald and Gleadow (1990) for displacements across faults along the Mt. Allen–Mt.Doorly ridge system of the Transantarctic Mountains.

A different approach exploits the shear heating by fault movements for dating this event. Such heating might be sufficiently intense and thus reset the apatite fission-track clock. Tagami *et al.* (1988b) collected apatite and zircon from three traverses across the Median Tectonic line, which is considered as a strike-slip fault

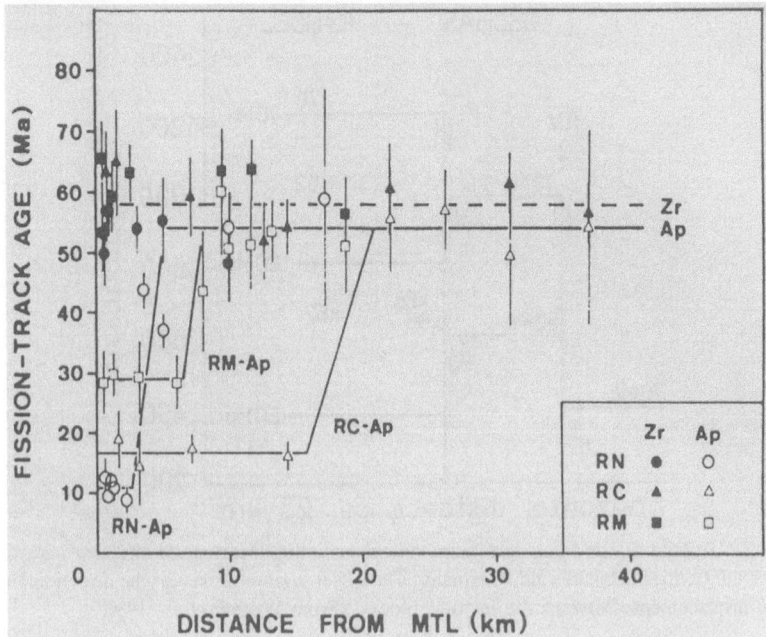

Figure 7.13. Zircon (Zr) and apatite (Ap) fission-track ages versus distance from three traverses across the Median Tectonic Line, Japan. The lowering of the apatite ages towards the fault is caused by shear heating. The apatite ages of samples collected near the fault date the last active fault movements. (From Tagami *et al.*, 1988b.)

of Tertiary age, in order to unravel the thermo-tectonic evolution of the Ryoke Belt, southwest Japan. The zircon ages turned out to be reasonably consistent, averaging 58 Ma in the three traverses. On the other hand, the apatite ages decrease along the traverses from about 54 Ma down to 10 Ma as the tectonic line is approached (Figure 7.13). Track-length measurements indicate that the rejuvenation of the low apatite fission-track ages near the fault was caused by complete, thermal resetting. According to Tagami *et al.* (1988b), the shear heating of the mylonites caused sufficiently high temperatures to wipe out the previous track record in apatite (>105°C) but did not affect the zircon fission-track system (<240°C). Consequently, the apatite fission-track ages from near the fault, date the last active movements between 29 and 11 Ma ago for different sections of the Median Tectonic line.

Bar *et al.* (1974) attempted to date faults by measuring epidote fission-track ages of mineralization veins along fault planes. The ages on epidote from faults and fissures in the Sinai Peninsula revealed early Cretaceous and Miocene tectonic phases.

7.5. Thermal Evolution of Sedimentary Basins

During the last decade, much attention has been paid to the thermal evolution of sedimentary basins (N. D. Naeser and McCulloh, 1989). This growing interest has

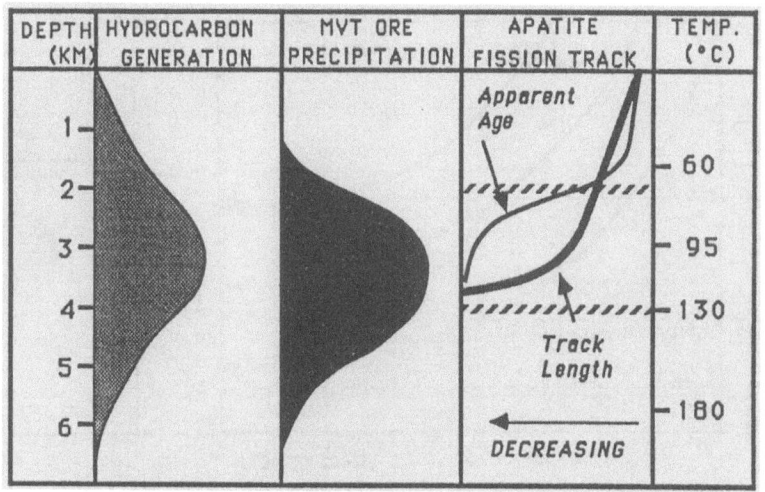

| DEPTH (KM) | HYDROCARBON GENERATION | MVT ORE PRECIPITATION | APATITE FISSION TRACK | TEMP. (°C) |

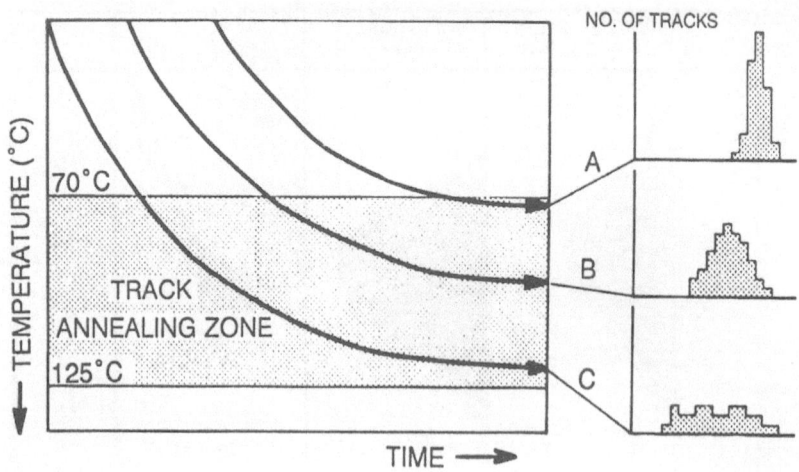

Figure 7.15. Hypothetical *T–t*-paths during burial and the resulting length distributions of confined fission tracks in apatite. (After Gleadow *et al.*, 1983.)

But it must be kept in mind that these two detrital minerals are less informative than apatite because an understanding of their track length behaviour during annealing is still lacking. Therefore, one has to rely solely on their fission-track ages, i.e., if they are older or younger than the stratigraphic age, in order to place upper limits on the palaeo-temperature. Such high temperatures were rarely reached in hydrocarbon prospective basins. In samples that have never been deeply buried and, consequently, remained closed, post-depositional fission-track systems, the fission-track ages may also give information of the source regions for single detrital grains (McGoldrick and Gleadow, 1977; Hurford *et al.*, 1984b; Hurford and Canter, 1991; Cerveny *et al.*, 1988; Yim *et al.*, 1985).

In routine applications of the apatite fission-track system as a hydrocarbon exploration tool, Green *et al.* (1989a) recommend an equally spaced sampling of drill cores down-hole to the bottom of the partial annealing zone (PAZ). Of course, the precise sampling interval depends on the geothermal gradient, the lithology, and the specific problem to be solved. Typically, a 400 m spacing would result in 6 to 10 samples per well. The present-day geothermal gradient should be well known as a base against which the palaeotemperatures derived from the track analysis can be compared. Most suitable samples (cores or cuttings) are sandstones and coarse siltstones, since they usually contain apatite as a minor detrital constituent. Normally 1–2 kg of sample material is sufficient. If present, detrital zircon and sphene should also be separated from the samples. Due to varying initial age values and chemical composition among single grains, the dating must be grain discrete, which is usually achieved with the external detector technique.

7.5.1. *Tejon Oil Field*

The early studies on the thermal history of sedimentary basins exploited the variation of apatite fission-track ages with bore-hole depth (Naeser, 1979). An illustrative example of this kind is the Tejon Oil Field area in the southern San Joaquin Valley, California (Briggs *et al.*, 1981; N. D. Naeser *et al.*, 1989). The valley contains a thick sequence of Cainozoic sediments. A seismically still active fault divides the area into the structurally extremely depressed, rapidly subsiding Basin Block and the less depressed, more slowly subsiding Tejon Block. At least since the early Miocene, the Tejon Block has a relatively higher tectonic position and lower burial rate (4 km Eocene-Holocene sediments) than the adjacent Basin Block (7.6 km Miocene-Holocene sediments). It is geologically believed that the sediments within both blocks are presently at the maximum depth of burial, but the duration of the burial is shorter for the Basin Block than for the Tejon Block. The different burial history of the sedimentary rocks of both blocks suggests that they have experienced a different thermal history. The present average geothermal gradient is about 22°C/km beneath the entire Tejon area. In order to investigate this thermo-tectonic difference between the two blocks, fission-track dating of apatite and zircon was carried out. The detrital grains were recovered mainly from sandstone collected from drill cores and cuttings as well as some outcrops. The age results are plotted against sampling depths in Figure 7.16. It appears that in depths shallower than 3 km, the apatite fission-track ages (*ca.* 60 Ma) are the same in both blocks. All these ages are older than the stratigraphic ages of the rocks from which the respective samples were collected. At greater depths, the apatite fission-track ages decrease steadily to zero, namely between 3 and 4 km in the Tejon Block and 4 and 5.5 km in the Basin Block. In terms of present temperature, the zero age is reached at *ca.* 115°C in the Tejon Block and 140°C in the Basin Block. Zircons were dated from four drill-hole samples at actual temperatures of up to 170°C. Their fission-track ages around 80 Ma do not vary noticeably with depth.

For interpreting the observed apatite fission-track ages in the Tejon Oil Field, one has to distinguish between the part in the age-versus-depth profile (Figure 7.16) where the ages do not vary with depth in contrast to were they do decrease with depth. For the upper, shallow part of such profiles, Naeser (1979) considers the fission-track ages as original detrital ages of source rocks unaffected by any post-depositional track annealing during subsidence and burial. Within this zone, the apatite fission-track age is higher than the stratigraphic age of the rock. On the other hand, in the lower part, the fission tracks were subjected to annealing at the progressively higher temperatures of the PAZ during burial. Within this zone, the apatite fission-track ages gradually become lower than the stratigraphic age of the rock. Such an age-versus-depth profile is to be expected for subsiding basins where the temperatures are presently at maximum (curve b in Figure 5.16). The significant difference of 25°C between the temperatures where zero age is observed in the two adjacent blocks, indicates that the blocks have different thermal histories, namely that the sediments in the Basin Block resided near their present temperature and depth for a shorter duration than those of the Tejon Block. From the known annealing behaviour of tracks in apatite, these durations are estimated as 10^6 and 10^7 a, respectively, under a geothermal regime similar to the present one. Since the zircon fission-track ages on the drill-hole samples turned out to be slightly older than those on the outcropping Eocene sandstone and,

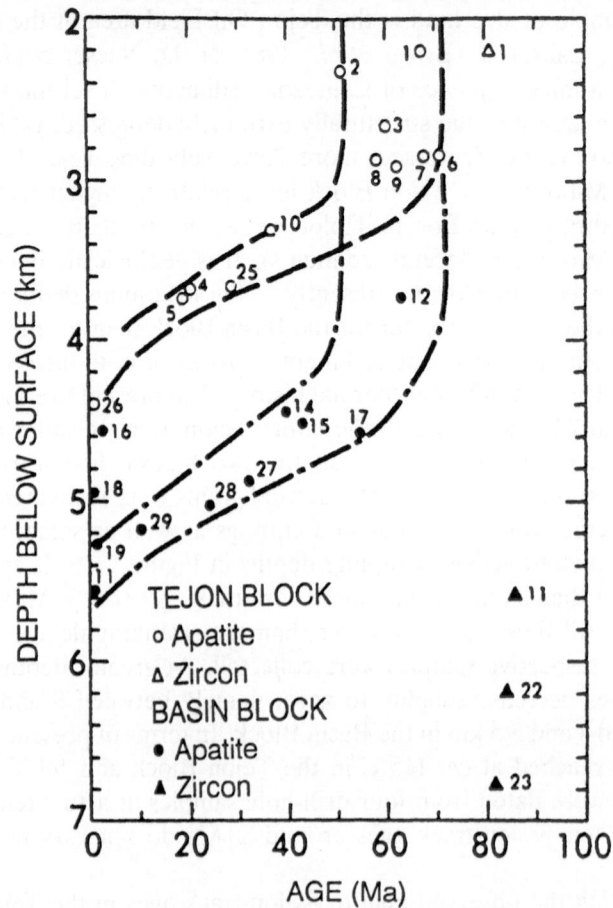

Figure 7.16. Fission-track ages of apatite and zircon versus drill-hole depth in the Tejon Oil Field, California. Although the present geothermal gradient is the same (22°C/km) in both blocks, the apatite fission-track ages reach zero at shallower depth and lower temperature in the Tejon Block (dashed lines) than in the Basin Block (dot-dashed lines) indicating the different burial history of the two blocks. (After Briggs *et al.*, 1981.)

since the ages do not change with depth, even the deepest samples have not been near their present temperatures long enough for significant track-annealing in zircon. As to the provenance of these sediments, the zircon fission-track ages suggest that the youngest Sierra Nevada intrusive complexes were probably a major source.

7.5.2. Otway Basin

An improved approach using several temperature-sensitive fission-track param-eters has been used by Gleadow *et al.* (1983) and Green *et al.* (1989a) for decipher-ing the thermal history of the hydrocarbon-bearing Otway Basin, south-eastern Australia. This basin contains 3–4 km thick fluviatile sediments of volcano-clastic origin. From fission-track dating of zircon and sphene, it has been inferred (Glea-dow and Duddy, 1981) that most of this early Cretaceous volcanogenic detritus was derived from contemporaneous volcanism. Geologically, the burial history of the Otway Basin seems to be typical of extensional basins found on passive continental margins with progressive heating since early Cretaceous and constant temperatures for approximately the last 20 Ma. Four drill holes with rock tempera-tures of up to 125°C were selected for sampling. The present geothermal gradient of approximately 30°C/km is uniform throughout the basin.

The apatite *fission-track ages* range between 120 Ma which corresponds to the stratigraphic age of the Otway Group and zero (Figure 4.11). Their variation with drill-hole temperature is similar to the above-mentioned example of the Tejon Oil Field. At shallow levels, including the surface, the ages ($\approx$120 Ma) do not change with depth and temperature. At temperatures between 60°C and 125°C, the ages are progressively reduced to zero. The ages around 120°C date the Cretaceous volcanism that acted as a source of the detrital apatite and, closely correspond to the stratigraphic age of the Otway Group. The apatites in the deeper part of the wells suffered post-depositional progressive track fading due to long residence times in the PAZ. Their fission-track ages are clearly younger than the age of the Otway Group deposition. Within each apatite sample, the *distribution of the single grain ages* is also indicative of the thermal history. To begin with, one might expect that all cogenetic apatite grains from the same sample experience identical track annealing and, consequently, have the same fission-track age. However, the annealing behaviour depends on the apatite composition, particularly on the Cl/F ratio (Figure 4.10). As this ratio varies for the Otway Group apatites, a correlation of the fission-track ages among the single grains from the sample results. Since this age-distribution was observed to characteristically change with down-hole temperature, it probably reflects to some degree the thermal history of the samples. *Track length* is the primary variable sensitive to annealing. The mean length of confined tracks in Otway apatites shows a progressive reduction with increasing well temperature from 14–15 μm below 50°C to zero at *ca.* 125°C – a temperature range which marks the PAZ. But more diagnostic than the mean length is the *shape of the track-length distribution*. The length distributions maintain, at shallow levels, a narrow symmetric form, but as the mean is progressively reduced down hole, the distributions become increasingly broad (Figure 7.17). Similarly, the distribution of projected track lengths changes systematically with bore-hole depth and temperature (Figure 7.18). As outlined in Section 5.4, the analysis of track length distribution reveals a wealth of information on the past *T–t*-path. In short, the information derived from all of the mentioned variables sensitive to track annealing, are in accord with the above-mentioned relatively simple thermal evol-ution of the Otway Basin.

As discussed in detail by Green *et al.* (1989a), the temperature-sensitive pa-rameters of fission tracks in apatite observed in the Otway Basin, may be applied to other sedimentary basins in order to reconstruct their thermal histories. How-ever, the analysis of other basins may be more difficult due to problems such as the mixed provenance of the detrital grains and complex thermal evolution. Re-cently, Green *et al.* (1989b) presented a methodology of modelling the apatite

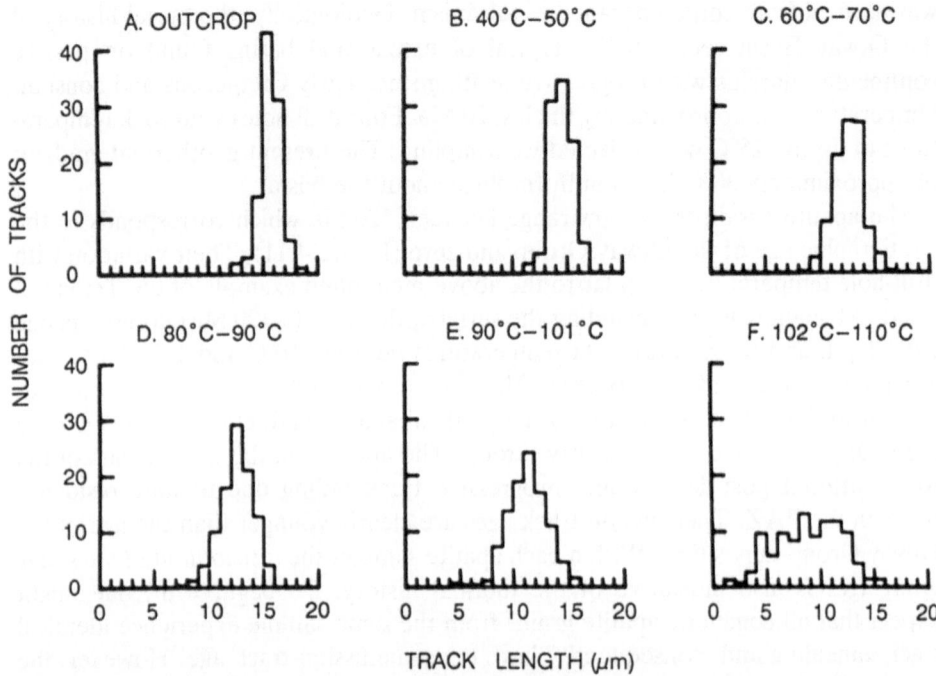

Figure 7.17. Length distributions of confined fission tracks (normalized to a total of 100 tracks) in apatite samples from different down-hole temperatures in Otway Group, Australia. While with increasing temperature the shape characteristically changes, the mean shifts to lower values. (After Green *et al.*, 1989.).

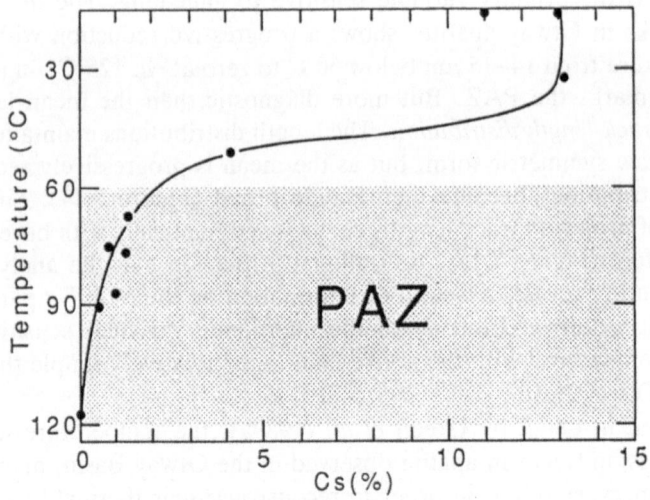

Figure 7.18. Relationship between projected track length (expressed as the fraction c_s of tracks with length $>10\ \mu m$) and down-hole temperature for apatites from the Otway Basin, Australia. The length parameter c_s characteristically changes with temperature. (From Wagner *et al.*, 1989b.)

fission-track age and the distribution of confined track lengths for various assumed geological *T–t*-paths. The comparison of models predicted with actually observed fission-track data will certainly facilitate the elimination of unlikely thermal evolutions. One of the major advantages the apatite fission-track analysis has over other methods of assessing the maturation of hydrocarbons, is that it also yields, in addition to palaeo-temperatures, information on the age of thermal events.

7.6. Age and Thermal History of Ore Deposits

Rock fluids play a major role among the geological processes which form ore deposits. The content and migration of such fluids are largely controlled by thermal evolution. For a given deposit, there is usually no lack of theories about its formation. But what are mostly lacking are time and temperature constraints for testing these theories. Fission-track analysis may yield such *T–t*-constraints for the host rocks and even for the ore deposit itself. Owing to the track-annealing characteristics of the commonly used track-recording minerals, under geological conditions, the fission-track technique essentially applies to the temperature region below 250°C – a temperature region which otherwise is not well covered – and thus supplements other radiometric techniques such as K-Ar dating, that apply to higher temperatures. This thermo-chronometric potential of fission tracks has, amazingly, only been sporadically exploited with respect to the formation of ore deposits and still awaits broad recognition. Hitherto, only a few pertinent fission-track studies exist in the literature, dealing either with base and precious metal deposits in the western United States or some Mississippi Valley-type (MVT) lead-zinc deposits.

7.6.1. *Mineralization ages*

Most of the numerous mineralizations in the mountainous western part of North America are associated with intensive intrusive and volcanic activity since the Cretaceous. The mineral belts contain rich deposits of precious metals, such as gold and silver, and base metals, such as copper, lead, uranium, molybdenum, tungsten, and others. Some mineralizations are directly related to igneous and hydrothermal processes. Others occur in altered rocks and were caused by migrating fluids which became mobilized by intrusive and extrusive heat sources. The establishment of a timeframe for the igneous activity as well as for the mineralizations – apart from its scientific interest – is of great economic importance for the exploration of mineral resources in these areas. With this scope in mind, several fission-track studies have been carried out in Colorado (Bookstrom *et al.*, 1987; Bryant *et al.*, 1981; Cunningham *et al.*, 1977; Lipman *et al.*, 1976; Naeser *et al.*, 1979a), Wyoming (Zielinski and Naeser, 1977), Washington (Ludwig *et al.*, 1981; Weiland *et al.*, 1980) and Utah.

One example is the Marysvale area in Utah, near the north-eastern end of a broad belt of Tertiary igneous rocks that extends from southern Nevada to central

Utah. The Marysvale area is a composite volcanic centre made up of numerous local volcanoes that were active in the late Oligocene and Miocene. The igneous activity was accompanied by various mineralizations, including alunite, gold, silver, copper, uranium, molybdenum, and native sulphur. The chronology of the volcanism and mineralization of this complex area has been largely deciphered by fission-track dating of zircon and apatite, as well as by K-Ar dating (Steven *et al.*, 1979; Cunningham *et al.*, 1984). Concordant fission-track ages for zircon and apatite indicate the absence, whereas discordant ages reveal later thermal effects. Three thermal episodes were detected at the Indian Creek stock in the Marysvale area (Cunningham *et al.*, 1984). The monzonite stock was probably emplaced 23–24 Ma ago. A sample from within the stock shows that the heat associated with a later, nearby magmatic event completely reset the zircon fission-track age to 19.3 ± 1.0 Ma. A subsequent, nearby magmatic event associated with hydrothermal activity did not affect the tracks in zircon, but partially reset the apatite fission-track system in the same sample to 12.8 ± 2.1 Ma. In the central mining area of Marysvale, the intrusive igneous activity began ≈ 23 Ma ago and continued for several million years (Cunningham *et al.*, 1982) (Figure 7.19). There, the hydrothermal uranium mineralization was contemporaneous with the last igneous activity. This was substantiated by direct fission-track dating of the uranium mineralization. Quartzes in the hydrothermal veins were used as natural external detectors for fission tracks from adjacent pitchblende and gave a fission-track age of 16.5 ± 4.3 Ma in accordance with other geochronological data. This example demonstrates that the age of mineralization in a mining area is a primary factor in determining *when* and *in association with which rock* an ore deposit has formed.

7.6.2. *Thermal history of MVT deposits*

The Mississippi-Valley-type deposits are among the world's largest ore-deposits and represent the most economically important lead and zinc resources. They typically occur as replacement bodies in carbonate rocks that have never been deeply buried in the past. Mineralogically, they predominantly consist of galena and sphalerite. Mainly based on the observation of fluid inclusions trapped in crystals, it is generally believed that the MVT deposits form around 50–150°C from metalliferous brines as they migrate through the carbonates. The sources of the metals, the driving mechanism for the fluids, as well as the cause and time for the sulfide precipitation, are still controversial. As MVT deposits always occur within large sedimentary basins, it is best to consider these ore-forming processes in the wider framework of the low-temperature evolution of the basins.

In view of the coincidence of the temperatures estimated for the MVT ore formation with those for track annealing in apatite (Figure 7.14), it seems promising to study the thermal history of the deposits by means of the apatite fission-track system. Especially if the ore deposition was associated with thermal pulses, then this should be detected by apatite fission-track analysis. As apatite occurs in the host rock, it also seems possible to trace spatial temperature gradients around the deposit.

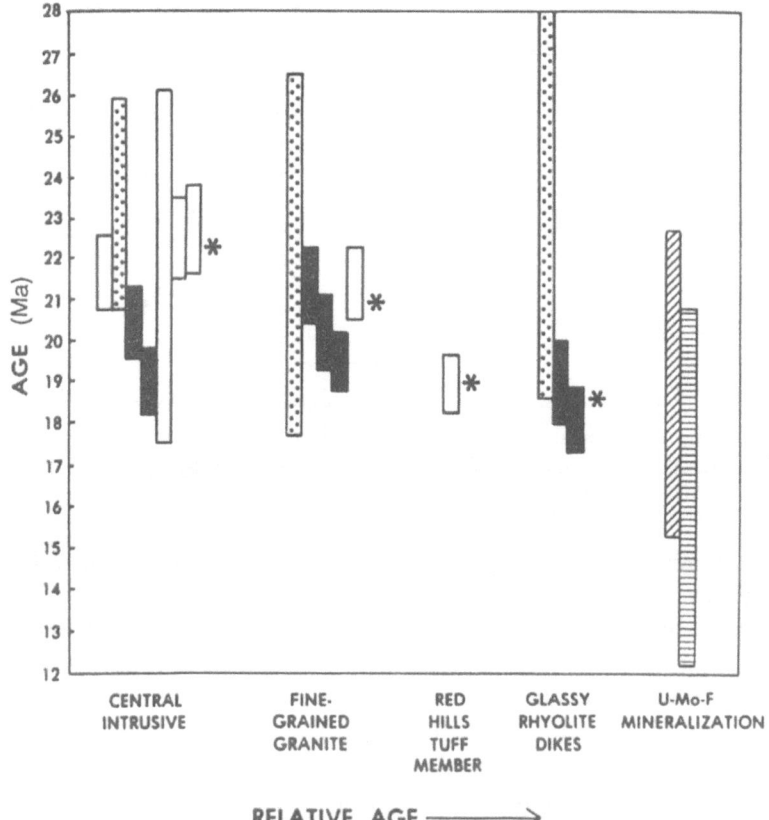

Figure 7.19. Radiometric ages versus relative geological ages of igneous rocks and uranium mineralization in the Central mining area, Utah; open pattern = K-Ar age; solid pattern = zircon fission-track age; stippled pattern = apatite fission-track age; diagonal pattern = lead-uranium isotopic age; horizontal pattern = fission-track age from quartz adjacent to pitchblende; * = mean age. (After Cunningham *et al.*, 1982.)

This potential has only recently been realized and practically tested by Arne *et al.* (1990). Apatites from three districts have been analyzed, including Viburnum Trend, Missouri, Pine Point, Canada and Lennard Shelf, Australia. In all three districts, the apatite fission-track ages of the sedimentary host rocks appear to be lower than the respective stratigraphic ages. This indicates appreciable post-sedimentary track annealing – a conclusion which is also supported by track-length measurements. Based on the apatite fission-track evidence, it seems that the three MVT ore deposits experienced a similar thermal history. After the sedimentation of the host carbonates, the areas underwent heating during shallow subsidence. The regional peak temperatures of 80–110°C, as indicated by the apatites in the host-rocks are generally similar to those derived from the fluid-inclusion studies. The question of whether the heating was solely due to subsidence or was enhanced by increased geothermal gradients is still open. The fact that no decrease of the apatite fission-track ages towards the ore bodies is observed, implies the association

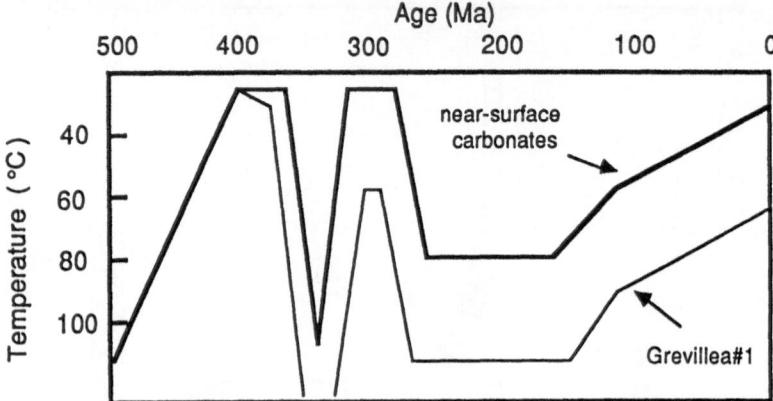

Figure 7.20. Schematic temperature-time paths derived from apatite fission-track analysis of samples from near-surface Devonian carbonates and from the drill hole Grevillea#1 in the MVT-ore district Lennard Shelf, Australia. (From Arne *et al.*, 1989.)

of the MVT mineralization with broad regional heating rather than with discrete hot injection pulses into cooler rocks.

In terms of apatite fission-track analysis, the best studied MVT district is the recently discovered lead-zinc deposit Lennard Shelf in north-western Australia (Arne *et al.*, 1989). It occurs in middle to late Devonian carbonates of the Canning Basin which were deposited during the Palaeozoic over a Precambrian basement. The geological evolution of the northern Canning Basin involved middle Devonian to early Carboniferous sedimentation, followed by a short period of uplift and by a long period of subsidence with sedimentation from late Carboniferous until the Triassic. As the last tectonic event, uplift and erosion affected the northern margin of the Canning Basin beginning in the early Jurassic. The fluid-inclusion evidence points to minimum temperatures of up to 115°C for the MVT-sulfide precipitation in the Devonian carbonates, of course, without giving any information about the timing. The samples taken for apatite fission-track analysis involved Devonian carbonates obtained from shallow exploration drill holes in and around the lead-zinc mineralization, as well as outcropping rocks from the Precambrian granitic basement and post-Devonian sediments (Arne *et al.*, 1989). The apatite fission-track ages of the Devonian carbonates generally range between 250 and 370 Ma and are significantly lower than the respective stratigraphic ages of 360–380 Ma. Together with the track-length data (mean confined track lengths fall between 12 and 13 μm), this indicates that track annealing occurred in post-Devonian times and that these fission-track ages are not event ages but complex ages (Section 5.5). It seems important to point out that the fission-track ages as well as the mean track lengths are similar, regardless of the sample's distance to the contact of the ore-mineralization. The apatite fission-track ages of the Precambrian granites range between 300 and 400 Ma and appear to be slightly older than those of the Devonian carbonates. According to the fission-track analysis, the Devonian carbonates underwent subsidence with a phase of elevated temperatures around 100°C during the late Devonian/early Carboniferous (Figure 7.20). After a short time of uplift and cooling, the carbonates were progressively buried and heated during the late Carboniferous and Triassic periods with maximum temperatures around 70–80°C experienced by samples presently at the surface. Uplift and cooling began during early Jurassic with a total of 1.5 km uplift since then. In order to reconcile the fluid-inclusion homogenization temperatures (in excess of 100°C) with the fission track derived *T–t*-constraints, the MVT mineralization in the carbonates needs to be assigned to the late Devonian/early Carboniferous rather than to the Triassic high-temperature phase. The lack of any decrease of the fission-track ages towards the contact of the ore bodies implies that the sedimentary host rocks as a whole reached high regional temperatures and, thus, supports a causal relationship between elevated ambient rock temperatures and the ore precipitation from migrating hot metalliferous fluids.

7.7. Meteoroid Impacts

Our solar system is populated by innumerable celestial bodies of various sizes. In decreasing order of size – and increasing order of number – these are the planets, asteroids, comets, meteoroids, and cosmic dust particles. Relative to the Earth, they travel with at least 11.2 km/s speed, which is the Earth's gravitational escape velocity. Some of these bodies move along highly elliptical trajectories and, thus, may collide with the Earth. The frequent fall of meteorites verify collisions with extraterrestrial bodies as a common phenomenon. The impact of extraterrestrial bodies with masses much larger than tons, may have catastrophic consequences for the Earth. Due to their large masses, such bodies are only partially slowed down, or not at all, by the atmosphere and reach the Earth's surface with their full cosmic velocity. During impact, their kinetic energy is instantaneously transformed into high shock-wave pressures (up to some hundred GPa) and temperatures (up to several thousand °C) by a process which is known as shock-wave metamorphism. Virtually all of the meteoroid's material is vaporized. The rocks of the target area suffer various degrees of shock-wave metamorphism involving effects such as vaporization, melting, phase transformations, heating, transport, brecciation, and fracturing. The intensity of these effects generally decreases sideward and downward from the point of impact. The term impactites applies to rocks intensively affected by these various shock-wave metamorphic processes. Impact craters ('astroblems') are the typical geological structures that witness past collisions with huge extraterrestrial bodies. Apart from these geologically visible scars left by the meteoroid impacts, there may also be dramatic consequences for the climate and life on Earth. The mass extinction of numerous species that occurred at the Cretaceous/Tertiary boundary is believed to have been caused by such a scenario.

The important role which the phenomenon of meteoroid impact plays in the geological and palaeontological evolution, has only been fully realized during the last one or two decades. One of its most important aspects is the *timing of impact events*. When and how often do impacts occur? Are they random or periodic? Fission-track dating can contribute to answering these questions. During impact, the temperatures become sufficiently high in order to reset the fission-track clock in various minerals. Beyond it, temperatures may exceed the melting points of silicates and impact glasses are formed that can be directly dated with fission tracks. Another application of fission-track dating is the reconstruction of the *thermal evolution of impact structures*. The *tektite problem* had been a scientific puzzle for a long time. The hypothesis of their terrestrial origin in connection with giant impact events can be and has been tested by simultaneous fission-track dating of both tektite strewn fields and possible impact sites.

7.7.1. *Age of impact structures*

During impact, very intense but instantaneous shock-wave heating takes place in the target rocks. After the passage of the shock wave, prolonged post-shock, rest-temperatures are reached. These temperatures depend on the level of shock-wave intensity (Stöffler, 1973). As an example, for shock-wave pressures in the 40 to 50 GPa range, rest temperatures between 500 and 1000°C can be assumed for the rocks. Such temperatures are sufficient to anneal fission tracks. Since the temperatures radially decrease away from the point of impact, one can expect, for the target area, a temperature field with all transitions from complete to partial annealing of fission tracks in different minerals. The question as to what extent the instantaneous thermo-barometric effects of the shockwave itself – in comparison to the prolonged thermic effects after the shockwave has passed – affect the fission-track systems, has not yet been thoroughly investigated. Experimental evidence by Fleischer *et al.* (1965b, 1974) suggests that strong shock waves can directly reset the fission-track systems. Anyway, for the present argument, it is sufficient to note that post-shock temperatures in the target rocks are high enough to reset the fission-track clock, allowing the impact to be dated. Most commonly, the highly temperature-sensitive apatite fission-track system is selected for impact dating (Wagner, 1977; Omar *et al.*, 1987), but sphene has also been successfully used (Wagner, 1977; Storzer and Wagner, 1977). The complete resetting of the fission-track systems needs to be established by track-length analysis. Minerals with reset fission-track systems should exhibit narrow track-length distributions with long mean lengths similar to those described for the *formation age type* (Section 5.5.1). In large craters that are formed in silicate rocks, glasses (Figure 7.21) are frequently observed which solidified from impact melts. Fission-track ages of such impact glasses can be directly related to the impact event (Fleischer *et al.* 1965c). However, as glasses are sensitive to track fading at ambient temperatures, their fission-track ages may be too low. The presence of track fading can be readily recognized by comparative track diameter analysis on spontaneous and induced fission tracks (Storzer and Wagner, 1969).

In fact, most of the impact glasses older than 10^4 years show some degree of fading of fossil fission tracks. Such glasses require the application of track-size and/or plateau-correction techniques (Section 5.7). So far, 20 out of *ca.* 200 known impact structures have been dated with fission tracks. The locations of these sites are shown in Figure 7.22. At some of these locations, such as the Libyan Desert and Muong Nong, only impact glass occurs and no impact structures are visible.

Figure 7.21. Impact glass from suevite Otting, Nördlinger Ries.

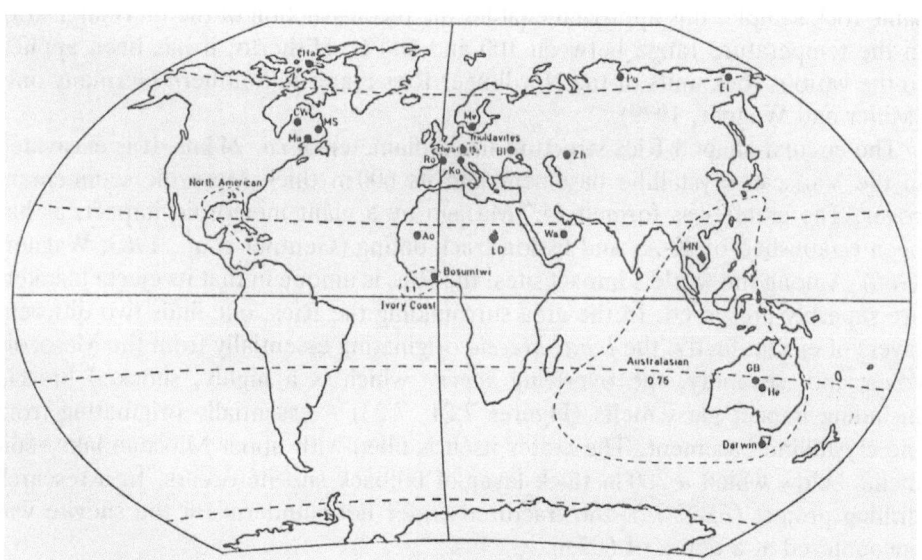

Figure 7.22. Terrestrial impact structures dated with fission tracks (compare Table 7.3). Also shown are the tektite strewn-fields with their mean fission-track ages (in Ma).

In nearly all cases of the dated sites, impact glasses were used as fission-track recording material. Most of the glasses showed fading phenomena of their fossil fission-track record and, thus, age correction procedures need to be applied. The fission-track ages of the dated impact sites are given in Table 7.3. The ages cover a wide time-range from 4 ka up to 300 Ma. As far as independent age information – K-Ar dates or geological estimates – are available, the agreement is generally good. There is no doubt that fission-track dating is excellently suited for dating impact events and that this potential has not yet been fully exploited.

7.7.2. *Thermal evolution of the Ries impact crater*

In addition to the dating of impacts, fission-track studies can also reveal the thermal history of the rocks which were affected by the impact. This approach uses the degree of track annealing by the impact's heat as a geothermometer. In principle, one has to distinguish three possibilities with regard to the effect which an impact has on the pre-existing fission-track systems in the target rocks: firstly, the temperatures are raised, but not high enough to cause any fission-track annealing. Secondly, the temperatures become high enough to cause partial track fading. Thirdly, the temperatures become sufficiently high in order to cause complete track annealing. If the annealing characteristics of the fission-track systems under consideration are known and valid assumptions on the duration of heating can be made, the first possibility allows the assessment of an upper temperature limit, the second possibility allows the evaluation of the actual temperature reached and the third possibility allows – in addition to the determination of the impact's age – the assessment of a minimal temperature reached during the impact. When using different fission-track systems, such as apatite, zircon, and sphene from the same rock sample, this approach enables the reconstruction of the thermal history in the temperature range between 100 and 500°C. Hitherto, it has been applied to the various rock units of the Nordlinger Ries crater in Southern Germany only (Miller and Wagner, 1979).

The circular-shaped Ries structure has a diameter of *ca.* 24 km. It is excavated in the Variscan crystalline basement and its 600 m thick Mesozoic sedimentary cover. The crater was formed 14.7 Ma ago by a giant meteoroid impact, as has been established by K-Ar and fission-track dating (Gentner *et al.*, 1963; Wagner, 1966). Among the world's impact sites, the Ries is unique in that its ejecta blankets are superbly preserved. In the area surrounding the Ries, one finds two different layers of ejecta, firstly, the *bunte breccia* originating essentially from the Mesozoic cover and, secondly, the overlying *suevite* which is a highly, shocked breccia including impact glass melts (Figures 7.21, 7.23) – essentially originating from the crystalline basement. The crater itself is filled with upper Miocene lake sediments below which a 300 m thick layer of fallback suevite occurs. In a research drilling project (FBN 73), the fractured crater floor underneath the suevite was encountered at a depth of 602 m.

Outcrop samples as well as cores from the 1206 m deep drill hole were used for fission-track analysis. Apatite and, wherever possible, sphene and zircon fractions

Table 7.3. Fission-track ages of impact structures (compare Figure 7.22 for locations)

Structure	Material	Fission-track age (Ma)	Corr. techn.	Reference
Henbury/He	Glass	0.0042 ± 0.0019		Storzer and Wagner (1977)
Wabar/Wa	Glass	0.0064 ± 0.0026		Storzer (1971)
		0.0064 ± 0.0025		Storzer and Wagner (1977)
Köfels/Kö	Glass	0.008 ± 0.006		Storzer et al. (1971)
		0.0089 ± 0.0029		Storzer and Wagner (1977)
Muong Nong/MN	Glass	0.45 ± 0.78		Fleischer et al. (1965c)
		0.70 ± 0.01		Storzer (1971)
		0.71 ± 0.07	Size	Storzer and Wagner (1977)
Darwin	Glass	0.37 ± 0.76		Fleischer et al. (1969a)
		0.72 ± 0.04		Gentner et al. (1969a)
		0.71 ± 0.01		Gentner et al. (1973)
		0.74 ± 0.04	Size	Storzer and Wagner (1977)
Bosumtwi	Glass	1.04 ± 0.20		Gentner et al. (1967)
		0.32 ± 1.48		Fleischer et al. (1965c)
				Fleischer and Price (1967)
		1.03 ± 0.02		Storzer (1971)
		1.04 ± 0.11	Size	Strozer and Wagner (1977)
Zhamanshin/Zh	Glass	1.07 ± 0.05	Plateau	Storzer and Wagner (1977)
		1.07 ± 0.06	Plateau	Storzer and Wagner (1979)
		0.81 ± 0.16		Florenski et al. (1979)
Aouelloul/Ao	Glass	0.16 ± 0.50		Fleischer et al. (1965c)
				Fleischer and Price (1967)
		0.46 ± 0.61		Wagner (1966)
		3.25 ± 0.50	Size	Storzer and Wagner (1977)
Elgygytgyn/El	Glass	4.52 ± 0.11	Plateau	Storzer and Wagner (1979)
Ries	Glass	15.2 ± 1.6		Fleischer et al. (1965c)
		14.8 ± 0.7		Wagner (1966)
		14.6 ± 0.6		Gentner et al. (1967)
		14.6 ± 0.6		Storzer and Gentner (1970)
		14.7 ± 0.3	Plateau	Storzer and Wagner (1977)
	Sphene	14.8 ± 0.4		Wagner (1977)
	Apatite	14.7 ± 0.4		Wagner (1977)
Haughton/Ha	Apatite	22.4 ± 1.4		Omar et al. (1987)
Libyan Desert/LD	Glass	37.5 ± 2.4		Fleischer et al. (1965c)
		32.1 ± 3.0		Wagner (1966)
		26.6 ± 1.3		Gentner et al. (1969a)
		28.5 ± 2.3		Storzer (1971)
		29.4 ± 0.5	Plateau	Storzer and Wagner (1977)
Popigai/Po	Glass	38.9 ± 2.2	Size	Komarov and Raikhlin (1976)
		30.5 ± 1.2	Plateau	Storzer and Wagner (1977)
Manson	Apatite	61 ± 9		Hartung et al. (1986)
Mistastin/Ms	Glass	39.6 ± 4.4	Size	Storzer and Wagner (1977)
Mien/Me	Glass	92 ± 6	Size	Storzer and Wagner (1977)
			Size	Storzer and Wagner (1977)

Table 7.3. Continued

Structure	Material	Fission-track age (Ma)	Corr. techn.	Reference
Boltysh/Bl	Glass	100 ± 10	Size	Komarov and Raichlin (1976)
Goses Bluff/GB	Apatite +Zircon	134 ± 6		Milton and Sutter (1987)
	Apatite	<175 ± 4		Tingate (1988)
Manicouagan/Ma	Glass	208 ± 25		Fleischer *et al.* (1969d)
		200 ± 30	Size	Storzer (1971)
			Size	Storzer and Wagner (1977)
Rochechouart/Ro	Glass	210 ± 100		Storzer (1971)
		206 ± 39	Size	Storzer and Wagner (1977)
	Apatite	198 ± 25	Size	Storzer and Wagner (1977)
Clear Water/CW	Glass	290 ± 30		Fleischer *et al.* (1969d)
		287 ± 43	Size	Storzer (1971)

were separated from the suevites and crystalline rocks (gneisses, amphibolites). The resulting fission-track ages can be clearly divided into three groups in accordance with the three above-mentioned possibilities of track fading in the pre-existing fission-track systems. The first group (ages >150 Ma) consists of zircon and sphene ages around 300 Ma and one apatite age of 151 Ma. These ages are believed to be representative of the basement rocks whose fission-track systems were thermally not disturbed by the impact. The fact that the fission-track age of the apatite is lower than those of sphene and zircon reflects the pre-impact thermal history of the basement and is of no concern in the context of the present discussion. The second group consists of apatite, sphene, and zircon ages between 30 and 130 Ma. These ages are believed to represent fission-track systems which were partially annealed during the impact event. If the ages of the first group (>150 Ma) are

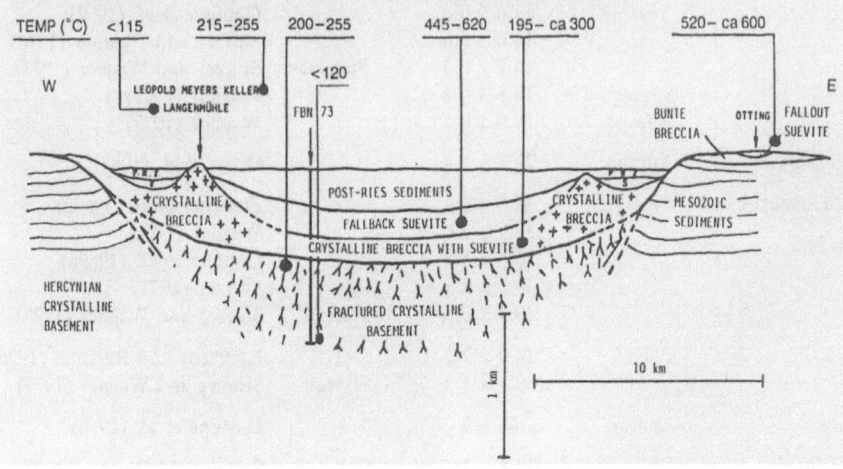

Figure 7.23. Schematic cross-section of the Ries crater with various rock units and their temperatures caused by the impact. These temperatures have been assessed by fission-track analysis of apatite, sphene, and zircon. (After Miller and Wagner, 1979.)

taken as the fission-track systems which remained unaffected by the impact, then the percentage of track fading for the affected systems in the second group can be calculated from known annealing characteristics. Taking into account the known annealing characteristics and assuming reasonable durations of annealing between 1 h and 1000 years for different rock units, one can derive the effective annealing temperatures during the impact event. Finally, the third group consists of apatite and sphene fission-track ages around 14.7 Ma which represent completely impact-reset fission-track systems. The resulting temperature picture for the various rock units of the Ries crater during and immediately after the impact is summarized in Figure 7.23. Due to complete track annealing in apatite and sphene, the suevite i.e., the fallout as well as the fallback suevite, experienced – apart from melting – the hottest temperatures of 500 to 600°C. These high temperatures are confirmed by the inverse magnetization of the suevite. On the other hand, the crater floor stayed relatively cool (<250°C) during the impact.

Two interesting observations seem worth mentioning. Firstly, no direct relationship was observed between the independently determined shock-wave intensity in single constituents within the suevite breccia and the degree of fission-track fading in apatite. From this, it is concluded that most of the track fading is due to annealing at the residual temperatures which became uniformly distributed throughout each suevite unit. Secondly, it is interesting to note that the impact caused a reversed palaeo-thermal gradient for some time in the upper crust, which is evidenced by the fission-track analysis (Figure 7.23). The establishment of a reverse palaeo-thermal gradient might become a useful criterion to distinguish an impact from a volcanic hypothesis for structures whose origin is still under debate.

7.7.3. *Tektites*

Tektites have puzzled geologists ever since they were first described in 1787 by Joseph Mayer in Bohemia. These natural glasses are geographically limited to restricted areas, so-called *strewn fields*. Four tektite strewn fields are known (Figure 7.22): North America (bediasites from Texas) Georgia tektites) Czechoslovakia (moldavites), Ivory Coast (Ivory Coast tektites, and south-eastern Asia as well as Australia (indochinite, javanites, philippinites, billitonites, australites). Tektites are characterized by a relatively uniform chemical composition (e.g., 70–80% SiO_2, 11–16% Al_2O_3, 0.5–1.7% Na_2O), the almost total absence of water (<0.01%), and the small amount of inclusions which indicate very high melting temperatures (<1500°C). All these properties are distinctly different from the other group of natural glasses, which are the obsidians.

These properties and the fact that tektites are not chemically related to the lithology of the region where they are found, made tektites appear as geologically strange objects to which various modes of origin were assigned. Without going into any detail, the two main hypotheses which were most intensely discussed concern terrestrial against lunar origin. In the hypothesis of terrestrial origin, the tektites are thought to have been formed as a consequence of a giant meteoroid's impact on the Earth's surface. During the impact, the target rocks were melted

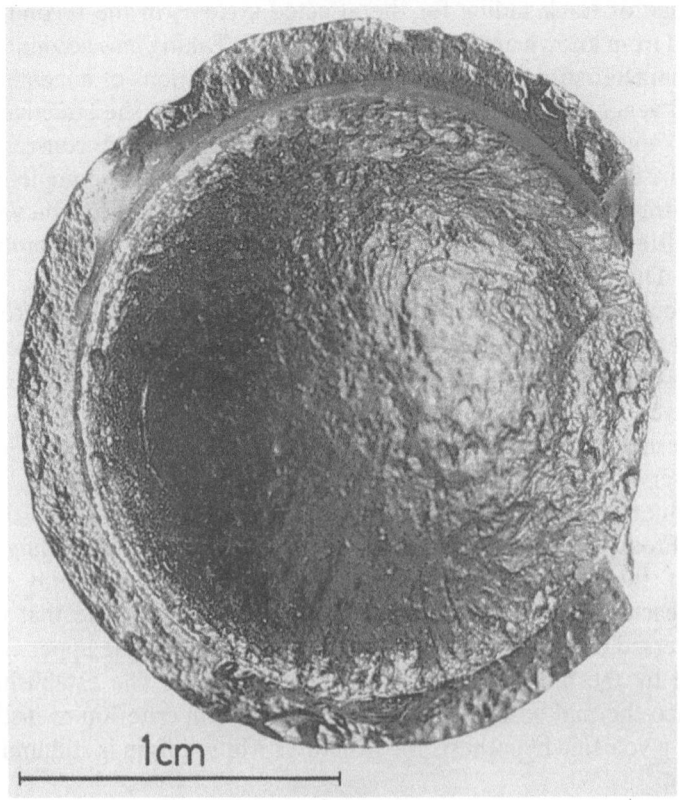

Figure 7.24. Australite; the aerodynamic button-shape with core and flange which is caused by remelting as the tektite reentered the atmosphere.

and vaporized. These silicates melted and the vapours moved several hundred to a thousand kilometers – probably above the dense atmosphere – before they solidified as tektite glass and fell back to the Earth's surface (Figure 7.24). In this scenario, the ages of the tektites should be the same as the impact crater from which they were derived. It is at this point that fission-track dating steps in as a criterion for the validity of the terrestrial hypothesis.

Tektite glass is excellently suited for fission-track dating. The virtual absence of inclusions, the homogeneous distribution and optimal content (1–4 $\mu g/g$) of uranium, the easy etchability, and the unequivocal recognition of the tracks make it an almost ideal material for fission-track recording and dating (Figure 6.5). It is also for this reason that tektites belong to the earliest samples extensively used for fission-track dating (Fleischer and Price, 1964b; Fleischer *et al.*, 1965c; Wagner, 1966; Gentner *et al.*, 1967, 1969a,b; Durran and Hancock, 1970).

These early fission-track studies were able to confirm previous K-Ar dating results (Gentner and Zähringer, 1960) which indicated that the four tektite-strewn fields have distinct ages (Figure 7.22). Furthermore, due to the absence of cosmic ray-induced fission tracks, Fleischer *et al.* (1965a) were able to set an upper limit of 300 years for a hypothetical cosmic ray-exposure time of tektites – a finding which supports their terrestrial origin.

In most tektite specimens, the fossil-track record shows some fading which may be caused by partial track annealing at ambient temperatures. Therefore, the fission-track ages of most tektites require some correction using the track-size or plateau-correction procedures (Section 7.7). In fact, the necessity of correcting partially lowered fission-track ages was originally recognized for australites by Storzer and Wagner (1969). By fission-track dating, the association of microtektites recovered from deep-sea cores in the Caribbean Sea and the Atlantic, Indian, and Pacific oceans with the tektite strewn fields on the continents, has also been proven (Gentner *et al.*, 1970; Glass *et al.*, 1973). Furthermore, it has been shown that two of the strewn fields were simultaneously formed with nearby impact craters, namely the Czechoslovakian moldavites (14.7 ± 0.4 Ma, Wagner, 1966; Storzer and Wagner, 1977) with the Nördlinger Ries (14.7 ± 0.3 Ma) crater in southern Germany, and the Ivory Coast tektites (1.08 ± 0.10 Ma) with the Bosumtwi crater (1.04 ± 0.11 Ma) in Ghana, supporting – at least in these two cases – the terrestrial impact hypothesis of tektite formation. In the meantime, this genetic relationship between tektites and source crater rocks was substantiated by chemical and isotopic evidence. However, for the remaining strewn fields in North America (34.6 ± 0.7 Ma, Storzer and Wagner, 1977), south-eastern Asia (0.693 ± 0.025 Ma, Storzer and Wagner, 1980), and Australia (0.830 ± 0.028 Ma), the source craters are not yet recorded, although several potential impact sites have been identified, e.g., the Canadian craters Mistastin, Manicouagan, and Clear Water Lake, as well as the Siberian craters Elgygytgyn, Popigai, and Zhamanshin. Fission-track dating (Table 7.3) disproved all these speculative assignments (Fleischer *et al.*, 1969d; Storzer and Wagner, 1977). Although the fission-track age of Darwin glass (0.74 ± 0.04 Ma) is similar to the huge Australasian strewn field, the Darwin crater in Tasmania, owing to its small size (*ca.* 1 km), probably has to be disregarded as the source for those tektites. There have been reports of postulating multiple tektite formation events in the Australasian region (Fleischer *et al.*, 1969c; Kashka-rov *et al.*, 1985; Storzer, 1985; Storzer and Wagner, 1980) – an issue which needs further attention.

A point of special interest is the fission-track dating of the aerodynamic flanges found on some australites (Figure 7.24). These flanges have probably been shaped by remelting due to frictional heating as the already solidified tektite glass re-entered the atmosphere. The hypothesis of the terrestrial formation of tektites requires that the flanges and the cores of the australites have the same age. In order to test its validity, the flanges and cores of four australites from the Nullabor Plains were dated separately (Storzer and Wagner, 1969). After correcting for track fading, the fission-track ages of both flanges and cores, correspond to each other for all four specimens.

7.8. Sea-Floor Spreading

According to the concept of sea-floor spreading, the oceanic crust is formed in a dynamic volcanic regime at the axial zone of the mid-ocean ridges. The newly created crust moves on top of the diverging lithosperic plates away from the ridge axis. Therefore, the age of the basaltic oceanic crust should increase with distance from the axial zone and reveal the spreading rates. The general validity of this concept has been proven by applying the magnetic time scale to the magnetized basalts. Continuous spreading with rates typically in the order of cm/a, has been found in this way. But the resolution of the magnetic time scale is, on average, limited to one or more Ma. Consequently, it does not disclose any information on the spreading behaviour within shorter time spans. Does the crust accrete gradually or episodically? Are there short volcanic outbursts with long periods of quiescence in between? How wide is the active volcanic zone? Since, for very young (i.e., less than several 10^5 years) deep-sea basalts, K-Ar dating is inappropriate, due to various difficulties such as excess argon and potassium exchange with seawater, the fission-track method seems to be the only suitable candidate left. This potential was realized more than two decades ago by Fleischer *et al.* (1968b).

The material most suitable for fission-track dating is basaltic glass which occurs on the margins of pillow lavas. The glass forms when the surface of the hot extruding lava is quenched by the cold seawater. Since, thereafter, the glasses spent all their geological history at ambient sea-bottom temperatures around 4°C, no track fading can be expected and, thus, the fission-track age should give the time of glass formation (Fisher, 1968; Fleischer *et al.*, 1968b; Aumento, 1969). Unfortunately, this assumption turned out to be too optimistic. As demonstrated by Selo and Storzer (1981)) even at low sea-floor temperatures, the fission tracks may fade to some degree due to the low thermal stability of tracks in basaltic glasses (Appendix B). Consequently, correction procedures need to be applied in order to obtain the true formation ages (Section 5.7) of the glasses. Another problem of basaltic glass is its low uranium content, mostly between 0.1 and 0.3 μg/g – yielding spontaneous fission-track densities as low as a few tracks/cm^2 and even less in young samples (Figure 6.1). Therefore, large scanning areas and ample scanning time are required in order to count the few tracks under the microscope. Large scanning areas can be conveniently obtained by repeated re-grinding (>20 μm off), repolishing, and re-etching of the sample. The statistics of a fission-track age which is based on only a few spontaneous fission tracks, may seem to be very poor but, for young ages, a large relative error is still small in absolute terms. Because, in altered basalts, over geological time, uranium becomes enriched (up to four times) due to the interaction with seawater, care must be taken to separate only remnants of fresh unweathered glass for fission-track dating (Storzer and Selo, 1976). Weathering and devitrification of the basaltic deep-sea glass restrict its use for fission-track dating to the last 50 Ma.

In their original study, Fleischer *et al.* (1968b) reported fission-track ages of between 10 and 30 ka for basaltic glass from the Mid-Atlantic Ridge at 45°N. In accordance with the concept of sea-floor spreading, the ages were observed to increase with distance from the ridge axis, revealing a local spreading rate of 2.5 cm/a. One sample at 24 km distance from the Median valley, with a much younger age (17 ka) than the general trend, was interpreted to originate from renewed volcanic activity in older terranes well removed from the axis of the ridge. According to Aumento (1969), who supplemented that study with ages of

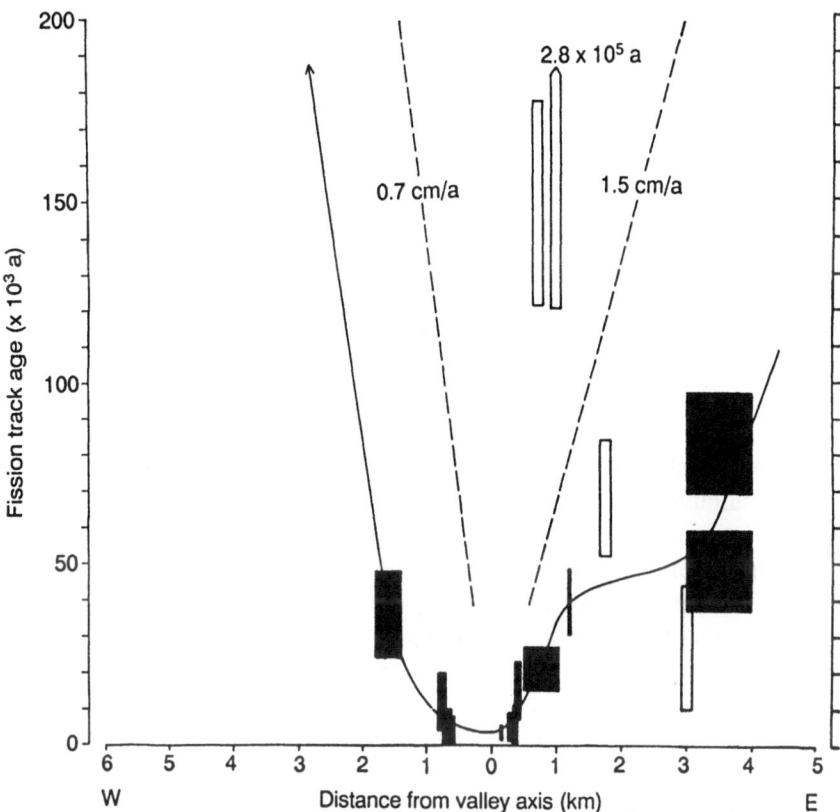

Figure 7.25. Fission-track ages of basaltic glass from the sea floor of the Mid-Atlantic Ridge at 37°N. The ages (in ka) are plotted against the sampling distance (in km) from the rift axis in the Median Valley. The samples (full squares) from the Median Valley increase with distance. The inverse slope of the curve gives the spreading rate. The data indicate an asymmetric, episodic spreading behaviour. The open squares represent samples from the intersection of the valley with a transform fault. (After Storzer and Selo, 1976.)

up to 16 Ma at 136 km off the ridge axis, the Median Valley floor is younger than 0.1 Ma. He also inferred from the data a sudden increase in the spreading rate.

In a study of the rift-valley axis situated more to the south (37°N) Storzer and Selo (1976) reported fission-track ages between 3 and 600 ka, which again increase with distance from the axis (Figure 7.25). Since, in this section, very young fission-track ages (less than a few 10 ka) are restricted to an axial zone only a few km wide, the new crust seems to form essentially within this narrow zone of the Median Valley, with flank volcanism and dike injection playing only a negligible role. Near the centre of recent volcanic activity spreading rates of up to 9 cm/a were found. They decrease with distance from the valley axis, namely to 0.7 cm/a towards the west and 1.5 cm/a towards the east. This indicates an acceleration of sea-floor spreading during the last few 10 ka and an episodic crustal accretion. The fission-track ages also disclose, for the last 10^5 years, an asymmetric crustal accretion pattern of the two diverging plates with slower spreading to the west than to the east. This detailed record of sea-floor spreading, based on fission-track evidence, contrasts favourably, with the palaeo-magnetically derived average rate of 2.2 cm/a over the last 0.69 Ma for this section of the Mid-Atlantic Ridge.

In view of this great potential which fission-track dating has towards the better understanding of the phenomenon of sea-floor spreading, it is surprising that only a very limited number of case studies have been performed so far. It is probably the tedious and time-consuming effort of assessing very low track densities that has prevented widespread application. The introduction of microcomputer-based counting and processing systems might improve the technique's attraction.

7.9. Archaeological Application

Not all fission-track recording materials which are useful in a geological context, are equally suitable for archaeologic applications. In fact, only a few of them are of practical interest for archaeology. Apart from their availability in archaeologic sites, the limitation is essentially due to the low number of fission tracks which accumulate during man's comparatively short presence on earth. After all, one has to bear in mind that the half-life of spontaneous uranium fission is about 10^{16} years. In order to obtain sufficient precision in age, at least a hundred or, preferably, several hundred fission tracks are required. Therefore, large-size samples with high uranium contents are desirable (Figure 6.1).

Counting low track densities, say a few tracks per cm^2, is tedious and time-consuming. Another obstacle that may prevent fission-track dating is the presence of a background of spurious 'tracks'. Etch pits which resemble fission tracks can develop along dislocations, fluid, and gas inclusions as well as in microlites. Spurious etch pits are a common phenomenon, but this does not matter too much if the fission-track density is comparably high. However, in the case of low fission-track densities the genuine tracks might easily be overlooked against a higher background of spurious 'tracks' (Figure 6.6).

Because of these various difficulties, fission-track dating is rarely applied in archaeology. Support by fission-track analysis is most commonly requested for the dating of man-made glasses, ceramics, and heated stones. Other suitable archaeological questions are related to tephrochronology if hominid-bearing layers are intercalated with datable volcanic layers. Generally speaking, in archaeological dating the fission-track method cannot compete with radiocarbon or thermoluminescence dating. But it is excellently suited for specific materials, especially obsidian and zircons. Probably its most potential applications fall in the Palaeolithic and Quaternary periods beyond the limited age-range covered by radiocarbon dating, i.e., beyond 30 ka, which is otherwise difficult to date. A further application of fission-track analysis is the geological provenance of archaeological raw materials.

One material well suited for the archeological application of the fission-track method is glass. Among the natural glasses, obsidian is of particular interest. Owing to its remarkable hardness, obsidian was highly valued for making sharp tools and weapons in prehistoric times and, thus, was widely traded. The uranium in obsidian is often homogeneously distributed and normally amounts to a few $\mu g/g$, but may reach up to 20 $\mu g/g$. With such uranium contents, fission-track ages as low as a few thousand years can be measured without too much difficulty. Artificial glasses – with the exception of *uranium-glass* to which uranium was deliberately added as a pigment – contain little uranium, commonly around 1 $\mu g/g$. This low content renders artificial glasses, impractical for fission-track dating, despite their large counting areas.

Minerals found in archaeological samples such as ceramics and burnt stones, are usually tiny and, therefore, only those with a high uranium content are of interest. Such minerals are zircon, monazite, sphene, and, to a lesser degree, epidote and apatite. They occur as accessories in granites, gneisses, volcanic rocks, and sandstones which may be associated with fireplaces. Although attempts have been undertaken to date apatite-bearing teeth or bones, so far these biogenic materials have proved to be unsuitable for fission-track dating because of the extremely small size of the crystals. Calcite crystals extracted from narrow cavities of bones found in African hominid-bearing breccias, turned out to be free of fossil fission tracks. This was ascribed to complete track fading at ambient temperatures (Macdougall and Price, 1974).

7.9.1. *Man-made glass*

Due to their low uranium content, only a few artificial glasses have been dated with the fission-track method. Nevertheless, man-made glasses can be dated, provided the archaeological relevance of the age justifies the long hours that are required for track counting. For instance, a glass-shard collected from mortar in the wall of a Gallo-Roman bath at Chassenon near Limoges was dated. Its uranium content was 3 $\mu g/g$ and fission-track dating gave AD 150 which agrees closely with the known time of the bath's construction. Unfortunately, the counting of a total of 29 fossil fission tracks on an area of 25 cm^2, demanded 100 h of scanning time and yielded a counting precision of only 20%. Yabuki *et al.* (1973) reported fission-track ages of several thousand years for Iranian glass vessels, but again with low precision. These examples demonstrate one of the principal problems when applying fission-track dating to young artifacts: the method works but is impractical. In order to circumvent the wearisome task of scanning large areas, it was suggested to introduce automatic track-counting devices (Carpenter, 1972). However, many glasses contain numerous tiny bubbles which only the experienced human eye can distinguish from genuine tracks.

Also glassy *metallurgical slags* are interesting objects for fission-track dating, provided they have sufficient uranium and a low background of bubble and mineral inclusions. In the glass phases of ancient lead and copper slags, up to 20 $\mu g/g$ homogeneously, distributed uranium was observed (Elitzsch *et al.*, 1983). The dating of bloomery slags has also been proposed (Scott, 1976), but not yet exercised.

Potentially, the simple presence of fossil fission tracks allows one to establish

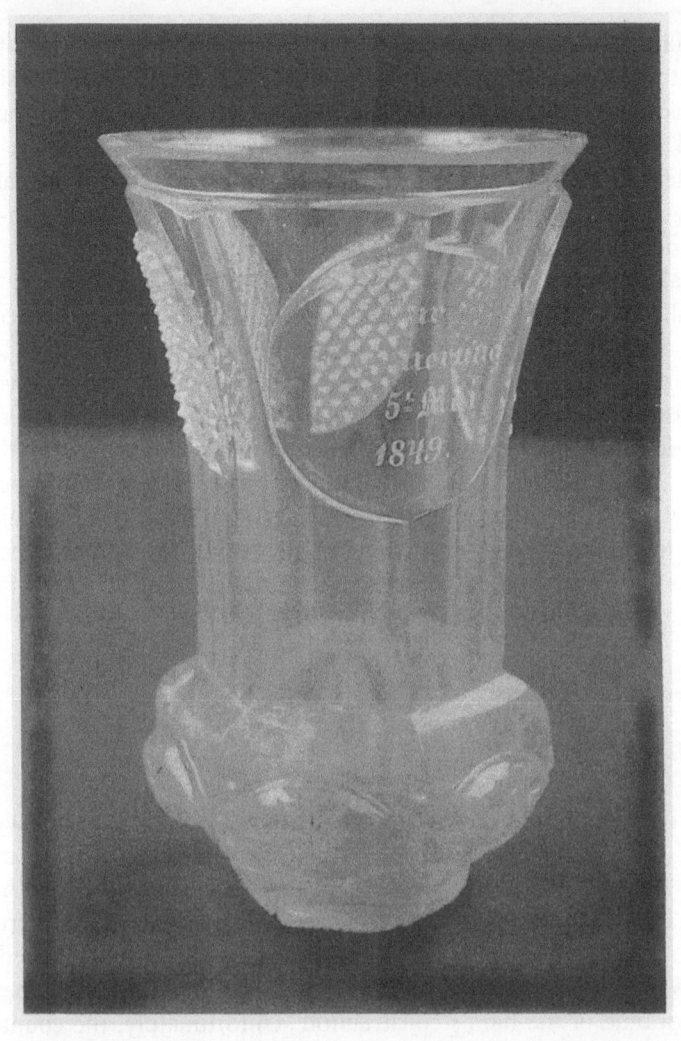

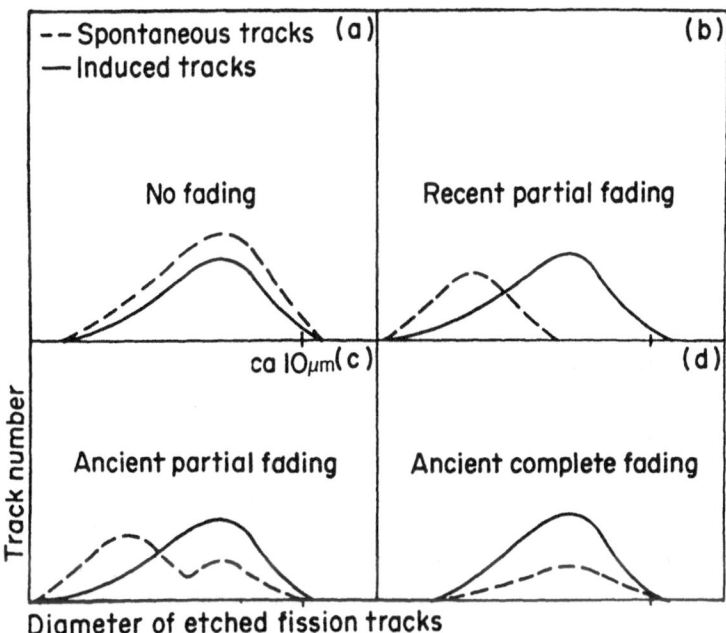

Figure 7.27. Schematic illustration of typical size distributions of spontaneous and induced fission-track etch pits in glass with different types of thermal history. (After Wagner, 1978.)

7.9.2. *Obsidian*

Obsidian is usually associated with young, acidic volcanism, especially around the Pacific, the Mediterranean, the Near East, and East Africa. In all these areas, prehistoric man collected large and compact lumps of obsidian in order to make hard, sharp objects, such as knives, saws, and arrow heads. Because of its restricted geographical occurrence, obsidian raw material as well as artifacts were widely traded. During production and use of the artifacts, for some unknown reason, the material was occasionally subjected to heat treatment which resulted in track annealing. One conceivable reason is that fire was used in the quarrying or splitting of the hard obsidian blocks. Nevertheless, fission-track dating may give the answers to both the questions of *geographic provenance* of the material and the *age of manufacture*. The *geological* fission-track age of unheated artifacts and their uranium content may be specific for a certain source area and, thus, by matching the artifacts with potential source areas, one may establish ancient trade routes. On the other hand, if the artifacts were sufficiently heated by prehistoric man, the fission-track age dates the moment of the artifact's manufacture or use.

In order to establish if the fission-track clock in an obsidian artifact was partially or completely reset by heat treatment, one has to study track sizes. An obsidian sample that has not undergone any track fading after its solidification, shows similar size distributions for both the fossil and induced fission tracks (Figure 7.27a). However, this is rarely met in geological and archaeological specimens. Commonly, the fossil fission tracks have undergone partial fading which is indicated by their smaller size (Figure 7.27b or c). First of all, it is known that some fading of fission tracks in natural glasses occurs at ambient surface temperatures, particularly when heated by exposure to the Sun. In addition, partial track fading may have been caused by the heat of subsequent volcanic events or ancient man's activities resulting typically – but not necessarily – in a bimodal size-distribution (Figure 7.27c). Also, the complete removal of fission tracks in obsidian, which requires only 300–400°C, could be easily achieved by ancient man. The spontaneous fission tracks formed after a strong heating event – although being low in number – will again have the same size distribution as the induced ones (Figure 7.27d).

All these hypothetical cases have been met in reality. Fleischer *et al.* (1965g) dated a knife made of obsidian. The knife had been excavated at Gambles Cave, Elmenteita. Its wilted shape revealed that, at some stage, it had been exposed to temperatures which were almost high enough to melt it. Since such temperatures are far above track-retention temperatures, all previous fission tracks must have been annealed at that time. Indeed fission-track dating gave an age of 3700 ± 900 a.

A similar application was reported by Watanabe and Suzuki (1969) who separated obsidian fragments embedded in various ceramics from Japan. The fission-track ages of between 5080 ± 400 and 520 ± 110 a revealed when the pottery was fired.

In both cases, the archaeological evidence or context indicated that the obsidian samples had been heated sufficiently for complete track loss. However, in most instances, obsidian artifacts seem to exhibit no signs of heating. Even such artifacts may turn out to be feasible for fission-track dating because they may have been exposed to relatively mild heating (<400°C) which is sufficient for partial or complete track removal but which is otherwise undetectable.

Examples of this kind were presented by Miller and Wagner (1981) for various Pre-Columbian obsidians from South America. Three samples were analyzed from the site El Inga near Quito, Ecuador, where numerous obsidian flakes were found and which probably represents an ancient place of tool working. All three samples gave different fission-track ages, namely 1.1 Ma (O75C), 0.14 Ma (O75D), and 2060 years (O75E) (Table 7.4). Track-size measurements indicated that the two oldest obsidian flakes, were partially annealed, as evidenced by the smaller sizes of fossil compared with induced tracks. On the other hand, in sample O75E, the mean size of the fossil tracks was identical within 2% to that of the induced tracks (Figure 7.28). Consequently, the young age of 2060 years, which is based on 70 fossil tracks, marks the time of strong heating resulting in complete track removal.

When applying the plateau correction method (Section 5.7) to the less annealed sample, O75C, an age of 1.72 Ma was found. Since all three obsidian flakes from the El Inga site have a very similar petrographic appearance and uranium content (10.7–11 µg/g), it may be concluded that they all originate from the same geological source, which is 1.72 Ma old, and from which the raw material was brought to El Inga where the artifacts were worked 2060 years ago.

Another site of interest is Cerro la Tefa, Columbia. Two obsidian artifacts were analyzed from this site. The specimen O80A bears a complex fission-track record. Track-size analysis revealed a bimodal distribution of fossil tracks consisting of two sub-populations, one with a high track density but small mean size (size

Table 7.4. Fission-track data of South American obsidian artifacts and geological sources (Miller and Wagner, 1981)

Samples	Measured age (Ma)	Plateau-corr. age (Ma)	Uranium content ($\mu g/g$)
Artifacts:			
El Inga, Ecuador			
O75C	1.1	1.72	11.0
O75D	0.15		
O75E	0.00206		
Palmar, Ecuador			
O94A	0.16		5.42
O94B	0.16		
O94E	0.14	0.25	5.57
O94M	0.15	0.29	5.87
O94O	0.16		5.92
O94D	1.31	2.01	6.91
O94L	0.98		6.54
O94T	1.25	1.89	6.58
La Esperanza, Ecuador			
O78A	0.22	0.21	6.7
Alofa, Bolivia			
O79A	3.5	6.4	11.2
Cerro la Tefa, Columbia			
O80A	0.26	2.0 (size corr.)	11.6
	0.0026	0.0034	
O80C	0.0037		
Geological occurrences:			
Mullumica, Ecuador			
M1	0.19		8.4
M2	0.21		8
Macusani, Peru			
O3	3.4	6.15	15.2
Quito, Ecuador			
O41		1.81	9.7
O42	0.68	1.78	9.7
Lipez, Bolivia			
O46	2.7	5.09	9.7
	3.8	5.15	

reduction $d/d_0 = 0.36$), and one with a low track density but large size ($d/d_0 = 0.89$) (Figure 7.29). The bimodal size distribution indicates an ancient heating event with strong partial track annealing, resulting in the small-sized sub-population. But the fission tracks which accumulated subsequently, also experienced fading – although to a lesser degree – represented by the slightly reduced sub-population (50× amplified in the semi-circle of Figure 7.29). Applying the size correction method to this post-event sub-population, yields a corrected fission-track age of 3400 years. This age probably dates the working of the obsidian artifact. The same age is found for another, completely annealed artifact (O80C) from the same site. The application of the size-correction method to the smaller

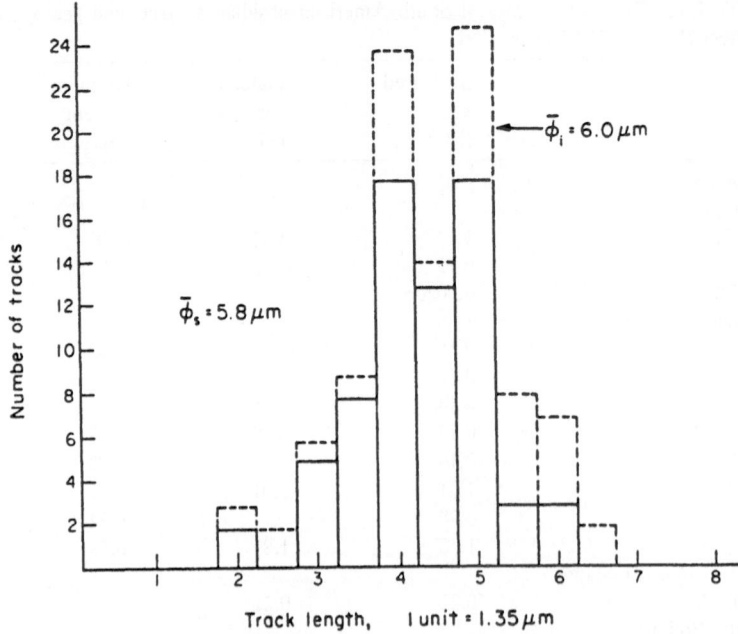

Figure 7.28. Size-distributions of fossil (solid line) and induced (dashed line) fission-track etch pits (major axis of elliptical openings) in obsidian flake O75E from El Inga, Ecuador. The glass was etched for 5 min at 23°C in 16% HF. $\bar{\phi}_s$ and $\bar{\phi}_i$ are the mean sizes for spontaneous and induced fission-tracks, respectively. (From Miller and Wagner, 1981.)

sub-population gives a corrected age of 2 Ma, which represents the geological age of the obsidian source material. Thus, it is possible to retrieve both the geological and the archaeological ages from one sample.

As has been first demonstrated by Suzuki (1970) for Japanese obsidian, the geological age and the uranium content – as analyzed by fission-track analysis – might help to identify the source of the raw material. Durrani *et al.* (1971b) were able to show that the obsidian artifacts from the Franchthi caves, Peloponnesus, originated from Milos Island. This dates the earliest known sea trade of obsidian back to the Mesolithic period. Other Mediterranean geological obsidian occurrences were characterized in terms of fission-track age and uranium content by Bigazzi *et al.* (1971a), Wagner *et al.* (1976), and Wagner and Weiner (1987). Together with chemical analyses, these data enable the identification of the raw material sources for obsidian artifacts in that region. A further example of this kind has been reported from Carpathian obsidian sources by Bigazzi *et al.* (1990).

A number of obsidian flakes from Palmar, an Ecuadorean Coastal site, have been analyzed to determine whether they were produced from one or several raw material sources, and whether the sources could be identified (Miller and Wagner, 1981). By measuring their uranium content and fission-track age, two groups were recognized: one group with uranium contents below 6 μg/g and fission-track ages around 0.15 Ma, and the other group with uranium contents above 6 μg/g and fission-track ages around 1.1 Ma (Table 7.4). From track-size analysis, it was realized that all investigated flakes exhibited partial track-fading. The plateau-

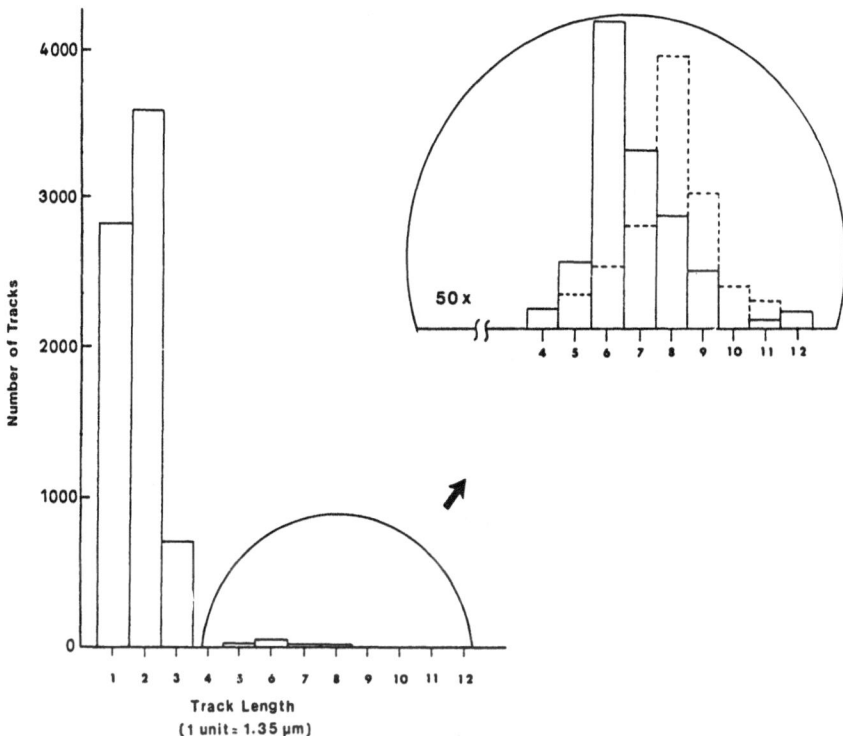

Figure 7.29. Size-distributions of fossil (solid line) and induced (dashed line) fission-track etch pits (major axis in case of elliptical pits) in obsidian flake O80A from Cerro la Tefa, Columbia. The glass was etched for 5 min at 23°C in 16% HF. Part of the histogram is shown magnified 50×. (From Miller and Wagner, 1981.)

corrected fission-track ages for the two groups are 0.27 Ma and 1.96 Ma, respectively. These results clearly establish the subdivision of these flakes into two different material sources. As to the geographical identification of these sources, some of the available geological obsidian occurrences in South America, too, were analyzed with the fission-track method. None of these sources matches, in terms of fission-track age and uranium content, the flakes from the Palmar site (Table 7.4).

7.9.3. *Fired stones and baked soils*

Mineral inclusions in heated rocks, baked soils, fired bricks, and ceramics can be used for fission-track dating of a firing event. In order to completely reset the fission-track clock and to remove all previously stored tracks, the firing temperatures have to exceed the track-retention temperatures. The explicit temperatures depend on the type of mineral and the duration of heating, but are generally above 500°C at a one-hour firing duration (Appendix B). Attainment of complete track annealing in antiquity can be established by track-length measurements analogous to the track-diameter measurements in glasses (Figure 7.27). So far,

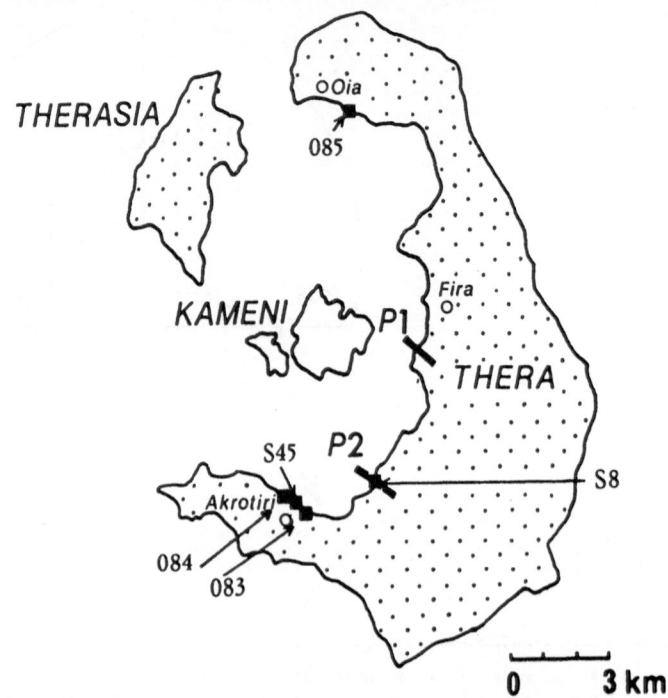

Figure 7.30. Thera (Santorini) island in the Agean Sea, Greece. The old ring islands, Thera and Therasia (dotted), surround the late-Minoan Santorini caldera and post-Minoan Kameni islands. Sample locations (O = glass, S = zircon) are marked. P1 and P2 are the locations of the schematic profiles in Figure 7.31. (After Seward *et al.*, 1980.)

this potential of the fission-track technique has not been fully exploited. The main problem is to find enough fossil fission tracks to make an age determination with acceptable precision. Since the inclusions typically have a size in the order of 100 μm, at least several hundred grains are required. The method is of particular interest for the Palaeolithic beyond 10 ka. The dating of zircons separated from baked soils and pottery found at various Japanese localities was attempted by Nishimura (1971), who reported fission-track ages of between 700 and 2300 years. Unfortunately, in his revised age table (Nishimura, 1972), nothing was mentioned about age precision, which was probably very low. Similarly, Nishimura and Tosi (1976) presented fission-track ages for baked earth and pottery from the proto-urban settlement Shar-i Sokhta, Iran. From each sample (*ca.* 10 kg), several hundred zircon grains were separated and 60 to 85 fossil fission tracks were counted. The calculated fission-track ages place the samples into the 4th and 3rd millennium BC, which is archaeologically reasonable. A very successful performance of dating fired mineral grains is the age determination of Peking man (*Homo erectus pekinensis*) who used fire (Guo Shilun *et al.*, 1980, 1990). Of the 17 layers within the Choukoutien cave near Peking, where Peking man was discovered, layers 3 to 11 carry remains from human fossils and activities, including firing ashes. In these ashes, collected from layers 4 and 10, several hundred grains of sphene were found ranging in size between 50 to 300 μm. Track-length criteria

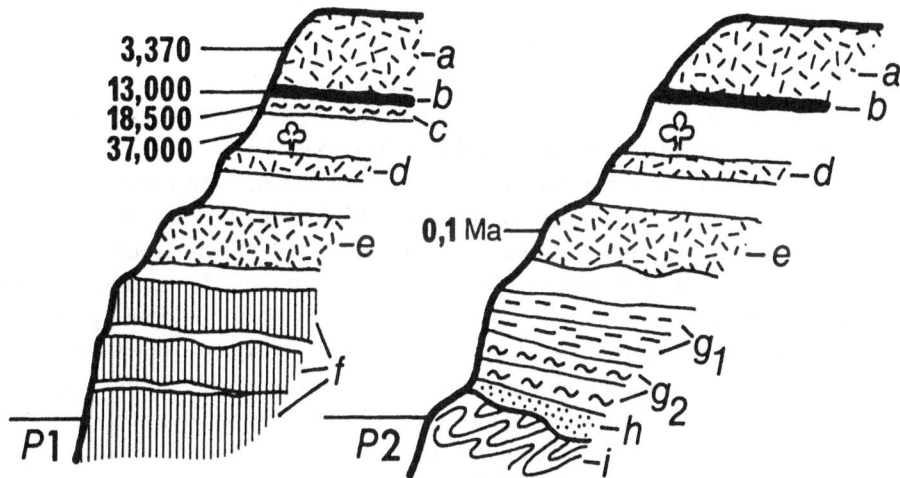

Figure 7.31. Schematic profiles at localities P1 and P2 in Figure 7.30. Age numbers are radiocarbon ages in profile P1, and fission-track age (glass in layer e) in profile P2; a = upper pumice, b = palaeosol, c = ignimbrite, d = middle pumice, e = lower pumice, f = lava, g_1 = red scoriae, g_2 = ignimbrites, h = white vitric tuff; i = phyllite basement. (After Seward *et al.*, 1980.)

were used in order to select only those grains whose fission-track clocks were completely reset by the fire. Altogether, *ca.* 100 grains passed this test. Fission-track ages of 306 ± 56 ka (layer 4) and 462 ± 45 ka (layer 10) were determined.

7.9.4. *Palaeolithic marker horizons*

A different type of fission-track application in archaeology and palaeo-anthro-pology is the dating of artifact- and hominid-bearing stratigraphic sequences that contain datable tuffaceous marker horizons. This approach is closely connected with tephrochronology (Section 7.1). Since tephra layers are instantaneously for-med and extend over wide areas, they can be stratigraphically related to the over- and underlying sediments with their fossil and cultural remains. In this way, the absolute time marks of these horizons yield age brackets for intercalated sedi-ments. For fission-track dating of tephra layers, either the glassy phases (pumice, glass shards, obsidian) or uranium-rich minerals, such as zircon, sphene, and apatite, can be used.

Sampling has to be very carefully performed, since some glasses and minerals might be older detrital contaminants rather than primary phases of the volcanic eruption. For instance, on the island of Thera in Greece (Figure 7.30), the tuff from the basal layer (layer h in Figure 7.31) of the caldera fillings contains grains from the crystalline basement as well as primary zircon grains (Seward *et al.*, 1980). The two zircon populations could be clearly distinguished under the binocular: The primary zircon population tended to be subhedral to euhedral and generally colourless, while the detrital zircon population was anhedral, generally rounded, and slightly pink. The primary zircons gave a fission-track age of 1.05 ± 0.16 Ma, whereas the detrital population gave 13.9 ± 1.2 Ma. For various volcanic glasses from the overlying pumice layer (e in Figure 7.31) a mean fission-track age of 0.1 ± 0.02 Ma – after applying the plateau-correction technique – was calculated. The composite, chronology – including fission-track as well as radiocarbon ages – of the caldera-filling layers on Thera is shown in Figure 7.31.

The question as to whether a grain is primary or detrital can often be settled by dating single grains. This has been firmly established for the KBS tuff at Lake Turkana (formerly Rudolph), Kenya, by Gleadow (1980) who found a mean fission-track age of 1.87 ± 0.04 Ma for its primary zircons. In addition to the primary zircons, Gleadow (1980) observed two detrital populations, one with single crystal ages ranging between 24 and 27 Ma – probably from some older volcanic source – and another with ages around 500 Ma, probably originating from a crystalline basement. Also, other important hominid-bearing sites in East Africa have been dated with fission tracks, among them Bed I at the Olduvai Gorge, Tanganyika, with 2.03 ± 0.28 Ma (Fleischer *et al.*, 1965f), the Hadar formation, Ethiopia, with 2.58 ± 0.28 Ma (Aronson *et al.*, 1976), the Cinder Tuff of the Middle Awash, Ethiopia, with 3.93 ± 0.28 Ma (Hall *et al.*, 1984), and tuffs from the Buluk Member, Bakate Formation, Kenya (Hurford and Watkins, 1987). In all these cases, the fission-track ages supplemented K-Ar dates which are often controversial due to excess argon or to argon loss.

Etching Conditions for the Revelation of Fission Tracks

Several reports have been published on the experimental etching conditions used for fission-track revelation in a wide variety of materials. An extensive list is found in Fleischer *et al.* (1975) which also served as a basis for the table of etchants given below. Nevertheless, the original publications have been consulted as far as possible and new information has been added. The table is confined to natural terrestrial materials capable of being dated with the fission-track method. Relevant information which helps to select the optimal etching conditions can be found in Section 2.3.3. For a correct interpretation of fission-track ages, it is essential to define the etching conditions precisely and to quote them in the publication.

Note that often slightly different etching conditions are applied to the same material by various authors. Among other reasons, this may be due to variations in chemical composition of the material studied. Hence, the etchants listed below only serve as a general guide.

Material	Etchant	Temp (°C)	Time	Reference
Apatite	0.5% (≈0.1 N) to 65% HNO_3	20–30	10–80 s	Fleischer and Price (1964d)
				Wagner and Reimer (1972)
Autunite	10% HCl	23	10–30 s	Fleischer and Price (1964d)
Barite	70% HNO_3	100	3 h	Fleischer and Price (1964d)
Barysilite	CH_3COOH (acetic acid)	23	5–10 s	Haack (1973)
Bastnaesite	20% HCl	155	20–150 min	Fleischer and Naeser (1972)
Benitoite	HF (40%):HNO_3 (65%) (ratio 1:1)	230	5 min	Haack (1973)
Beryl	KOH (aq)	150	9 h	Fleischer and Price (1964d)
Calcite	HCO_2H (formic acid)	23	1 min	Sippel and Glover (1964)
				MacDougall and Price (1974)
Chlorite	48% HF	23	10 min	Debeauvais et al. (1964)
	48% HF	60	35–40 min	Sharma et al. (1977)
Clinopyroxene				
– augite	KOH (aq)	220	1 min	Fleischer and Price (1964d)
	HF (48%):H_2SO_4 (80%):H_2O (2:1:4)	23	5–20 min	Crozaz et al. (1970)
– diopside	3g NaOH, 2g H_2O	boiling	85–95 min	Lal et al. (1968)
	3g NaOH, 2g H_2O	boiling	75–85 min	Lal et al. (1968)
Epidote				
– allanite	50N NaOH	150	2–60 min	Naeser et al. (1970)
– epidote	50N NaOH	150	1–3 h	Naeser and Dodge (1970)
		boiling		Saini et al. (1978)
	25N NaOH followed by 48% HF	150	2 h	Chakranarayan and Powar (1982)
	HF:HCl (1:1)	23	15 min	Lal and Waraich (1983)
– zoisite (tanzanite)	50N NaOH	23	20–25 min	Naeser and Saul (1974)
		140	0.5–2 h	
Feldspar				
– microcline and orthoclase	5g KOH:1g H_2O	190	80 min	Fleischer et al. (1975)
	48% HF	23	1–10 s	Fleischer and Price (1964d)
– plagioclase	KOH (aq)	210	15–30 min	Fleischer et al. (1965d)
	6g NaOH:4g H_2O	boiling	3–26 min	Lal et al. (1968)
Fluorite	98% H_2SO_4	23	10 min	Fleischer and Price (1964d)

	Etchant	Temp (°C)	Time	Reference
Garnet				
– almandine, pyrope and spessartine	50N NaOH	140 or boiling	5 min–2 h	Haack and Gramse (1972)
– andradite and grossularite	75N NaOH	boiling	1–6 h	Haack and Gramse (1972)
Glass				
– basaltic and andesitic glass	5–20% HF	23	1–5 min	Fleischer et al. (1969d)
	24% HBF:5% HNO$_3$:0.5% CH$_3$COOH	23	10–50 min	MacDougall (1971a)
– pumice	5% HF	23	500 s	Fleischer et al. (1965f)
– obsidian	48% HF	23	60 s	Fleischer and Price (1964b)
– pitchstone	48% HF	23	15 s	Storzer (1970)
– tektite	48% HF	23	15–45 s	Fleischer and Price (1964c)
Gypsum	5% HF	23	5–10 s	Fleischer and Price (1964d)
Glauconite	HF:H$_2$SO$_4$	23		Srivastava et al. (1983)
Hornblende	48% HF	23–60	5–60 s	Fleischer and Price (1964d)
Hübnerite	5g NH$_4$Cl:5g Na$_4$P$_2$O$_7$: 5 ml H$_3$PO$_4$:20 ml H$_2$O	boiling	110 min	Haack (1973)
Kyanite	40% HF	21	2 h	Saini et al. (1983)
Mica				
– biotite	20% HF	23	1–2 min	Price and Walker (1962c)
– lepidolite	48% HF	23	3–70 s	Fleischer and Price (1964d)
– muscovite	48% HF	23	10–40 min	Price and Walker (1962c)
– phlogopite	48% HF	23	1–5 min	Price and Walker (1962c)
Microlite	HF (48%):HNO$_3$ (65%) (1:1)	23	6 min	Haack (1973)
Mimetite	HNO$_3$ (65%):H$_2$O (1:2)	23	3 s	Haack (1973)
Monazite	35% HCl	boiling	45 min	Shukoljukov and Komarov (1970)
Olivine	1 ml H$_3$PO$_4$ (98%):40 g EDTA: 1 g oxalic acid:100 ml H$_2$O + NaOH till pH = 8 (WN-solution)	boiling	2–6 h	Krishnaswami et al. (1971)
Orpiment	1% KOH	50	1–2 min	Jakupi et al. (1982)
	1% NaOH	50	2 min	Jakupi et al. (1990)

Material	Etchant	Temp (°C)	Time	Reference
Orthopyroxene				
– bronzite	6g NaOH:4g H_2O	boiling	35–50 min	Lal et al. (1968)
	48% HF	23	3 min	Dorling et al. (1974)
– enstatite and hypersthene	6g NaOH:4g H_2O	boiling	30–55 min	Lal et al. (1968)
Prehnite	$HF:H_2SO_4:HCl:H_2O$			Koul et al. (1985)
Pyromorphite	$HNO_3(65\%):H_2O$ (1:2)	23	5 s	Haack (1973)
Quartz				
– quartz	KOH (aq)	150	3 h	Fleischer and Price (1964d)
	48% HF	23	24 h	Fleischer and Price (1964d)
	60% NaOH	boiling	20–40 min	Khan and Ahmad (1976)
– tridymite	10% HF	23	1 h	Fleischer et al. (1965d)
Scheelite	$NH_4Cl:HCl:H_2O$	boiling	1–2 min	Haack (1973)
	6.25N NaOH	90	95 min	Fleischer et al. (1975)
Sphene	HCl (conc)	90	0.5–1.5 h	Naeser and Faul (1969)
	4g NaOH:2 gH_2O ($\approx$50N)	130	20–30 min	Calk and Naeser (1973)
	$HF:HNO_3:HCl:H_2O$ (1:2:3:6)	20	1–30 min	Naeser and McKee (1970)
Stibiotantalite	HF (48%):HNO_3 (65%) (1:1)	23	6 min	Gleadow (1978)
Thorite	H_3PO_4 (85%)	250–300	1 min	Haack (1970)
Topaz	KOH (aq)	150	100 min	Fleischer et al. (1965d)
Tourmaline	KOH (aq)	220	20 min	Fleischer and Price (1964d)
	48% HF	30	15–20 min	Fleischer and Price (1964d)
Vanadinite	HNO_3 (65%):H_2O (1:2)	23	1 s	Nand Lal et al. (1977b)
Vermiculite	48% HF	23	5–10 s	Haack (1973)
	24% HF	30	3–5 min	Fleischer et al. (1975)
Vesuvianite	75N NaOH	boiling	30–45 min	Sharma et al. (1979)
				Haack (1976a)
zeolite				
– chabazite	3g KOH:4g H_2O	boiling	30–60 min	Koul et al. (1983)
– heulandite	10 ml Aq. Reg.:1 ml HF (48%).	23	30 s	Fleischer et al. (1975)
– stilbite	1% HF	23	1 min	Fleischer et al. (1975)
Zircon	H_3PO_4	375–500	1 min	Fleischer et al. (1964c)
	20g NaOH:5g H_2O ($\approx$100N)	220	15 min–4 h	Naeser (1969)
	HF (48%):H_2SO_4 (98%) (1:1)	150–180	2–9 h	Krishnaswami et al. (1974)
	8g NaOH:11.5g KOH (eutect.)	200–220	2–70 h	Gleadow et al. (1976)
	NaOH:KOH:LiOH:H_2O (6:14:1) (eutect.)	200	2–5 h	Zaun and Wagner (1985)

Annealing Properties of Fission Tracks in Minerals and Glasses

Numerous reports exist in the literature of annealing experiments on fission tracks in a large variety of terrestrial materials. In order to review the annealing characteristics and to facilitate access to the original works, this appendix compiles, in tabular form, most results obtained in these studies together with the corresponding references. The data presented here, are not the original experimental data because most authors graphically reported their data as lines of equal track-density reduction in Arrhenius diagrams rather than numerically. Furthermore, the contour lines result from the interpolation and extrapolation of the original experimental data.

The data presented here were graphically read as accurately as possible from the line which marks the 50% reduction in fission-track density. In order to allow better comparability between the various experiments and materials, the annealing characteristics are presented as temperatures for the same annealing durations. The following annealing durations were chosen: 1 h, 1 year, 10^2, 10^4, 10^6, and 10^8 years. Since most experiments were carried out for annealing durations of less than a year, the long-time durations are obviously calculated by extrapolation.

The equation to be used for extrapolating the annealing data from laboratory to geologic timescales needs some consideration. Commonly, it is assumed that fission-track annealing obeys first-order reaction kinetics according to the equation

$$\ln t = E/(kT) + C,$$

with t = annealing time, E = activation energy (eV), k = Boltzmann's constant, ($= 8.616 \times 10^{-5}$ eV/K), T = annealing temperature (K), and C = specific constant.

For apatite, it has been shown by Green et al. (1988) that the assumption of first-order kinetics is an erroneous oversimplification and that there is actually no good reason why it should be expected at all to describe fission-track annealing (Section 4.2.5). Hence, extrapolations to the geologic timescale based on first-order kinetics, may lead to large errors. In fact, it remains questionable whether the contours of equal annealing degrees in the Arrhenius diagram need to be straight rather than curved (Laslett et al., 1987). In the latter case, extrapolation from laboratory geologic timescales could result in considerable errors in prediction. Nevertheless, Laslett et al. (1987) demonstrated that a straight-line, narrow fanning model gives a convenient and statistically satisfactory description of their apatite-annealing data, but stressed that "extrapolation of the fitted model outside the range of the experimental data should be treated with caution".

Before accepting the geologic predictions from the preferred model, the extrapolations need to be tested against valid geologic constraints. Indeed, such comparison between extrapolated and geologically observed annealing temperatures have revealed gross systematic discrepancies in the cases of zircon, sphene, epidote, and garnet. These differences may be due to various reasons, among them – apart from wrong model assumptions – insufficiently designed and performed annealing experiments, as well as parameters other than solely time and temperature that affect the annealing behaviour.

In view of this complex situation and uncertaintiesa linear annealing equation within the Arrhenius diagram

$$\ln t = C_1/T + C_2,$$

with C_1 and C_2 being empirical specific constants, has been adopted for extrapolating the annealing temperatures from short experimental to long geologic durations (Table B.1.). Of the various equal track-density reduction contours, only the 50% line has been selected, since it seems that this contour line is affected less than other lines by errors introduced in the extrapolation (Green *et al.*, 1985; Laslett *et al.*, 1988) and probably closely represents the effective track-retention temperatures. Owing to various difficulties (Chapter 4), the extrapolated annealing temperature for geologic timescales presented in this appendix are *not meant to be applied directly to geologic situations* without any further thought.

Table B.1. Fission-track retention temperatures of terrestrial materials based on annealing experiments and long-time extrapolations. The (fixed) temperatures are given for the reduction degree $\rho/\rho_0 = 0.5$ at which 50% of the density ρ_0 freshly induced fission tracks is erased that was present prior to annealing. Owing to insufficient information for some of the reported annealing data no extrapolated values can be calculated.

| Material | 50% retention temperature (°C) | | | | | | Reference |
Locality	1 h	1 a	10^2 a	10^4 a	10^6 a	10^8 a	
(a) *Minerals*							
Allanite							
Kasipatnam	690	460	380	310	260	220	Saini *et al.* (1975)
Amber							
Baltic	110	80	65	50	40	25	Uzgiris and Fleischer (1971)
Apatite							
Sljudjanka	336	220	174	138	108	80	Wagnet (1968)
unknown	300						Seitz *et al.* (1970)
Eldora	330	220	175	140	110	80	Naeser and Faul (1972a)
Flockenbach	340	225	175	140	115	85	Wagner (1972a)
Liebersbach	330	240	200	165	140	120	Wagner and Reimer (1972)
Durango	335	225	180	145	115	90	Märk *et al.* (1973)
Borra Mine	350	210	166	125	90	60	Koul (1979)
unknown	330	240	205	170	145	125	Nagpaul *et al.* (1974b)
Durango	323	227	186	153	123	95	Watt and Durrani (1985)
Aragonite							
unknown	140						Fleischer *et al.* (1968a)
Autunite							
unknown	40						Fleischer and Price (1964d)
Biotite							
unknown	390	225	165	120	80	50	Nagpaul *et al.* (1974b)
unknown						100	Saini (1978)/Nagpaul (1982)
Bytownite							
unknown	750						Fleischer *et al.* (1965d)
Calcite							
unknown	350						Fleischer *et al.* (1968a)
Chlorite							
Nellore	450	260	190	145	110	70	Sharma *et al.* (1977)
Diopside							
unknown	850						Fleischer *et al.* (1968a)
Epidote							
Gateway	680	625	600	580	560	540	Naeser *et al.* (1970)
Gabi	650	580	545	510	480	455	Reimer (1972)
SW Africa	620	515	470	430	390	360	Haack (1976c)
Sausar	653	400	320	250	210	160	Saini *et al.* (1978)
Garnet							
Rüdenau-Nord ($An_{41}Gr_{56}Al_3$)	665	545	495	455	415	380	Haack and Potts (1972)
Bhunas ($An_8Sp_{31}Al_{57}Py_3$)	620	495	445	400	360	325	Nand Lal *et al.* (1977a)
unknown						350	Saini (1978)/Nagpaul (1982)
Hornblende							
unknown	590						Fleischer and Price (1964d)
unknown	590						Fleischer *et al.* (1968a)
unknown	710						Maurette (1970b)
unknown						145	Saini (1978)/Nagpaul (1982)

Table B.1. Continued

Material Locality	50% retention temperature (°C) 1 h	1 a	10^2 a	10^4 a	10^6 a	10^8 a	Reference
Hypersthene							
unknown	475						Fleischer *et al.* (1968a)
unknown	560						Maurette (1970b)
Kyanite							
Himalaya	370						Saini *et al.* (1983)
Monazite							
Aktschatai	300						Shukolyukov *et al.* (1970)
Muscovite							
Renfrew Cty.	520						Fleischer *et al.* (1964b)
unknown	650						Maurette (1970b)
unknown	600						Berzina *et al.* (1967)
Kuhara	530						Mehta and Rama (1969)
Kodarma	560	350	285	230	185	150	Barman and Nandi (1978)
unknown						170	Saini (1978)/Nagpaul (1982)
Olivine							
Red Sea	510	370	320	280	240	205	Fleischer *et al.* (1965b)
Orpiment							
Alshar	<260	<190	<165	<140	<115	<95	Todorovic *et al.* (1980)
Phlogopite							
Mahatsinjo	450	270	215	170	130	100	Maurette *et al.* (1964)
Neyyur	535	315	250	195	150	115	Parshad *et al.* (1978)
Gardiner	560	300	235	175	125	85	Koul *et al.* (1983)
Quartz							
unknown	1030						Fleischer *et al.* (1968a)
unknown	780						Haack (1972)
Sphene							
Magnet Cove							
(HCl etch)	610	490	440	390	350	320	Naeser and Faul (1969)
(NaOH etch)	730	580	530	480	430		Calk and Naeser (1973)
unknown	600	480	430	380	340	310	Nagpaul *et al.* (1947b)
75-14	730						Gleadow (1978a)
Australia	640						Watt and Durrani (1985)
unknown						340	Saini (1978)/Nagpaul (1982)
Stibiotantalite							
Alto Ligonha	510						Haack (1970)
Tanzanite							
Ally Mine	675						Naeser and Saul (1974)
Tourmaline							
Pisangan	560	385	320	270	225	190	Nand Lal *et al.* (1977b)
Vermiculite							
Kasipatnam	480	290	220	170	130	95	Sharma *et al.* (1979)
Vesuvianite							
Ombujongupa	615	450	385	330	285	250	Haack (1976c)
Zeolite (chabazite)							
Vagar	400						Koul *et al.* (1983)
Mersey Valley	490	230	155	100	55	10	Koul *et al.* (1986)
Zircon							
Indochina	700	550	480	420	370	330	Fleischer *et al.* (1965b)
Alcoota	700	520	450	390	340	300	Krishnaswami *et al.* (1974)
Sri Lanka	620						Nishida and Takashima (1975)

Table B.1. Continued

Material Locality	50% retention temperature (°C) 1 h	1 a	10^2 a	10^4 a	10^6 a	10^8 a	Reference
Japan	700						Nishida and Takashima (1975)
India	630						Nishida and Takashima (1975)
Formosa	660						Nishida and Takashima (1975)
Koto	670						Tagami *et al.* (1990)
(b) *Glasses*							
Obsidians							
Japan	390	245	190	145	110	80	Suzuki (1970)
Japan	450	260	210	160	116	65	Suzuki (1973)
Irukamura	400	205	165	130	90	50	Suzuki (1973)
Columbia	205						Storzer (1978)
Tachylite							
Rossberg	215						Storzer (1978)
Pitchstones							
Tisens	390	230	175	130	95	50	Storzer (1970)
Auer	354						Storzer (1978)
San Lugano	360						Storzer (1978)
Basaltic deep-sea glasses							
N. Atlantic	275	200	160	130	100	80	Aumento (1969)
N. Atlantic	290	170	130	90	60	85	Fleischer *et al.* (1968b)
DSDP	170	90	60	30	10	−10	MacDougall (1976)
	235	124	84	51	24	±0	Storzer and Selo (1978)
Impact glasses							
Cl. Water Lake	220	110	70	40	15	−10	Fleischer *et al.* (1968d)
Manicouagan	205	155	130	105	85	70	Fleischer *et al.* (1969d)
LDG 699-2	460	280	220	180	140	105	Storzer and Wagner (1971)
LDG-L.A	410						Storzer (1978)
Riesglas	210						Storzer (1978)
Darwinglass	375						Storzer (1978)
Tektites							
australite 4	360	210	160	120	88	55	Storzer and Wagner (1969)
australite 9	305						Storzer (1978)
australite 14	295						Storzer (1978)
bediasite	360	195	140	100	65	40	Durrani and Khan (1970)
bediasite	365	205	155	115	80	50	Storzer and Wagner (1971)
moldavite	300						Storzer (1978)
Cuba	310						Storzer (1978)
Lee County	370						Storzer (1978)
DG-A2	365						Storzer (1978)
NMNH 2345	370						Storzer (1978)
East Corner	380						Storzer (1978)
Artificial glasses							
U-gl. Sarat.	220	140	110				Wagner *et al.* (1975)
U-gl. Grpl.	220	140	110				Wagner *et al.* (1975)
U-gl. 1844	230	90	45				Damm (1973)
micr. slide	223						Storzer (1978)
NBS-SRM	235	149	110				Reimer *et al.* (1972)

APPENDIX C

Fundamentals of Error Calculation in Fission-Track Dating

In the calculation of a fission-track age, both the track-density ratio R ($= \rho_s/\rho_i$) and the fluence ϕ can be regarded as independent variables. The quadratic error on the age (t) will thus be given by

$$s_t^2 = (\delta t/\delta R)^2 s_R^2 + (\delta t/\delta \phi)^2 s^2 \phi, \qquad (C.1)$$

where s_R and s_ϕ are the errors on R and ϕ, respectively. The age equation (3.10) can be written as

$$t = k_1 \ln(k_2 R\phi + 1), \qquad (C.2)$$

with

$$k_1 = 1/\lambda_\alpha \text{ and } k_2 = (\lambda_\alpha/\lambda_f) QGI\sigma.$$

Taking into account Equation (C.2), one obtains for s_t, after calculating the partial derivatives in Equation (C.1),

$$s_t = \frac{k_1 k_2}{k_2 R\phi + 1} \sqrt{\phi^2 s_R^2 + R^2 s_\phi^2}$$

and the relative error of the age is then given by

$$s_t/t = K\sqrt{(s_R/R)^2 + (s_\phi/\phi)^2}, \qquad (C.3)$$

with

$$K = \frac{k_2 R\phi}{(k_2 R\phi + 1)\ln(k_2 R\phi + 1)}.$$

The factor K can be expressed as a function of time which yields (Johnston et al., 1979):

$$K = k_1/t(1 - e^{-t/K_1}) \qquad (C.4)$$

Substitution of the exact value of K, as calculated from Equation (C.4), in the error equation (C.3), is only worthwhile for samples with Precambrian fission-track ages; for younger samples, the vlaue of K can be put equal to 1 ($K = 0.955$ for $t = 600$ Ma).

248

The equation for calculating the error on the track-density ratio depends upon the dating procedure that is used. If the induced tracks are counted on a sample mount which does not already contain etched spontaneous tracks (non-subtraction method), the quadratic error on R is given by

$$s_R^2 = (\delta R / \delta \rho_s)^2 s_{\rho_s}^2 + (\delta R / \delta \rho_i)^2 s_{\rho_i}^2 \qquad (C.5)$$

where s_{ρ_s} and s_{ρ_i} are the errors on ρ_s and ρ_i, respectively.

After calculating the partial derivatives, one obtains

$$s_R = \sqrt{(1/\rho_i)^2 s_{\rho_s}^2 + (\rho_s/\rho_i^2)^2 s_{\rho_i}^2}$$

and, for the relative error.

$$s_R / R = \sqrt{(s_{\rho_s}/\rho_s)^2 + (s_{\rho_i}/\rho_i)^2}. \qquad (C.6)$$

If the induced tracks are counted on a sample mount which does already contain etched spontaneous tracks, the density ρ_{s+i} is determined and ρ_i has to be derived by subtraction ($\rho_i = \rho_{s+i} - \rho_s$). The basic equation for the calculation of the error on R becomes (Bigazzi *et al.*, 1986):

$$s_R^2 = (\delta R / \delta \rho_s)^2 s_{\rho_s}^2 + (\delta R / \delta \rho_{s+i})^2 s_{\rho_{s-i}}^2, \qquad (C.7)$$

which, after calculation of the partial derivatives, yields

$$s_R = \sqrt{\left(\frac{\rho_{s+i}}{(\rho_{s+i} - \rho_s)^2}\right)^2 s_{\rho_s}^2 + \left(\frac{\rho_s}{(\rho_{s+i} - \rho_s)^2}\right)^2 s_{\rho_{s+i}}^2}$$

and for the relative error on R one obtains

$$s_R / R = \frac{\rho_{s+i}}{\rho_{s+i} - \rho_s} \sqrt{(s_{\rho_s}/\rho_s)^2 + (\rho_{\rho_{s-i}}/\rho_{s+i})^2}. \qquad (C.8)$$

References

Abdullaev, Kh., Gorbachev, S. K., Perelygin, V. P., and Tretjakova, S. P. (1966): The determination of the geological ages of mica: muscovite, phlogopite and biotite with uranium fission tracks, (Russ.), Dubna P 3-2961, 1–10.

Agar, S. M., Cliff, R. A., Duddy, I. R., and Rex, D. C. (1989): Accretion and uplift in the Shimanto Belt, SW Japan, *J. Geol. Soc. London* **146**, 893–896.

Ahrens, T. J., Fleischer, R. L., Price, P. B., and Woods, R. T. (1970): Erasure of fission tracks in glasses and silicates by shock waves, *Earth Planet. Sci. Lett.* **8**, 420–426.

Albrecht, D., Armbruster, P., Spohr, R., and Roth, M. (1982): Small angle neutron scattering from oriented latent nuclear tracks, *Radiation Effects.* **65**, 145–148.

Albrecht, D., Armbruster, P., Spohr, R., Roth, M., Schaupert, K., and Stuhrmann, H. (1984): Small angle scattering from oriented latent nuclear tracks, *Nucl. Instr. Meth.* **B2**, 702–705.

Albrecht, D., Armbruster, P., Spohr, R., Roth, M., Schaupert, K. J, and Stuhrmann, H. (1985): Investigation of heavy ion produced defect structures in insulators by small angle scattering, *Appl. Phys. A* **37**, 37–46.

Albrecht, D., Balanzat, E., and Schaupert, K. (1986): X-ray small angle scattering investigation of high energy Ar-tracks in mica, *Nucl Tracks Radiat. Meas.* **11**, 93–94.

Ali, A. and Durrani, S. A. (1977): Etched-track kinetics in isotropic detectors, *Nucl. Track Detection* **1**, 107–121.

Amin, Y. M. (1988): Effect of damage on the etching and X-ray diffraction properties of zircon, *Nucl. Tracks Radiat. Meas.* **15**, 119–123.

Andriessen, P. A. M. (1990): Anomalous fission track apatite ages of the Precambrian basement in the Hunnedalen region, south-western Norway, *Nucl. Tracks Radiat. Meas.* **17**, 285–291.

Arias, C., Bigazzi, G., and Bonadonna, F. P. (1981): Size corrections and plateau age in glass shards, *Nucl. Tracks* **5**, 129–136.

Arne, D. C., Green, P. F., Duddy, I. R., Gleadow, A. J. W., Lambert, I. B., and Lovering, J. F. (1989): Regional thermal history of the Lennard Shelf, Canning Basin from apatite fission track analysis: implications for the formation of Pb–Zn ore deposits, *Austral. J. Earth Sci.* **36**, 495–513.

Arne, D. C., Green, P. F., and Duddy, I. R. (1990): Thermochronologic constraints on the timing of Mississippi Valley-type ore formation from apatite fission track analysis, *Nucl. Tracks Radiat. Meas.* **17**, 319–323.

Aronson, J. L., Schmitt, T. J., Walter, R. C., Taieb, M., Tiercelin, J. J., Johanson, D. C., Naeser, C. W., and Nairn, A. E. M. (1976): New geochronologic and paleomagnetic data from the hominid-bearing Hadar Formation of Ethiopia, *Nature* **267**, 323–327.

Aschenbach, J., Fiedler, G., Schreck-Köllner, H., and Siegert, G. (1974): Special glasses as energy detectors for fission fragments, *Nucl. Instr. Meth.* **116**, 389–396.

Aumento, F. (1969): The Mid-Atlantic ridge near 45°N. Fission track and ferro-manganese chronology, *Canad. J. Earth Sci.* **6**, 1431–1440.

Baard, J. H., Zijp, W. L., and Nolthenius, H. J. (1989): *Nuclear Data Guide for Reactor Metrology*, Kluwer Acad. Publ., Dordrecht.

Banks, N. G. and Stuckless, J. S. (1973): Chronology of intrusion and ore deposition at Ray, Arizona: Part II, Fission-track ages, *Econ. Geol.* **68**, 657–664.

Baptista, Z. N. R., Mantovani, M. S. M. and Ribeiro, F. B. (1981): Contribuicao para a determinacao da constante de fissao espontanea do Uranio, *Ann. Acad. Brasil Cienc.* **53**, 437–441.

Bar, M., Kolodny, Y., and Bentor, Y. K. (1974): Dating faults by fission track dating of epidotes – an attempt, *Earth Planet. Sci. Lett.* **22**, 157–162.

Barman, T. R. and Nandi, K. (1978): Fission track dating of mica from Kodarma, Bihar, eastern India, U.S. Geol. Surv. Open-File Report 78-701, 24–26.

Baumhauer, H. (1894): *Die Resultate der Aetzmethode*, Leipzig.

Bean, C. P., Doyle, M. V. and Entine, G. (1970): Etching of submicron pores in irradiated mica, *J. Appl. Phys.* **141**, 1454–1459.

Belenky, S. N., Shorokhvatov, M. D. and Etenko, A. V. (1983): Measurement of the characteristics of spontaneous fission of 238U and 236U, *At. Energ.* **55**, 528–530.

Belyaev, A. D., Bahromi, I. I., Beresina, N. V., Bikbova, Z. S., Volkova, N. I., Gorevoi, A. A., Kogan, V. I., Muminov, A. I., Pikul, V. P., and Usmandiarov, A. M. (1980a): Critical angles for fission fragment registration in some solid state track detectors, *Nucl. Tracks* **4**, 49–52.

Belyaev, A. D., Bahromi, I. I., Beresina, N. V., Bikbova, Z. S., Gorevoi, A. A., Kogan, V. I., Muminov, A. I., Pikul, V. P., and Usmandiarov, A. M. (1980b): Relations of fission-fragment ranges to mica crystal orientation, *Nucl. Tracks* **4**, 53–56.

Belyaev, A. D., Bikbova, Z. S., Gayshan, V. L., Khabibullaev, P. K., Kogan, V. I., Pikul, V. P., and Usmandiarov, A. M. (1988): The ion track etching rate in crystal track detectors, *Nucl. Tracks Radiat. Meas.* **14**, 369–372.

Benjamin, M. T., Johnson, N. M., and Naeser, C. W. (1987): Recent rapid uplift in the Bolivian Andes: Evidence from fission track dating, *Geology* **15**, 680–683.

Benton, E. V. (1970): On latent track formation in organic nuclear charged paricle track detectors, *Radiation Effects* **2**, 273–280.

Bertagnolli, E., Keil, R., and Pahl, M. (1983): Thermal history and length distribution of fission tracks in apatite: part I, *Nucl. Tracks* **7**, 163–177.

Bertel, E., Märk, T. D., and Pahl, M. (1977): A new method for the measurement of the mean etchable fission track length and of extremely high fission track densities in minerals, *Nucl. Track Detection* **1**, 123–126.

Berzina, I. G., Berman, I. B., and Zlotova, I. M. (1966): Age determination of micas with uranium fission tracks (Russ.), *Izv. Akad. Nauk. SSSR. Ser. Geol.* **31**, 10–25.

Berzina, I. G., Vorob'eva, I. V., Geguzin, Ya. E., and Zlotova, I. M. (1967): Annealing of tracks of fragments from spontaneous fission of uranium in glasses and mica crystals, *Soviet Phys. Dokl.* **11**, 1105–1107.

Bhandari, N., Bhat, S. G., Lal, D., Rajagopalan, G., Tamhane, A. S. J., and Venkatavaradan, V. S. (1971): Fission fragment tracks in apatite: recordable track lengths, *Earth Planet. Sci. Lett.* **13**, 191–199.

Bigazzi, G. (1967): Length of fission tracks and age of muscovite samples, *Earth Planet. Sci. Lett.* **3**, 434–438.

Bigazzi, G. (1981): The problem of the decay constant λ_f of ^{238}U, *Nucl. Tracks* **5**, 35–44.

Bigazzi, G., Bonadonna, F. P, Belluomini, G., and Malpieri, L. (1971a): Studi sulle ossidiane italiane. IV. Datazione con il metodo delle trace di fissione, *Boll. Soc. Geol. It.* **90**, 469–480.

Bigazzi, G., Cattani, M., Cordani, U. G., and Kawashita, K. (1971b): Comparison between radiometric and fission track ages of micas, *An. Acad. Brasil. Ciênc.* **43**, 633–638.

Bigazzi, G., Ferrara, G., and Innocenti, F. (1972): Fission track ages of gabbros from Northern Apennines ophiolites, *Earth Planet. Sci. Lett.* **14**, 242–244.

Bigazzi, G., Bonadonna, F. P., Maccioni, L., and Pecorini, G. (1976): Research on Monte Arci (Sardinia) subaerial volcanic complex using the fission track method, *Boll. Soc. Geol. It.* **95**, 1–16.

Bigazzi, G., Bonadonna, F., and Neto, J. C. H. (1986): Contribution to statistics in fission track counting, *Nucl. Tracks Radiat. Meas.* **11**, 123–136.

Bigazzi, G., Hadler Neto, J. C., Norelli, P., Osorio Araya, A. M., Paulino, R., Poupeau, G., and Stella de Navia, L. (1988): Dating of glass: the importance of correctly identifying fission tracks, *Nucl. Tracks Radiat. Meas.* **15**, 711–714.

Bigazzi, G., Marton, P., Norelli, P., and Rozloznik, L. (1990): Fission track dating of Carpathian obsidians and provenance identification, *Nucl. Tracks Radiat. Meas.* **17**, 391–396.

Bodu, R., Bouzigues, H., Morin, N., and Pfiffelmann, J. P. (1972): On the existence of anomalous isotopic abundances in uranium from Gabon, *Compt. Rend. Acad. Sci., Paris* **D275**, 1731–1732.

Boellstorff, J. (1976): The succession of late Cenocoic volcanic ashes in the Great Plains: A progress report in *Stratigraphy and Faunal Sequence Meade County, Kansas*; Guidebook 24th Ann. Meet. Midwestern Friends of the Pleistocene, May 1976, Guidebook Series 1, Kansas Geol. Survey.

Boellstorff, J. and Steineck, L. (1975): The stratigraphic significance of fission-track ages on volcanic ashes in the marine late Cenozoic of Southern California, *Earth Planet. Sci. Lett.* **27**, 143–154.

Boellstorff, J. and Te Punga, M. T. (1977): Fission-track ages and correlation of middle and lower Pleistocene sequences from Nebraska and New Zealand, *NZ J. Geol. Geophys.* **20**, 47–58.

Bohannon, R. G., Naeser, C. W., Schmidt, D. L., and Zimmermann, R. A. (1989): The timing of uplift, volcanism and rifting peripheral to the Red Sea: A case for passive rifting? *J. Geophys. Res.* **94**, 1683–1702.

Bookstrom, A. A., Naeser, C. W., and Shannon, J. R. (1987): Isotopic age determinations, unaltered and hydrothermally altered igneous rocks, North-Central Colorado mineral belt, *Isochron/West* **49**, 13–20.

Brandl, W. (1991): Spaltspurdatierung an Vesuvian aus dem Westrand der Böhmischen Masse, Bayerischer Wald im Umfeld der Kontinentalen Tiefbohrung (KTB) und Ausheilexperimente mit Vesuvian-Spaltspuren, Diplomarbeit, Univ. Heidelberg.

Briggs, N. D., Naeser, C. W., and McCulloh, T. H. (1981): Thermal history of sedimentary basins by fission-track dating, *Nucl. Tracks* **5**, 235–237.

Brill, R. H. (1965): Applications of fission-track dating to historic and prehistoric glasses, *Archaeometry* **7**, 51–57.

Brill, R. H., Fleischer, R. L., Price, P. B., and Walker, R. M. (1964): The fission track dating of man-made glasses, *J. Glass Studies* **6**, 151–155.

Brown, R. W., Rust, D. J., Summerfield, M. A., Gleadow, A. J. W., and De Wit, M. C. J. (1990): An early Cretaceous phase of accelerated uplift on the south-western margin of Africa: evidence from apatite fission track anlysis and the off-shore sedimentary record, *Nucl. Tracks Radiat. Meas.* **17**, 339–350.

Bryant, B. and Naeser, C. W. (1980): The significance of fission track-ages of apatite in relation to the tectonic history of the Front and Sawatch Ranges, Colorado, *Geol. Soc. Amer. Bull.* **91**, 156–164.

Bryant, B., Marvin, R. F., Naeser, C. W., and Mehnert, H. H. (1981): Ages of igneous rocks in the South Park-Breckenridge region, Colorado, and their relation to the tectonic history of the Front Range uplift, Geol. Surv. Prof. Paper 1199 C, 15–35.

Burchart, J. (1972): Fission-track age determinations of accessory apatite from the Tatra Mountains, *Earth Planet. Sci. Lett.* **15**, 418–422.

Burchart, J. (1981): Evaluation of uncertainties in fission-track dating: some statistical and geochemical problems, *Nucl. Tracks* **5**, 87–92.

Burchart, J. and Reimer, G. M. (1972): Effect of ionic solution on fission track stability in apatite, *Trans. Amer. Nucl. Soc.* **15**, 129–130.

Burchart, J., Dakowski, M., and Galazka, J. (1975): A technique to determine extremely high fission track densities, *Bull. Acad. Polon. Sci., Ser. Sci. Terr.* **23**, 1–7.

Burchart, J., Butkiewicz, T., Dakowski, M., and Galazka-Friedman, J. (1979): Fission track retention in minerals as a function of heating time during isothermal experiments: a discussion, *Nucl. Tracks* **3**, 109–117.

Calk, L. C. and Naeser, C. W. (1973): The thermal effect of a basalt intrusion on fission tracks in quartz monzonite, *J. Geol.* **81**, 189–198.

Carbonnel, J. P. and Poupeau, G. (1969): Premiers éléments de datation absolue par traces de fission des basaltes de l'Indochine méridionale, *Earth Planet. Sci. Lett.* **6**, 26–30.

Carlson, W. D. (1990): Mechanisms and kinetics of apatite fission-track annealing. *Amer. Mineral.* **75**, 1120–1139.

Carpena, J. (1985): Tectonic interpretation of an inverse gradient of zircon fission-track ages with respect to altitude: alpine thermal history of the Gran Paradiso basement, *Contrib. Mineral. Petrol.* **90**, 74–82.

Carpenter, S. B. (1972): Quantitative applications of the nuclear track technique, *Microscope* **20**, 175–182.

Carpenter, S. B. (1984): Standard reference materials: calibrated glass standards for fission track use, Nat. Bur. Stand. Spec. Publ. 260-92.

Carpenter, S. B. and Reimer, G. M. (1974): Standard reference materials: calibrated glass standards for fission track use, Nat. Bur. Stand. Spec. Publ. 260-49.

Cerveny, P. F., Naeser, N. D., Zeitler, P. K., Naeser, C. W., and Johnson, N. M. (1988): History of the uplift and relief of the Himalaya during the past 18 million years: evidence from fission-track ages of detrital zircons from sandstones of the Siwalik Group, in K. L. Kleinspehn and C. Paola (eds), *New Perspectives in Basin Analysis*, Springer-Verlag, New York, pp. 43–61.

Chadderton, L. T. (1988): On the anatomy of a fission fragment track, *Nucl. Tracks Radiat. Meas.* **15**, 11–29.

Chadderton, L. T. and Torrens, I. McC. (1969): *Fission Damage in Crystals*, Methuen, London.

Chadderton, L. T., Biersack, J. P., and Koul, S. L. (1988): Discontinuous fission tracks in crystalline detectors, *Nucl. Tracks Radiat. Meas.* **15**, 31–40.

Chakranarayan, A. B. and Powar, K. B. (1982): A new etching technique for developing fission tracks in epidote, *Nucl. Tracks* **6**, 193–195.

Clarke, A. C. W. V. and Carter, A. (1987): Handling of counting data for fission track dating, *Nucl. Tracks Radiat. Meas.* **13**, 105–110.

Coates, D. A. and Naeser, C. W. (1984): Map showing fission-track ages of clinker in the Rochelle Hills, Southern Campbell and Weston Counties, Wyoming, U.S. Geol. Surv. Map I-1462, Miscellaneous Investigations Series.

Comer, J. B., Naeser, C. W., and McDowell, F. W. (1980): Fission-track ages of zircon from Jamaican bauxite and Terra Rossa, *Econ Geol.* **75**, 117–121.

Cowan, G. A. and Adler, H. H. (1976): The variability of the natural abundance of ^{235}U, *Geochim. Cosmochim. Acta* **40**, 1487–1490.

Crough, S. Th. (1983): Apatite fission-track dating of erosion in the eastern Andes, Bolivia, *Earth Planet. Sci. Lett.* **64**, 396–397.

Crowley, K. D. (1985): Thermal significance of fission-track length distributions, *Nucl. Tracks* **10**, 311–322.

Crowley, K. D. (1986): Neutron dosimetry in fission-track analysis, *Nucl. Tracks Radiat. Meas.* **11**, 237–243.

Crowley, K. D. and Kuhlman, St. L. (1988): Apatite thermochronometry of Western Canadian Shield: Implications for origin of the Williston Basin, *Geophys. Res. Lett.* **15**, 221–224.

Crowley, K. D., Ahern, J. L., and Naeser, C. W. (1985): Origin and epeirogenic history of the Williston basin: evidence from fission track analysis of apatite, *Geology* **13**, 620–623.

Crowley, K. D., Naeser, C. W., and Babel, C. A. (1986): Tectonic significance of Precambrian apatite fission-track ages from the midcontinent United States, *Earth Planet. Sci. Lett.* **79**, 329–336.

Crozaz, G., Haack U., Hair, M., Maurette, M., Walker, R. J., and Woolum, D. (1970): Nuclear track studies of ancient solar radiations and dynamic lunar surface processes, *Proc. Apollo 11 Lunar Sci. Conf.* 3, Pergamon, New York, pp. 2051–2080.

Cunningham, C. G., Naeser, C. W., and Marvin, R. F. (1977): New ages for intrusive rocks in the Colorado mineral belt, U.S. Geol. Surv. Open-File Report 77-573.

Cunningham, C. G., Ludwig, K. R., Naeser, C. W., Weiland, E. K., Mehnert, H. H., Steven, T. A., and Rasmussen, J. D. (1982): Geochronology of hydrothermal uranium deposits and associated igneous rocks in the Eastern source area of the Mount Belknap volcanics, Marysvale, Utah, *Econ. Geol.* **77**, 453–463.

Cunningham, C. G., Steven, T. A., Campbell, D. L., Naeser, C. W., Pitkin, J. A., and Duval, J. S. (1984): Multiple episodes of igneous activity, mineralization, and alteration in the western Tushar mountains, Utah, in *Igneous Activity and Related Ore Deposits in the Western and Southern Tushar Mountains, Marysvale Volcanic Field, West-Central Utah*, U.S. Geol. Surv. Prof. Paper 1299-A, 1–21.

Dakowski, M. (1978): Length distributions of fission tracks in thick crystals, *Nucl. Track Detection* **2**, 181–189.

Dakowski, M., Burchart, J., and Galazka, J. (1974): Experimental formula for thermal fading of fission tracks in minerals and natural glasses, *Bull. Acad. Polon. Sci., Ser. Sci. Terre* **22**, 11–17.

Damm, G. (1973): Untersuchungen zum Ausheilverhalten von Kernspaltspuren in Gläsern und einige Anwendungen der Spaltspurmethode, Diplomarbeit, Univ. Köln.

Danis, A., Onescu, M., and Sandru, P. (1969): Determination of geologic age by the fission track method (Rumanian), *St. Cerc. Geol. Geofiz. Geogr.*, *Seria Geofiz.* **7**, 181–189.

Dartyge, E., Duraud, J. P., and Langevin, Y. (1978): Thermal annealing of iron tracks in muscovite, labradorite and olivine, *Nucl. Tracks Suppl.* **1** (Proc. 9th Conf. SSNTD München), 395–399.

Dartyge, E., Duraud, J. P., Langevin, Y., and Maurette, M. (1981): New model of nuclear particle tracks in dielectric minerals, *Phys. Rev.* **B23**, 5213–5229.

Davie, I. W. and Durrani, S. A. (1978): Anisotropic track etching in olivine crystals using WN solution, *Nucl. Track Detection* **2**, 199–205.

Debeauvais, M., Maurette, M., Mory, J., and Walker, R. (1964): Registration of fission fragment tracks in several substances and their use in neutron detection, *Int. J. Appl. Rad. Isotopes* **15**, 289–299.

De Carvalho, H. G., Martins, J. C., De Souza, I. O., and Tavares, O. A. P. (1975): Spontaneous emission of heavy ions from uranium, Centro Brasileiro de Pesquisas Fisicas, Rio de Janeiro, Brazil, Preprint presented to the Academia Brasileira de Ciências.

De Carvalho, H. G., Martins, J. B., Medeiros, E. L., and Tavares O. A. P. (1982): Decay constant for the spontaneous-fission process in ^{238}U, *Nucl. Instr. Meth.* **197**, 417–426.

De Corte, F., Van den haute, P., De Wispelaere, A., and Jonckheere, R. (1991): Calibration of the fission-track dating method: Is Cu useful as an absolute thermal neutron fluence monitor? *Chem. Geol. (Isot. Geosci. Sect.)* **86**, 187–194.

Dodson, M. H. (1973): Closure temperature in cooling geochronological and petrological systems, *Contrib. Mineral. Petrol.* **40**, 259–274.

Dokka, R. K. (1984): Fission-track geochronologic evidence for Late Cretaceous mylonitization and Early Palaeocene uplift of the Northeastern Peninsular Ranges, California, *Geophys. Res. Lett.* **11**, 46–49.

Donelick, R. A. (1991): Crystallographic orientation dependence of mean etchable of fission track in apatite: An empirical model and experimental observations, *Amer. Mineral.* **76**, 83–91.

Donelick, R. A., Roden, M. K., Mooers, J. D., Carpenter, B. S., and Miller, D. S. (1990): Etchable length reduction of induced fission tracks in apatite at room temperature ($\approx 23°C$): crystallographic orientation effects ad 'initial' mean lengths, *Nucl. Tracks Radiat. Meas.* **17**, 261–265.

Dorling, G. W., Bull, R. K., Durrani, S. A., Fremlin, J. H., and Khan, H. A. (1974): Anisotropic etching of charged-particle tracks in crystals, *Radiation Effects* **23**, 141–143.

Duddy, I. R., Green, P. F., and Laslett, G. M. (1988): Thermal annealing of fission tracks in apatite, 3. Variable temperature behaviour, *Chem. Geol. (Isot. Geosci. Sect.)* **73**, 25–38.

Duraud, J. P. (1978): Enregistrement des traces d'ions lourds dans les silicates. Application à la détermination de l'abondance des éléments ultra-lourds dans le rayonnement cosmique solaire ancien, Thèse de Doctorat d'Etat, Orsay, Paris, 158 p.

Durrani, I. R. and Bull. R. K. (1987): *Solid State Nuclear Track Detection (Principles, Methods and Applications)*, Pergamon Press, Oxford.

Durrani, S. A. and Hancock, D. A. (1970): Effect of strain on fission-track ages of tektites, *Earth Planet. Sci. Lett.* **8**, 157–162.

Durrani, S. A. and James, K. (1988): The effects of irradiation temperature on the registration and annealing properties of nuclear tracks in crystals, *Nucl. Tracks Radiat. Meas.* **15**, 223–230.

Durrani, S. A. and Khan, H. A. (1970): Annealing of fission tracks in tektites: corrected ages of bediasites, *Earth Planet. Sci. Lett.* **9**, 431–445.

Durrani, S. A. and Khan, H. A. (1971): Ivory Coast microtektites: fission track ages and geomagnetic reversals, *Nature* **232**, 320–323.

Durrani, S. A., Khan, H. A., Taj, M., and Renfrew, C. (1971): Obsidian source identification by fission track analysis, *Nature* **233**, 242–245.

Durrani, S. A., Khan, H. A., Malik, S. R., Aframian, A., Fremlin, J. H., and Tarney, J. (1973): Charged-particle tracks in Apollo 16 lunar glasses and analogous materials, *Proc. 4th Lunar Sci. Conf., Geochim. Cosmochim. Acta, Suppl.* **4**, 2291–2305.

Durrani, S. A., Khazal, K. A. R., Malik, S. R., Fremlin, J. H., and Hendry, G. L. (1975): Thermoluminescence and fission track studies of the Oklo fossil reactor materials. in *The Oklo Phenomenon*. Proc. of Symp. Oklo phenomenon, IAEA-SM-204/6, Vienna, pp. 207–222.

Edwards, J. (1967): A comparison between etched fission track densities on internal and external glass surfaces after neutron irradiation, *Geophys. J. Roy. Astron. Soc.* **13**, 541–543.

Elitzsch, C., Pernicka, E., and Wagner, G. A. (1983): Thermoluminescnce dating of archaeometallurgical slags, *PACT J.* **9**, 271–286.

Emma, V. and Lo Nigro, S. (1975): Decay constant for spontaneous fission of ^{238}U and ^{232}Th, *Nucl. Instr. Meth.* **128**, 355–357.

Emmermann, R. (1986): Das deutsche Kontinentale Tiefbohrprogramm, *Geowiss.i.u. Zeit.* **4**, 19–33.

Endo, K. and Doke, T. (1973): Calibration of plastic nuclear track detectors for identification of heavy charged nuclei using fission fragments, *Nucl. Instr. Meth.* **111**, 29–37.

Faust, J. W., Jr. (1959): Etching of metals and semiconductors. In H. C. Gatos, (ed.), *The Surface Chemistry of Metals and Semiconductors*, Wiley, New York, pp. 151–173.

Fisher, E. D. (1968): Fission track ages of deep sea glasses, *Nature* **221**, 549–550.

Fitzgerald, P. G. and Gleadow, A. J. W. (1988): Fission-track geochronology, tectonics and structure of the Transantarctic Mountains in northern Victoria Land, Antarctica, *Chem. Geol. (Isot. Geosci. Sect.)* **73**, 169–198.

Fitzgerald, P. G. and Gleadow, A. J. W. (1990): New approaches in fission-track geochronology as a tectonic tool: examples from the Transantarctic Mountains, *Nucl. Tracks Radiat. Meas.* **17**, 351–357.

Fitzgerald, P. G., Sandiford, M., Barett, P. J., and Gleadow, A. J. W. (1986): Asymmetric extension associated with uplift and subsidence in the Transantarctic Mountains and Ross Embayment, *Earth Planet. Sci. Lett.* **81**, 67–78.

Fleischer, R. L. (1990): Where are nuclear tracks leading? Directions in track studies, Gen. Electr. Res. Dev. Center Techn. Inf. Ser. 90CRD212.

Fleischer, R. L. and Hart, H. R., Jr. (1972): Fission track dating: techniques and problems. in W. W. Bishop, D. A. Miller and S. Cole (eds), *Proc. Burg Wartenstein Conf. on Calibration of Hominoid Evolution*, Scott. Acad. Press, Edinburgh, pp. 135–170.

Fleischer, R. L. and Naeser, C. W. (1972): Search for ^{244}Pu tracks in Mountain Pass Bastnaesite, *Nature* **240**, 465.

Fleischer, R. L. and Price, P. B. (1963a): Charged particle tracks in glass, *J. Appl. Phys.* **34**, 2903–2904.

Fleischer, R. L. and Price, P. B. (1963b): Tracks of charged particles in high polymers, *Science* **140**, 1221–1222.

Fleischer, R. L. and Price, P. B. (1964a): Decay constant for spontaneous fission of U^{238}, *Phys. Rev.* **133B**, 63–64.

Fleischer, R. L. and Price, P. B. (1964b): Fission track evidence for the simultaneous origin of tektites and other natural glasses, *Geochim. Cosmochim. Acta* **28**, 755–766.

Fleischer, R. L. and Price, P. B. (1964c): Glass dating by fission fragment tracks, *J. Geophys. Res.* **69**, 331–339.

Fleischer, R. L. and Price, P. B. (1964d): Techniques for geological dating of minerals by chemical etching of fission fragment tracks, *Geochim. Cosmochim. Acta* **28**, 1705–1714.

Fleischer, R. L. and Price, P. B. (1967): Ages of impact glasses from the Ashanti and Auoelloul craters: a correction, *Geochim. Cosmochim. Acta* **31**, 2451–2452.

Fleischer, R. L., Price, P. B., and Symes, E. M. (1964a): On the origin of anomalous etch figures in minerals, *Amer. Mineral.* **49**, 794–800.

Fleischer, R. L., Price, P. B., Symes, E. M., and Miller, D. S. (1964b): Fission track ages and track annealing behaviour of some micas, *Science* **143**, 349–351.

Fleischer, R. L., Price, P. B., and Walker, R. M. (1964c): Fission track ages of zircons, *J. Geophys. Res.* **69**, 4885–4888.

Fleischer, R. L., Price, P. B., and Walker, R. M. (1964d): Fossil records of nuclear fission, *New Scientist* **21**, 406–408.

Fleischer, R. L., Price, P. B., Walker, R. M., and Hubbard, E. L. (1964e): Track registration in various solid-state nuclear track detectors, *Phys. Rev.* **133–5A**, 1433–1449.

Fleischer, R. L., Naeser, C. W., Price, P. B., Walker, R. M., and Maurette, M. (1965a): Cosmic ray exposure ages of tektites by the fission-track technique, *J. Geophys. Res.* **70**, 1491–1496.

Fleischer, R. L., Price, P. B., and Walker, R. M. (1965b): Effects of temperature, pressure, and ionization on the formation and stability of fission tracks in minerals and glasses, *J. Geophys. Res.* **70**, 1497–1502.

Fleischer, R. L., Price, P. B., and Walker, R. M. (1965c): On the simultaneous origin of tektites and other natural glasses, *Geochim. Cosmochim. Acta* **29**, 161–166.

Fleischer, R. L., Price, P. B., and Walker, R. M. (1965d): Solid-state track detectors: applications to nuclear science and geophysics, *Ann. Rev. Nuc. Sci.* **15**, 1–28.

Fleischer, R. L., Price, P. B., and Walker, R. M. (1965e): Application of fission tracks and fission track dating to anthropology, General Electr. Report N. 65-R1-3876 M, 1–12.

Fleischer, R. L., Price, P. B., Walker, R. M., and Leakey, L. S. B. (1965f): Fission track dating of bed I, Olduvai Gorge, *Science* **148**, 72–74.

Fleischer, R. L., Price, P. B., Walker, R. M., and Leakey, L. S. B. (1965g): Fission track dating of a mesolithic knife, *Nature* **205**, 1138.

Fleischer, R. L., Price, P. B., and Walker, R. M. (1965h): The ion explosion spike mechanism for formation of charged particle tracks, *J. Appl. Phys.* **36**, 3645–3652.

Fleischer, R. L., Price, P. B., Walker, R. M., and Hubbard, E. L. (1967): Criterion for registration in dielectric track detectors, *Phys Rev.* **156**, 353–355.

Fleischer, R. L., Price, P. B., and Walker, R. M. (1968a): Charged particle tracks: tools for geochronology and meteorite studies. in E. I. Hamilton and R. M. Farquhar, (eds), *Radiometric Dating for Geologists*, Wiley Interscience, London, pp. 417–435.

Fleischer, R. L., Viertl, J. R. M., Price, P. B., and Aumento, F. (1968b): Mid-atlantic ridge: age and spreading rates, *Science* **161**, 1339–1342.

Fleischer, R. L., Price, P. B., Viertl, J. R. M., and Woods, R. T. (1969a): Ages of Darwin glass, Macedon glass, and Far Eastern tektites, *Geochim. Cosmochim. Acta* **33**, 1071–1074.

Fleischer, R. L., Price, P. B., and Walker, R. M. (1969b): Fission track dating and processes in the earth's interior. in S. Runcorn, (ed.), *The Application of Modern Physics to the Earth and Planetary Interiors*, Wiley Interscience, London, Part 6, Chap 1.

Fleischer, R. L., Price, P. B., and Woods, R. T. (1969c): A second tektite fall in Australia, *Earth Planet. Sci. Lett.* **7**, 51–52.

Fleischer, R. L., Viertl, J. R. M., and Price, P. B. (1969d): Age of the Manicouagan and Clearwater Lakes craters, *Geochim. Cosmochim. Acta* **33**, 523–527.

Fleischer, R. L., Viertl, J. R. M., Price, P. B., and Aumento, F. (1971): A chronological test of ocean bottom spreading in the north Atlantic, *Radiation Effects* **11**, 193–194.

Fleischer, R. L., Comstock, G. M., and Hart, H. R., Jr. (1972): Dating of mechanical events by deformation-induced erasure of particle tracks, *J. Geophys. Res.* **77**, 5050–5053.

Fleischer, R. L., Woods, R. T., Hart, H. R., Jr., Price, P. B., and Short, N. M. (1974): Effect of shock on fission track dating of apatite and sphene crystals from the Hardhat and Sedan underground nuclear explosions, *J. Geophys. Res.* **79**, 339–342.

Fleischer, R. L., Price, P. B., and Walker, R. M. (1975): *Nuclear Tracks in Solids; Principles and Applications*, University of California Press, Berkeley.

Flerov, G. N. and Petrzhak, K. A. (1940): Spontaneous fission of uranium, *J. Phys.* **3**, 275–280.

Flisch, M. (1986): Die Hebungsgeschichte der ostalpinen Silvretta-Decke seit der mittleren Kreide, *Bull. Ver. schweiz. Petroleum-Gl. u. Ing.* **53**, 23–49.

Florenski, P. V., Perelygin, V. P., Bazhenov, M. L., Lhagvasuren, D. D., and Stetsenko, S. G. (1979): Complex determination of the age of the meteoritic crater Zhamanshin (Russian), *Astronom. Vestnik* **13/3**, 178–186.

Friedlander, G., Kennedy, J. W., Macias, S. M., and Miller, J. M. (1981): *Nuclear and Radiochemistry*, Wiley, New York.

Galazka, J. and Burchart, J. (1976): Cumulative isothermal heating (CIH) as a method to correct thermally lowered fission track ages. I. Mathematical models, *Bull. Acad. Polon. Sci., Ser. Sci. Terre* **25**, 1–7.

Galbraith, R. F. (1981): On statistical models for fission track counts, *Math. Geol.* **13**, 471–438.

Galbraith, R. F. (1982): Statistical analysis of some fission-track counts and neutron-fluence measurements, *Nucl. Tracks* **6**, 99–107.

Galbraith, R. F. (1984): On statistical estimation in fission track dating, *Math. Geol.* **16**, 653–669.

Galbraith, R. F. (1986a): Allocation of grains in the population method of fission track dating, *Nucl. Tracks Radiat. Meas.* **11**, 201–203.

Galbraith, R. F. (1986b): Statistical analysis of C. W. Naeser's Fish Canyon zircon data, *Nucl. Tracks Radiat. Meas.* **11**, 295–300.

Galbraith, R. F. (1988): Graphical display of estimates having differing standard errors, *Technometrics* **30**, 271–281.

Galbraith, R. F. (1990): The radial plot: graphical assessment of spread in ages, *Nucl. Tracks Radiat. Meas.* **17**, 207–214.

Galbraith, R. F. and Green, P. F. (1990): Estimating the component ages in a finite mixture, *Nucl. Tracks Radiat. Meas.* **17**, 197–206.

Galbraith, R. F. and Laslett, G. M. (1985): Some remarks on statistical estimation in fission track dating, *Nucl. Tracks* **10**, 361–363.

Galliker, D., Hugentobler, E., and Hahn, B. (1970): Spontane Kernspaltung von ^{238}U und ^{241}Am, *Helv. Phys. Acta* **43**, 593–606.

Gatos, H. C. (ed.) (1959): *The Surface Chemistry of Metals and Semiconductors*, Wiley, New York.

Geguzin, Ya. E., Berzina, I. G., and Vorob'eva, I. V. (1968): The kinetics of healing of fission-fragment tracks in thin mica sheets, *Soviet Phys. Dokl.* **12**, 814–816.

Gentner, W. and Wagner, G. A. (1969): Altersbestimmungen an Riesgläsern und Moldaviten, *Geol. Bavarica* **61**, 296–303.

Gentner, W. and Zähringer J. (1960): Das Kalium-Argon-Alter von Tektiten, *Z. Naturforsch.* **15A**, 93–99.

Gentner, W., Lippolt, H. J. and Schaeffer, O. A. (1963): Das Kalium-Argon Alter der Gläser des Nördlinger Rieses und der böhmisch-märischen Tektite, *Geochim. Cosmochim. Acta* **27**, 191–200.

Gentner, W., Kleinmann, B., and Wagner, G. A. (1967): New K-Ar and fission track ages of impact glasses and tektites, *Earth Planet. Sci. Lett.* **2**, 83–86.

Gentner, W., Storzer, D., and Wagner, G. A. (1969a): New fission track ages of tektites and related glasses, *Geochim. Cosmochim. Acta* **33**, 1075–1081.

Gentner, W., Storzer, D., and Wagner, G. A. (1969b): Das Alter von Tektiten und verwandten Gläsern, *Naturwissenschaften* **56**, 255–260.

Gentner, W., Glass, B., Storzer, D., and Wagner, G. A. (1970): Fission track ages and ages of deposition of deep-sea microtektites, *Science* **168**, 359–361.

Gentner, W., Kirsten, T., Storzer, D., and Wagner, G. A. (1973): K-Ar and fission track dating of Darwin Crater glass, *Earth Planet. Sci. Lett.* **20**, 204–210.

Giger, M. (1991): Geochronologische und petrographische Studien an Geröllen und Sedimenten der Gonfolite Lombarda Gruppe (Südschweiz und Norditalien) und ihr Vergleich mit dem alpinem Hinterland, Dissertation, Univ. Bern.

Giger, M. and Hurford, A. J. (1989): Tertiary intrusives of the Central Alps: their Tertiary uplift, erosion, redeposition and burial in the south-alpine foreland, *Eclogae Geol. Helv.* **82/3**, 857–866.

Glass, B. P., Baker, R. N., Storzer, D., and Wagner, G. A. (1973): North American microtektites from the Caribbean Sea and their fission track age, *Earth Planet. Sci. Lett.* **19**, 184–192.

Gleadow, A. J. W. (1978a): Anisotropic and variable track etching characteristics in natural sphenes, *Nucl. Track Detection* **2**, 105–117.

Gleadow, A. J. W. (1978b): Comparison of fission track dating methods: effects of anisotropic etching and accumulated a-damage, *U.S. Geol. Surv. Open-File Report* **78–701**, 143–145.

Gleadow, A. J. W. (1980): Fission track age of the KBS tuff and associated hominid remains in northern Kenya, *Nature* **284**, 225–230.

Gleadow, A. J. W. (1981): Fission track dating methods: What are the real alternatives?, *Nucl. Tracks* **5**, 3–14.

Gleadow, A. J. W. and Brooks, C. K. (1979): Fission track dating, thermal histories and tectonics of igneous intrusions of East Greenland, *Contrib. Mineral. Petrol.* **71**, 45–60.

Gleadow, A. J. W. and Duddy, I. R. (1980): Early Cretaceous volcanism and the early break-up history of southeastern Australia: evidence from fission track dating of volcanoclastic sediments, *5th Int. Gondwana Symposium, Wellington 1980*, pp. 295–300.

Gleadow, A. J. W. and Duddy, I. R. (1981): A natural long-term track annealing experiment for apatite, *Nucl. Tracks* **5**, 169–174.

Gleadow, A. J. W. and Fitzgerald, P. G. (1987): Uplift history and structure of the Transantarctic Mountains: new evidence from fission track dating of basement apatites in the Dry Valleys area, southern Victoria Land, *Earth Planet. Sci. Lett.* **82**, 1–14.

Gleadow, A. J. W. and Lovering, J. F. (1975): Fission track dating methods. Dept. of Geology, School of Earth Sciences, University of Melbourne, Publication No. 3.

Gleadow, A. J. W. and Lovering, J. F. (1977): Geometry factor for external detectors in fission track dating, *Nucl. Track Detection* **1**, 99–106.

Gleadow, A. J. W. and Lovering, J. F. (1978a): Fission track geochronology of King Island, Bass Strait, Australia: relationship to continental rifting, *Earth Planet. Sci. Lett.* **37**, 429–437.

Gleadow, A. J. W. and Lovering, J. F. (1978b): Thermal history of granitic rocks from Western Victoria: A fission-track dating study, *J. Geol. Soc. Austral.* **25**, 323–340.

Gleadow, A. J. W., Hurford, A. J., and Quaife, R. D. (1976): Fission track dating of zircon: improved etching techniques, *Earth Planet Sci. Lett.* **33**, 273–276.

Gleadow, A. J. W., McKelvey, and Ferguson, K. U. (1984): Uplift history of the Transantarctic Mountains in the Dry Valleys area, Antarctica, from apatite fission track ages, *NZ J. Geol. Geophys.* **27**, 457–464.

Gleadow, A. J. W., Duddy, I. R., and Lovering, J. F. (1983): Fission track analysis: a new tool for the evaluation of thermal histories and hydrocarbon potential, *Austral. Petrol. Expl. Ass. J.* **23**, 93–102.

Gleadow, A. J. W., Duddy, I. R., Green, P. F., and Lovering, J. F. (1986): Confined fission track lengths in apatite: a diagnostic tool for thermal history analysis, *Contrib. Mineral. Petrol.* **94**, 405–415.

Gold, R. A., Armani, R. J., and Roberts, J. H. (1968): Absolute fission rate measurements with solid-state track recorders, *Nucl. Sci. Eng.* **34**, 13–32.

Gold, R., Roberts, J. H., Preston, C. C., McNeece, J. P., and Ruddy, F. H. (1984): The status of automated nuclear scanning systems, *Nucl. Tracks Radiat. Meas.* **8**, 187–197.

Green, P. F. (1980): On the cause of the shortening of spontaneous fission tracks in certain minerals, *Nucl. Tracks* **4**, 91–100.

Green, P. F. (1981a): A new look at statistics in the fission track dating, *Nucl. Tracks* **5**, 77–86.

Green, P. F. (1981b): 'Track-in-track' length measurements in annealed apatites, *Nucl. Tracks* **5**, 121–128.

Green, P. F. (1985): Comparison of zeta calibration baselines for fission-track dating of apatite, zircon and sphene, *Chem. Geol. (Isot. Geosc. Sect.)* **58**, 1–22.

Green, P. F. (1986): On the thermo-tectonic evolution of Northern England: evidence from fission track analysis, *Geol. Mag.* **123**, 493–506.

Green, P. F. (1988): The relationship between track shortening and fission track age reduction in apatite: combined influences of inherent instability, annealing anisotropy, length bias and system calibration, *Earth Planet. Sci. Lett.* **89**, 335–352.

Green, P. F. (1989): Thermal and tectonic history of the East Midlands shelf (onshore UK) and surrounding regions assessed by apatite fission track analysis, *J. Geol. Soc. London* **146**, 755–773.

Green, P. F. and Duddy, I. R. (1989): Some comments on palaeotemperature estimation from apatite fission track analysis, *J. Petrol. Geol.* **12**, 111–114.

Green, P. F. and Durrani, S. A. (1977): Annealing studies of tracks in crystals, *Nucl. Track Detection* **1**, 33–39

Green, P. F. and Durrani, S. A. (1978): A quantitative assessment of geometry factors for use in fission track studies, *Nucl. Track Detection* **2**, 207–213.

Green, P. F. and Hurford, A. J. (1984): Thermal neutron dosimetry for fission track dating, *Nucl. Tracks* **9**, 231–241.

Green, P. F., Bull, R. K., and Durrani, S. A. (1978): Particle identification from track etch-rates in minerals, *Nucl. Instr. Meth.* **157**, 185–193.

Green, P. F., Duddy, I. R., Gleadow, A. J. W., and Lovering, J. F. (1989a): Apatite fission-track analysis as a paleotemperature indicator for hydrocarbon exploration, in N. D. Naeser and Th. H. McCulloh (eds), *Thermal History of Sedimentary Basins*, Springer-Verlag, New York, pp. 181–195.

Green, P. F., Duddy, I. R., Gleadow, A. J. W., Tingate, P. R., and Laslett, G. M. (1985): Fission track annealing in apatite: track length measurements and the form of the Arrhenius plot, *Nucl. Tracks* **10**, 323–328.

Green, P. F., Duddy, I. R., Gleadow, A. J. W., Tingate, P. R., and Laslett, G. M. (1986): Thermal annealing of fission tracks in apatite: 1. A qualitative description, *Chem. Geol. (Isot. Geosci. Sect.)* **59**, 237–253.

Green, P. F., Duddy, I. R., and Laslett, G. M. (1988): Can fission track annealing in apatite be described by first-order kinetics?, *Earth Planet. Sci. Lett.* **87**, 216–228.

Green, P. F., Duddy, I. R., Laslett, G. M., Hegarty, K. A., Gleadow, A. J. W., and Lovering, J. F. (1989b): Thermal annealing of fission tracks in apatite: 4. Quantitative modelling techniques and extension to geological timescales, *Chem. Geol. (Isot. Geosci. Sect.)* **79**, 155–182.

Groeneveld, K.-O. E. (1988): Nuclear track formation related electron production and transport from ion penetration through solids, *Nucl. Tracks Radiat. Meas.* **15**, 51–60.

Grundmann, G. and Morteani, G. (1985): The young uplift and thermal history of the Central Eastern Alps (Austria/Italy). Evidence from apatite fission track ages, *Jb. Geol. B.-A.* **128**, 197–216.

Guo Shilun, Zhou Shuhua, Meng Wu, Zhang Pengfa, Hao Xiuhong, Liu Shunsheng, Zhang Feng, Hu Ruiying, and Liu Jinfa (1980): Fission track dating of Peking man, *Kexue Tongba* **25**, 770–772.

Guo Shilun, Liu Shunsheng, Sun Shengfen, Zhang Feng, Zhou Shuhua, Hao Xiuhong, Hu Ruiying, Meng Wu, Zhang Pengfa, and Liu Jinfa (1990): Age and duration of Peking man site by fission track method, Abstr. 15th Internat. Conf. on Particle Tracks in Solids, Marburg.

Gupta, M. L., Nagpaul, K. K., and Mehta, P. P. (1971a): Fission track ages of some Indian muscovites, *Canad. J. Earth Sci.* **8**, 1491–1495.

Gupta, M. L., Mehta, P. P., and Nagpaul, K. K. (1971b): Fission track ages of some Indian biotites, *Indian J. Pure Appl. Phys.* **9**, 466–469.

Haack, U. (1970): Fission track age of stibiotantalite from Alto Ligonha, Mozambique, *Contrib. Mineral. Petrol.* **29**, 183–185.

Haack, U. (1972): Systematics in the fisson track annealing of minerals, *Contrib. Mineral. Petrol.* **35**, 303–312.

Haack, U. (1973): Suche nach überschweren Transuranelementen, *Naturwissenschaften* **60**, 65–70.

Haack, U. (1976a): Experiences with dating garnet, epidote, vesuvianite (idocrase) and apatite by fission tracks, *N. Jb. Min. Abh.* **127**, 143–155.

Haack, U. (1976b): Rekonstruktion der Abkühlungsgeschichte des Damara-Orogens in Südwest-Afrika mit Hilfe von Spaltspur-Altern, *Geol. Rundschau* **65**, 967–1002.

Haack, U. (1976c): The stability of fission tracks in epidote and vesuvianite, *Earth Planet. Sci. Lett.* **30**, 129–134.

Haack, U. (1977a): Fission track dating of garnet and vesuvianite (idocrase). *N. Jb. Min. Abh.* **129**, 160–170.

Haack, U. (1977b): The closing temperature for fission track retention in minerals, *Amer. J. Sci.* **277**, 459–464.

Haack, U. (1983): Reconstruction of the cooling history of the Damara orogen by correlation of radiometric ages with geography and altitude, in H. Martin and F. W. Eder (eds), *International Fold Beds*, Springer-Verlag, Heidelberg, pp. 873–884.

Haack, U. K. and Gramse, M. (1972): Survey of garnets for fossil fission tracks, *Contrib. Mineral. Petrol.* **34**, 258–260.

Haack, U. K. and Potts, M. J. (1972): Fission track annealing in garnet, *Contrib. Mineral. Petrol.* **34**, 343–345.

Hadler, J. C., Lattes, C. M. G., Marques, A., Marques, M. D. D., Serra, D. A. B., and Bigazzi, G. (1981): Measurement of the spontaneous-fission disintegration constant of ^{238}U, *Nucl. Tracks* **5**, 45–52.

Haggerty, S. E., Raber, E., and Naeser, C. W. (1983): Fission tack dating of kimberlitic zircons, *Planet. Sci. Lett.* **63**, 41–50.

Hahn, O. and Strassmann, F. (1939): Über den Nachweis und das Verhalten der bei Bestrahlung des Urans mittels Neutronen enstehenden Erdalkalimetalle, *Naturwissenschaften* **27**, 11–15.

Hall, C. M., Walter, R. C., Westgate, J. A., and York, D. (1984): Geochronology, stratigraphy and geochemistry of Cindery Tuff in Pliocene hominid-bearing sediments of the Middle Awash, Ethiopia, *Nature* **308**, 26–31.

Hammerschmidt, K., Wagner, G. A., and Wagner, M. (1984): Radiometric dating on research drill core Urach III: a contribution to its geothermal history, *J. Geophys.* **54**, 97–105.

Hanna, G. C., Westcott, C. H., Lemmel, H. D., Leonard, B. R., Story, J. S., and Attree, P. M. (1969): Revision of values for the 2200 m/s neutron constants for four fissile nuclides, *At. Energ. Rev.* **7/4**, 3–92.

Harrison, T. M. (1985): A reassessment of fission-track annealing behavior in apatite, *Nucl. Tracks* **10**, 329–333.

Harrison, T. M., Armstrong, R. L., Naeser, C. W., and Harakal, J. E. (1979): Geochronology and thermal history of the Coast plutonic complex, near Prince Rupert, British Columbia, *Canad. J. Earth Sci.* **16**, 400–410.

Hartung, J. B., Izett, G. A., Naeser. C. W., Kunk, M. J., and Sutter, J. F. (1986): The Manson, Iowa, impact structure and the Cretaceous-Tertiary boundary event, *Lunar Planet. Sci.* **17**, 313–314.

Hashemi-Nezhad, S. R. and Durrani, S. A. (1981a): Correction of thermally-lowered fission-track ages of minerals and glasses: the importance of the prolonged etching factor, *Nucl. Tracks* 5, 101–111.

Hashemi-Nezhad, S. R. and Durrani, S. A. (1981b): Registration of alpha recoil tracks in mica: the prospects for alpha-recoil dating method, *Nucl. Tracks* 5, 189–205.

Hashemi-Nezhad, S. R. and Durrani, S. A. (1986): Determination of fission track age of uranium-bearing veins in a biotite sample from Bancroft, Canada, *Nucl. Tracks Radiat. Meas.* 12, 905–908.

Hejl, E. and Grundmann, G. (1989): Apatit-Spaltspurdaten zur thermischen Geschichte der Nördlichen Kalkalpen, der Flysch- und Molassezone, *Jb. Geol. B.-A.* 132, 191–212.

Hejl, E. and Wagner, G. A. (1990): Geothermische und tektonische Interpretation von Spaltspurdaten am Beispiel der Kontinentalen Tiefbohrung in der Oberpfalz, *Naturwissenschaften* 77, 201–213.

Hejl, E. and Wagner, G. A. (1991): Spaltspuren in Apatit und Zirkon: Schlüssel zur Niedertemperatur- und Hebungsgeschichte der Alpen, *Schweiz. Mineral. Petrogr; Mitteil.* 71, 63–71.

Henke, R. P. and Benton, E. V. (1971): On geometry of tracks in dielectric nuclear track detectors, *Nucl. Instr. Meth.* 97, 483–489.

Holden, N. E. (1981): The uranium half-lives: a critical review, Brookhaven, Nat. Lab. Inf. Rep. BNL-NCS-51320.

Holden, N. E. (1989): Total and spontaneous fission half-lives for uranium, plutonium, americium and curium nuclides, *Pure Appl. Chem.* 61.

Holden, N. E. and Holden, K. A. (1989): Re-examination of 2200 meter/second cross section experiments for neutron capture and fission standards, *Pure Appl. Chem.* 61, 1505–1510.

Holmes, P. J. (1962): Practical applications of chemical etching, in P. J. Holmes (ed.), *The electrochemistry of semiconductors*, Academic Press, London, pp. 329–377.

Huang, W. H. and Walker, R. M. (1967): Fossil a-particle recoil tracks: a new method of age determination, *Science* 5, 1103–1106.

Hurford, A. J. (1974): Fission track dating of a vitric tuff from East Rudolf, North Kenya, *Nature* 249, 236–237.

Hurford, A. J. (1977a): A preliminary fission track dating survey of Caledonian 'newer and last granites' from the Highlands of Scotland, *Scot. J. Geol.* 3, 271–284.

Hurford, A. J. (1977b): Fission track dates from two Galloway granites, Scotland, *Geol. Mag.* 114, 299–304.

Hurford, A. J. (1986a): Standardization of fission track dating calibration: results of questionnaire distributed by International Union of Geological Sciences Subcommission on Geochronology, *Nucl. Tracks Radiat. Meas.* 11, 329–333.

Hurford, A. J. (1986b): Cooling and uplift patterns in the Lepontine Alps South Central Switzerland and age of vertical movement on the Insubric fault line, *Contrib. Mineral. Petrol.* 92, 413–427.

Hurford, A. J. (1990): Standardization of fission track dating calibration: Recommendation by the Fission Track Working Group of the I.U.G.S. Subcommission on Geochronology, *Chem. Geol.* (*Isot. Geosci. Sect.*) 80, 171–178.

Hurford, A. J. and Carter, A. (1991): The role of fission-track dating in discrimination of provenance, *Geol. Soc. Spec. Publ.* 57, 67–78.

Hurford, A. J. and Gleadow, A. J. W. (1977): Calibration of fission track dating parameters, *Nucl. Track Detection* 1, 41–48.

Hurford, A. J. and Green, P. F. (1981a): A reappraisal of neutron dosimetry and uranium-238 λ_f values in fission track dating, *Nucl. Tracks* 5, 53–61.

Hurford, A. J. and Green, P. F. (1981b): Standards, dosimetry and the uranium-238 λ_f decay constant: a discussion, *Nucl. Tracks* 5, 73–75.

Hurford, A. J. and Green, P. F. (1982): A users' guide to fission track dating calibration, *Earth Planet. Sci. Lett.* 59, 343–354.

Hurford, A. J. and Green, P. F. (1983): The zeta age calibration of fission track dating, *Isotope Geosci.* 1, 285–317.

Hurford, A. J. and Hammerschmidt, K. (1985): $^{40}Ar/^{39}Ar$ and K/Ar dating of the Bishop and Fish Canyon tuffs: calibration ages for fission-track dating standards, *Chem. Geol.* 58, 23–32.

Hurford, A. J. and Hunziker, J. C. (1985): Alpine cooling history of the Monte Mucrone eclogites (Sesia-Lanzo Zone): fission track evidence, *Schweiz. Mineral. Petrogr. Mitt.* 65, 325–334.

Hurford, A. J. and Hunziker, J. C. (1989): A revised thermal history for the Gran Paradiso massif, *Schweiz. Mineral. Petrogr. Mitt.* 69, 319–329.

Hurford, A. J. and Watkins, R. T. (1987): Fission-track age of the tuffs of the Buluk Member, Bakate formation, Northern Kenya: a suitable fission-track age standard, *Chem. Geol. (Isot. Geosci. Sect.)* **66**, 209–216.

Hurford, A. J., Fitch, F. J., and Clarke, A. (1984): Resolution of the age structure of the detrital zircon populations of two Lower Cretaceous sandstones from the Weald of England by fission track dating, *Geol. Mag.* **121**, 269–277.

Hurford, A. J., Hammerschmidt, K., Green P. F., Carter, A., and Lux D.: Age of the Tardree rhyolite, Co. Antrim: evidence for the timing of lower Tertiary magmatism in northern Ireland (unpublished manuscript)

Hurford, A. J., Flisch, M., and Jäger, E. (1989): Unravelling the thermo-tectonic evolution of the Alps: a contribution from fission track analysis and mica dating, in M. P. Coward, D. Dietrich, and R. G. Parks (eds.), *Alpine Tectonics*, Geol. Soc. Spec. Publ. 45, pp. 369–398.

Hurford, A. J., Hunziker, J. C. and Stöckert, B. (1991): Constraints on the late thermotectonic evolution of the Western Alps: Evidence for episodic rapid uplift, *Tectonics* **10**, 758–769.

Irving, B. A. (1962): Chemical etching of semiconductors, in P. J. Holmes (ed.), *The Electrochemistry of Semiconductors*, Academic Press, London, pp. 257–287.

Ishimori, T., Ueno, K., Kimura, K., Akatsu, E., Kobayashi, Y., Akatsu, J., Ono, R., and Hoshi, M. (1967): The spontaneous fission of uranium-238, *Radiochim. Acta* **7**, 95–103.

Ito, H., Sorkhabi, R. B., Tagami, T., and Nishimura, S. (1989): Tectonic history of granitic bodies in the south Fossa Magna region, Central Japan: new evidence from fission-track analysis of zircon, *Tectonophysics* **166**, 331–344.

Ivanov, K. N. and Petrzhak, K. A. (1974): Probability of fission by 1.33 MeV-rays and spontaneous fission half-life for ^{238}U, *Soviet Atom Energy* **36**, 514.

Ivanov, M. P., Ter-Akopian, G. M., Fefilov, B. V., and Voronin, A. S. (1985): Study of ^{238}U spontaneous fission using a double ionizaion chamber, *Nucl. Instr. Meth.* **A234**, 152–157.

Izett, G. A. and Naeser, C. W. (1976): Age of Bishop Tuff of eastern California as determined by the fission track method, *Geology* **4**, 577–590.

Izett, G. A. and Naeser, C. W. (1986): Fission-track ages of air-fall tuffs in Miocene sedimentary rocks of the Española Basin, Santa Fe County, New Mexico, *U.S. Geol. Surv. Bull.* **1622**, 25–29.

Izett, G. A., Obradovich, J. D., Naeser, C. W., and Cebula, G. T. (1981): Potassium-argon and fission-track zircon ages of Cerro Toledo rhyolite tephra in the Jemez Mountains, New Mexico, Geol. Surv. Prof. Paper 1199-D, pp. 37–43.

Jacobs, J. (1991): Structural evolution and cooling history of the Heimefrontfjella Mountains (western Dronning Maud Land/Antarctica), *Ber. Polarforsch.* **97**, 1–141.

Jaffey, A. H., Flynn, K. F., Glendenin L. E., Bentley, W. C., and Essling, A. M. (1971): Precision measurements of half-lives and specific activities of ^{235}U and ^{238}U, *Phys. Rev.* **C4**, 1889–1906.

Jakupi, B., Kostica, A., Antanasijevic, R., Jovanovic, L., Todorovic, Z., Perelygin, V. P., and Stetsenko, S. G. (1982): On the geological age determination of auripigments from Alshara (Macedonia) by fission fragment track method, *Bull. Mus. d'Histoire Naturelle Belgrade Serie A* **37**, 135–143.

Jakupi, B., Bytyci, M., Todorovic, Z., Antanasijevic, R., Perelygin, V. P., and Stetsenko, S. G. (1990): Tracks of heavy charged paticles in some natural minerals, *Nucl. Tracks Radiat. Meas.* **17**, 55–57

James, K. and Durrani, S. A. (1988): The registration-temperature dependence of heavy-ion track-etch rates and annealing sensitivity in crystals: implications for cosmic ray identification and fission track dating of meteorites, *Earth Planet. Sci. Lett.* **87**, 229–236.

Johnsson, M. J. (1985): Late Paleozoic-Middle Mesozoic uplift rate, cooling rate and geothermal gradient for south-central New York State, *Nucl. Tracks* **10**, 295–301.

Johnson, N. M., McGee, V. E., and Naeser, C. W. (1979): A practical method of estimating standard error of age in the fission track dating method, *Nucl. Tracks* **3**, 93–99.

Kamp, P. J. J., Green, P. F., and White, S. H. (1989): Fission track analysis reveals character of collisional tectonics in New Zealand, *Tectonics* **8**, 169–195.

Kaneoka, I. and Suzuki, I. (1970): K-Ar and fission track ages of some obsidians from Japan, *J. Geol. Soc. Japan* **76**, 309–313.

Kase, M., Kikuchi, J., and Doke, T. (1978): Half-life of ^{238}U spontaneous fission and its fragment kinetics, *Nucl. Instr. Meth.* **154**, 335–342.

Kashkarov, L. L., Ganaeva, L. I., and Lavrukhina, A. K. (1985): Fission-track ages of Vietnam tektites, *Meteoritics* **20**, 679–680.

Kashukejev, N., Taneva, T., Ignatova, R., and Neova, I. (1970): Determination of geological age of some Bulgarian mica by method of the U fission fragment tracks (Russian), *Compt. Rendue Acad. Bulg. Sci.* **23**, 559–562.

Kasuya, M. (1987): Comparative study of Miocene fission-track chronology and magneto-biochronology, *Sci. Rep. Tohoku Universit, Sendai, 2nd Ser.* **5**, 93–106.

Kasuya, M. and Naeser, C. W. (1988): The effect of a-damage on fission track annealing in zircon, *Nucl. Tracks Radiat. Meas.* **14**, 477–480.

Katz, R. and Kobetich, E. J. (1968): Formation of etchable tracks in dielectrics, *Phys. Rev.* **170**, 401–405.

Katz, R. and Kobetich, E. J. (1970): Formation of particle tracks, *Radiation Effects* **3**, 169–174.

Khan, H. A. (1973a): An important precaution in the etching of solid state nuclear track detectors, *Nucl. Instr. Meth.* **109**, 515–519.

Khan, H. A. (1973b): On thermal annealing of latent damage trails, *Radiation Effects* **18**, 51–54.

Khan, H. A. (1977): Etching characteristics of muscovite mica and feldspar crystalline track detectors, *Radiation Effects* **32**, 49–53.

Khan, H. A. (1980a): Track registration and development efficiencies of solid state nuclear track detectors, *Nucl. Instr. Meth.* **173**, 43–54.

Khan, H. A. (1980b): Some important factors affecting the track registration characteristics of solid state nuclear track detectors, *Nucl. Instr. Meth.* **173**, 55–62.

Khan, H. A. and Ahmad, I. (1975): The role of pre-irradiation annealing in changing the track development characteristics of glass track detectors, *Nucl. Instr. Meth.* **131**, 89–92.

Khan, H. A. and Ahmad, I. (1976): Anisotropy in the track development properties of various crystallographic planes of natural quartz crystals., *Radiation Effects* **30**, 159–165.

Khan, H. A. and Durrani, S. A. (1972a): Efficiency calibration of solid state nuclear track detectors *Nucl. Instr. Meth.* **98**, 229–236.

Khan, H. A. and Durrani, S. A. (1972b): Prolonged etching factor in solid state track detection and its applications, *Radiation Effects* **13**, 257–266.

Khan, H. A. and Durrani, S. A. (1973a): Fission fragment energy spectrum determination with solid state nuclear track detectors, *Nucl. Instr. Meth.* **109**, 341–348.

Khan, H. A. and Durrani, S. A. (1973b): Measurement of spontaneous-fission decay constant of ^{238}U with a mica solid state track detector, *Radiation Effects* **17**, 133–135.

Kleeman, J. D. and Lovering, J. F. (1971): A determination of the decay constant for spontaneous-fission of natural uranium using fission track accumulation, *Geochim. Cosmochim. Acta* **35**, 637–640.

Koch, G. S. Jr. and Link, R. F. (1970): *Statistical Analysis of Geological Data.* Vol. 1, Wiley, New York.

Kohn, B. P. (1979): Identification and significance of a Late Pleistocene Tephra in Canterbury District, South Island, New Zealand, *Quaternary Research* **11**, 78–92.

Kohn, B. P. and Eyal, M. (1981): History of uplift of the crystalline basement of Sinai and its relation to opening of the Red Sea as revealed by fission track dating of apatites, *Earth Planet. Sci. Lett.* **52**, 129–141.

Kohn, B. P., Shagam, R., Banks, P. O., and Burkley, E. A. (1984): Mesozoic-Pleistocene fission track ages on rocks of the Venezuelan Andes and their tectonic implications, *Mem. Geol. Soc. Amer.* **162**. 365–384.

Komarov, A. N. and Raichlin, A. I. (1976): Comparison of fission-track and potassium-argon dating on impactites, *Akad. Nauk. SSSR (Geology)* **228**, 35–38.

Komarov, A. N., Skovorodkin, N. W., and Karapetyan, S. G. (1972): Determination of the age of natural glasses according to tracks of uranium fission fragments (Russian) *Geochimiya* **6**, 693–698.

Komarov, A. N., Schitkov, A. S., Ilupin, I. P., and Skovorodkin, N. V. (1973): Determination of the age of kimberlites from Yakutia with the zircon fission-track method (Russian), *Geolog. Rudn. Mestoroschd.* **15**(4), 75–79.

Koshimizu S. (1981): Fission track ages of pyroclastic flows in the Pliocene Ashoro Formation and the Plio-Pleistocene Ikeda formation developed in eastern Hokkaido Japan, *J. Fac. Sci. Hokkaido Univ. Ser. IV* **19**, 505–518.

Koshimizu, S. and Kim, C. W. (1987): Fission track dating of the Cenozoic formations in Central-Eastern Hokkaido, Japan, *J. Geol. Soc. Japan* **95**, 217–227.

Koul, S. L. (1979): On the fission track dating and annealing behaviour of accessory minerals of eastern Ghats (Andhra Pradesh, India), *Radiation Effects* **40**, 187–192.

Koul, S. L., Chadderton, L. T., and Brooks, C. K. (1983): East Greenland and the Faeroe Islands: a fission track study, *Kong. Danske Vidensk. Selskab Mat.-fys. Med.* **40**, 13.

Koul, S. L., Chadderton, L. T., and Sutherland, F. L. (1986): On the annealing behaviour of fission tracks in zeolites, *Nucl. Tracks Radiat. Meas.* **11**, 85–92.

Krishnaswami, S., Lal, D., Prabhu, N. R., and Tamhane, A. S. (1971): Olivines: revelation of tracks of charged particles, *Science* **174**, 287–291.

Krishnaswami, S., Lal, D., Prabhu, N., and MacDougall, D. (1974): Characteristics of fission tracks in zircon: applications to geochronology and cosmology, *Earth Planet. Sci. Lett.* **22**, 51–59.

Lakatos, S. and Miller, D. S. (1970): Water-pressure effect in fission track-annealing in an alpine muscovite, *Earth Planet. Sci. Lett.* **9**, 77–81.

Lakatos, S. and Miller, D. S. (1972): Evidence for the effect of water content on fission-track annealing in volcanic glass, *Earth Planet. Sci. Lett.* **14**, 128–130.

Lakatos, S. and Miller, D. S. (1983): Fission track analysis of apatite and zircon defines a burial depth of 4 to 7 km for lowermost upper Devonian, Catshill Mountain, New York, *Geology* **11**, 103–104.

Lal, D., Muralli, A. V., Rajan, R. S., Tamhane, A. S., Lorin, J. C., and Pellas, P. (1968): Techniques for proper revelation and viewing of etch-tracks in meteoritic and terrestrial minerals, *Earth Planet. Sci. Lett.* **5**, 111–119.

Lal N. and Waraich R. S. (1983): Comments on the paper 'A new etching technique for developing fission tracks in epidote', *Nucl. Tracks* **7**, 191.

Lambert, M., Levelut, A. M., Maurette M., and Heckman, H. (1970): Etude par diffusion des rayons X au petits angles de micamuscovite irradié par des ions d'argon, *Radiation Effects* **3**, 155–160.

Langevin, Y. and Duraud, J. P. (1982): Thermal stability and chemical reactivity of defects produced in silicates by high energy heavy ions, *Radiation Effects* **65**, 139–142.

Laslett, G. M., Kendall, W. S., Gleadow, A. J. W., and Duddy, I. R. (1982): Bias in measurement of fission-track length distributions, *Nucl. Tracks* **6**, 79–85.

Laslett, G. M., Gleadow, A. J. W., and Duddy, I. R. (1984): The relationship between fission track length and track density in apatite, *Nucl. Tracks* **9**, 29–38.

Laslett, G. M., Green P. F., Duddy, I. R., and Gleadow, A. J. W. (1987): Thermal annealing of fission tracks in apatite: 2. A quantitative analysis, *Chem. Geol.* (*Isot. Geosci. Sect.*) **65**, 1–13.

Leach, B. F., Naeser, C. W., and Ward, G. K. (1981): The ages of Pacific obsidians from fission track analysis, *NZ J. Archaeol.* **3**, 71–82.

Lehmann Chr. (1977): *Interaction of Radiation with Solids and Elementary Defect Production*, North-Holland, Amsterdam.

Lehtovaara, J. (1976): Apatite fission track dating of Finnish precambrian intrusives, *Ann. Acad. Sci. Fenn. Ser. A III* **117**.

Leme, M. P. T., Renner, C., and Cattani, M. (1971): Determination of the decay constant for spontaneous fission of ^{238}U, *Nucl. Instr. Meth.* **91**, 577–579.

Lewis, C. L. E. (1990): Thermal history of the Kunlun batholith, N. Tibet, and implications for uplift of the Tibetan plateau, *Nucl. Tracks Radiat. Meas.* **17**, 301–307.

Lindsey, D. A., Naeser, C. W. and Shawe, D. R. (1975): Age of volcanism, intrusion, and mineralization in the Thomas Range, Keg Mountain, and Desert Mountain, Western Utah, *J. Res. U.S. Geol. Surv.* **3**, 597–604.

Lindsey, D. A., Glanzman, R. K., Naeser, C. W., and Nichols, D. J. (1980): Upper Oligocene Evaporites in Basin Fill of Sevier Desert Region, western Utah, *Amer. Assoc. Petrol. Geol.* **65**, 251–260.

Lipman, P. W., Fisher, F. S., Mehnert, H. M., Naeser, C. W., Luedke, R. G., and Steven, T. A. (1976): Multiple ages of mid-Tertiary mineralization and alteration in the Western San Juan Mountains, Colorado, *Econ. Geol.* **71**, 571–588.

Lipman, P. W., Mehnert, H. H. and Naeser, C. W. (1986): Evolution of the Latir volcanic field, northern New Mexico, and its relation to the Rio Grande rift, as indicated by Potassium-Argon and fission track dating, *J. Geophys. Res.* **91**, 6329–6345.

Lorenz, I. (1984): Mineralogische Methoden zur Untersuchung von Edelsteinen und Perlen, Diplomarbeit, Univ. Heidelberg.

Lovell, L. C. (1958): Dislocation etch pits in apatite, *Acta Metall.* **6**, 775–778.

Ludwig, K. R., Nash, J. T., and Naeser, C. W. (1981): U-Pb isotope systematics and age of uranium mineralization Midnite Mine, Washington, *Econ. Geol.* **76**, 89–110.

Luyendyk, B. P. and Fisher, D. E. (1969): A new point on the sea-floor spreading curve: fission track age of magnetic anomaly 10, *Science* **164**, 1516–1517.

Macdougall, D. (1971a): Fission track dating of volcanic glass shards in marine sediments, *Earth Planet. Sci. Lett.* **10**, 403–406.

Macdougall, D. (1971b): Deep sea drilling: age and composition of an atlantic basaltic intrusion, *Science* **171**, 1244–1245.

Macdougall, D. (1976): Fission track annealing and correction procedures for oceanic basalt glasses, *Earth Planet. Sci. Lett.* **30**, 19–26

Macdougall, D. and Price, P. B. (1974): Attempt to date early South African hominids by using fission tracks in calcite, *Science* **185**, 943–944.

McDougall, I. and Watkins, R. T. (1985): Age of hominoid-bearing sequence at Buluk, northern Kenya, *Nature* **318**, 175–178.

McDowell, F. W. and Keizer, R. P. (1977): Timing of mid-Tertiary volcanism in the Sierra Madre Occidental between Durango City and Mazatlan, Mexico, *Geol. Soc. Am. Bull.* **88**, 1479–1487.

MacFadden, B. J., Johnson, N. M., and Opdyke, N. D. (1979): Magnetic polarity stratigraphy of the Mio-Pliocene mammal-bearing big sandy formation of western Arizona, *Earth Planet. Sci. Lett.* **44**, 349–364.

MacFadden, B. J., Campbell, K. E., Jr., Cifelli, R. L., Siles, O., Johnson, N. M., Naeser, C. W., and Zeitler, P. K. (1985): Magnetic polarity stratigraphy and Mammalian fauna of the Deseadan (Late Oligocene – Early Miocene) Salla beds of northern Bolivia, *J. Geol.* **93**, 228–250.

McGee, V. E., Johnson, N. M., and Naeser, C. W. (1985): Simulated fissioning of uranium and testing of the fission-track dating method, *Nucl. Tracks* **10**, 365–379.

McGoldrick, R. J. and Gleadow, A. J. W. (1977): Fission track dating of lower Palaeozoic sandstones at Tatong, North Central Victoria, *J. Geol. Soc. Austral.* **24**, 461–464.

McLane, V., Dunford, C. L., and Rose P. F. (1988): *Neutron Cross Sections, Vol. 2: Neutron Cross Section Curves*, Academic Press, San Diego.

Märk, E., Pahl, M., and Märk, T. D. (1972): Fission track annealing behaviour of apatite, *Trans. Amer. Nucl. Soc.* **15**, 126–127.

Märk, E., Pahl, M., Purtscheller, F., and Märk, T. D. (1973): Thermische Ausheilung von Uran-Spaltspuren in Apatiten, Alterskorrekturen und Beiträge zur Geothermochronologie, *Tschermaks Min. Petr. Mitt.* **20**, 131–154.

Märk, T. D., Märk, E., Bertel, E., and Pahl, M. (1977): Fission track measurements of the decay constant for spontaneous fission of ^{238}U, *Ber. Nat.-Med. Ver. Innsbruck* **64**, 7–11.

Märk, T. D., Vartanian, R., Purtscheller, F., and Pahl, M. (1981): Fission track annealing and application to the dating of Austrian sphene, *Acta Phys. Austriaca* **53**, 45–59.

Manley, K. and Naeser, C. W. (1977): Fission-track ages for tephra layers in Upper Cenozoic rocks, Española basin, New Mexico, *Isochron/West* **18**, 13–14.

Marvin, R. F., Hearn, B. C., Jr., Mehnert, H. H., Naeser, C. W., Zartman, R. E., and Lindsey, D. A. (1980): Late Cretaceous-Paleocene-Eocene igneous activity in North-Central Montana, *Isochron/West* **29**, 5–25.

Maurette, M. (1966): Etude des traces d'ions lourds dans les minréaux naturels d'origine terrestre et extra-terrestre, *Bull. Soc. Franç. Minér Crist.* **89**, 41–75.

Maurette, M. (1970a): Track formation mechanisms in minerals, *Radiation Effects* **3**, 149–154.

Maurette, M. (1970b): On some annealing characteristics of heavy ion tracks in silicate minerals, *Radiation Effects* **5**, 15–19.

Maurette, M., Pellas, P., and Walker, R. M. (1964): Etude des traces de fission fossiles dans le mica, *Bull. Soc. Franç. Minér. Crist.* **87**, 6–17.

Mehta, P. P. and Nagpaul, K. K. (1971): Fission track ages of mica belts and some other precambrian muscovites of India, *Pure Appl. Geophys.* **87**, 174–191.

Mehta, P. P. and Rama (1969): Annealing effects in muscovite and their influence on dating by fission track method, *Earth Planet. Sci. Lett.* **7**, 82–86.

Miller, D. K., Bagley, M., and Rao, M. N. (1968): Fission track ages of mica minerals, *Z. Naturforsch.* **23a**, 1093–1094.

Miller, D. S. (1968): Fission track ages on 250 and 2500 m.y. micas, *Earth Planet. Sci. Lett.* **4**, 379–383.

Miller, D. S. and Duddy, I. R. (1989): Early Cretaceous uplift and erosion of the northern Appalachian basin, New York, based on apatite fission track analysis, *Earth Planet. Sci. Lett.* **93**, 35–49.

Miller, D. S. and Jäger, E. (1968): Fission track ages of some alpine micas, *Earth Planet. Sci. Lett.* **4**, 375–378.

Miller, D. S. and Lakatos, S. (1983): Uplift rate of Adirondack anorthosite measured by fission-track analysis of apatite, *Geology* **11**, 284–286.

Miller, D. S. and Wagner, G. A. (1979): Age and intensity of thermal events by fission track analysis: The Ries impact crater, *Earth Planet. Sci. Lett.* **43**, 351–358.

Miller, D. S. and Wagner, G. A. (1981): Fission-track ages applied to obsidian artifacts from South America using the plateau-annealing and the track-size age-correction techniques, *Nucl Tracks* **5**, 147–155.

Miller, D. S., Wagner, G. A., and Jäger, E. (1978): Fission track ages on apatite, sphene and zircon of Bergell rocks from Central Alps and of Bergell boulders in Oligocene sediments, U.S. Geol. Surv. Open-File Report 78-701, pp. 297–299.

Miller, D. S., Duddy, I. R., Green, P. F., Hurford, A. J., and Naeser, C. W. (1985): Results of interlaboratory comparison of fission-track age standards: fission-track workshop 1984, *Nucl. Tracks* **10**, 383–391.

Miller, D. S., Eby, N., McCorkell, R., Rosenberg, P. E., and Suzuki, M. (1990): Results of interlaboratory comparison of fission track ages for the 1988 Fission Track Workshop, *Nucl. Tracks Radiat. Meas.* **17**, 237–245.

Miller, R. S. and Kahn J. S. (1962): *Statistical Analysis in the Geological Sciences*, Wiley, New York.

Milton, D. J. and Sutter, J. F. (1987): Revised age for the Gosses Bluff impact structure, Northern Territory, Australia, based on $^{40}Ar/^{39}Ar$ dating, *Meteoritics* **22**, 281–289.

Monnin, M. M. (1980): Visualization of latent damage trails, *Nucl. Instr. Meth.* **173**, 1–14.

Monnin M. M. and Blanford G. E. (1973): Detection of charged particles by polymer grafting, *Science* **181**, 743–744.

Moore, M. E., Gleadow, A. J. W., and Lovering, J. F. (1986): Thermal evolution of rifted continental margins: new evidence from fission tracks in basement apatites from southeastern Australia, *Earth Planet. Sci. Lett.* **78**, 255–270.

Morgan, D. V. and Chadderton, L. T. (1968): Fission fragment tracks in semiconducting layer structures, *Philos. Mag.* **17**, 1135–1143.

Morgan, D. V. and Van Vliet, D. (1970): Charged particle tracks in solids, *Contemp. Phys.* **11**, 173–193.

Morley, M. E., Gleadow, A. J. W., and Lovering, J. F. (1980): Evolution of the Tasmanian Rift: apatite fission track dating evidence from the southeastern Australian continental margin, *5*th Int. Gondwana Symposium, Wellington 1980, pp. 289–293.

Mughabghab, S. F. (1984): *Neutron Cross Sections. Vol 1, Neutron Resonance Parameters and Thermal Cross Sections, Part B: Z = 61 to 100*, Academic Press, New York.

Mughabghab, S. F., Divadeenam, M., and Holden, N. E. (1981): *Neutron Cross Sections, Vol. 1, Neutron Resonance Parameters and Thermal Cross Sections, Part A. Z = 1 to 60*, Academic Press, New York.

Naeser, C. W. (1967): The use of apatite and sphene for fission track age determinations, *Bull. Geol. Soc. Amer.* **78**, 1523–1526.

Naeser, C. W. (1969): Etching fission tracks in zircons, *Science* **165**, 388.

Naeser, C. W. (1971): Geochronology of the Navajo-Hopi diatremes, Four Corners area, *J. Geophys. Res.* **76**, 4978–4985.

Naeser, C. W. (1979): Fission-track dating and geologic annealing of fission tracks, in E. Jäger and J. C. Hunziker (eds), *Lectures in Isotope Geology*, Springer-Verlag, Heidelberg, pp. 154–169.

Naeser, C. W. and Dodge, F. C. W. (1969): Fission track ages of accessory minerals from granitic rocks of the Central Sierra Nevada batholith, California, *Bull. Geol. Soc. Amer.* **80**, 2201–2212.

Naeser, C. W. and Faul, H. (1969): Fission track annealing in apatite and sphene, *J. Geophys. Res.* **74**, 705–710.

Naeser, C. W. and Fleischer, R. L. (1975): The age of the apatite at Cerro de Mercado, Mexico: a problem for fission track annealing corrections, *Geophys. Res. Lett.* **2**, 67–70.

Naeser, C. W. and Forbes, R. B. (1976): Variation of fission track ages with depth in two deep drill holes, *Trans. Amer. Geophys. Union* **57**, 353.

Naeser, C. W. and Maldonado, F. (1981): Fission-track dating of the Climax and Gold Meadows Stocks, Nye County, Nevada, Geol. Surv. Prof. Papers 1199-E, pp. 45–47.

Naeser, N. D. and McCulloh, Th. H. (eds), (1989): *Thermal History of Sedimentary Basins*, Springer-Verlag, New York, pp. 157–180.

Naeser, C. W. and McKee, E. H. (1970): Fission-track and K-Ar ages of Tertiary ash-flow tuffs, north-central Nevada, *Geol. Soc. Amer. Bull.* **81**, 3375–3384.

Naeser, N. D. and Naeser, C. W. (1984): Fission-track dating, in W. C. Mahaney (ed.), *Quaternary Dating Methods*, pp. 87–100.

Naeser, C. W. and Saul, J. M. (1974): Fission track dating of tanzanite, *Amer. Mineralogist* **59**, 613–614.

Naeser, C. W., Dodge, F. C. W., and Kistler, R. W. (1968): Fission track ages of accessory minerals from rocks of the Sierra Nevada Batholith, *Ann. M. Geol. Soc. Amer., Mexico City*, 11–13 Nov. 1968, Abstract, 214.

Naeser, C. W., Engels, J. C., and Dodge, F. C. W. (1970): Fission track annealing and age determination of epidote minerals, *J. Geophys. Res.* **75**, 1579–1584.

Naeser, C. W., Kistler, R. W., and Dodge, F. C. W. (1971): Ages of coexisting minerals from heat-flow borehole sites, Central Sierra Nevada batholith, *J. Geophys. Res.* **76**, 6462–6463.

Naeser, C. W., Izett, G. A., and Wilcox, R. E. (1973): Zircon fission-track ages of Pearlette Family ash beds in Meade County, Kansas, *Geology* **1**, 187–189.

Naeser, C. W., Cunningham, C. G., Marvin, R. F., and Obradovich (1979a): Pliocene intrusive rocks and mineralization near Rico, Colorado. U.S. Geol. Surv. Open-File Report 79-1093.

Naeser, C. W., Gleadow, A. J. W., and Wagner, G. A. (1979b): Standardization of fission-track data reports, *Nucl. Tracks* **3**, 133–136.

Naeser, C. W., Izett, G. A., and Obradovich, J. D. (1980): Fission-track and K-Ar ages of natural glasses, *U.S. Geol. Surv. Bull.* **1489**, 1–31.

Naeser, C. W., Zimmerman, R. A., and Cebula, G. T. (1981): Fission-track dating of apatite and zircon: an interlaboratory comparison, *Nucl. Tracks* **5**, 65–72.

Naeser, C. W., Bryant, B. C., Crittenden, M. D., Jr., and Sorensen, M. L. (1983): Fission track dating in the Wasatch Mountains, Utah: an uplift study, *Geol. Soc. Amer. Memoir* **157**, 29–36.

Naeser, C. W., McKee, E. H., Johnson, N. M., and MacFadden, B. J. (1987): Confirmation of a late oligocene-early Miocene age of the Deseadan Salla beds of Bolivia, *J. Geol.* **95**, 825–828.

Naeser, N. D., Naeser, C. W., and McCulloh, Th. H. (1989): The application of fission-track dating to the depositional and thermal history of rocks in sedimentary basins in N. D. Naeser and Th. H. McCulloh (eds), *Thermal History of Sedimentary Basins*, Springer-Verlag, New York, pp. 157–180.

Naeser, N. D., Crowley, K. D., McCulloh, T. H., and Reaves, C. M. (1990): High temperature annealing of fission tracks in fluorapatite, Santa Fe Springs oil field, Los Angeles Basin, California, *Nucl. Tracks Radiat. Meas.* **17**, 424.

Nagpaul, K. K. (1982): Fission track geochronology of India in J. N. Goswami (ed.) *Nuclear Tracks*, Indian Academy of Sciences, Bangladore, pp. 53–65.

Nagpaul, K. K., Mehta, P. P., and Gupta, M. L. (1974a): Fission track ages of cogenetic minerals of the Nellore mica belt of India, *Pure Appl. Geophys.* **112**, 140–148.

Nagpaul, K. K., Mehta, P. P., and Gupta, M. L. (1974b): Annealing studies on radiation damages in biotite, apatite and sphene and corrections to fission track ages, *Pure Appl. Geophys.* **112**, 131–139.

Nand Lal, Nagpaul, K. K., and Sharma, K. K. (1976a): Fission track ages and uranium concentration in garnets from Rajasthan, India, *Bull. Geol. Soc. Amer.* **87**, 687–690.

Nand Lal, Saini, H. S., and Nagpaul, K. K. (1976b): Study of annealing versus etching behaviour of fission damage in biotite, *Radiation Effects* **29**, 161–163.

Nand Lal, Parshad, R., and Nagpaul, K. K. (1977a): Fission track annealing characteristics of garnet, *Lithos* **10**, 129–132.

Nand Lal, Parshad, R. and Nagpaul, K. K. (1977b): Fission track etching and annealing of tourmaline, *Nucl. Track Detection* **1**, 145–148.

Nishida, T. and Takashima, Y. (1975): Annealing of fission tracks in zircons, *Earth Planet. Sci. Lett.* **27**, 257–264.

Nishimura, S. (1971): Fission track dating of archaeological materials from Japan, *Nature* **230**, 242–243 (1972 revised: *Nature* **237**, 57).

Nishimura, S. (1972): On the value of the decay constant for spontaneous fission of ^{238}U, *J. Japan. Assoc. Min. Petrol. Econ. Geol.* **67**, 139–142 (in Japanese).

Nishimura, S. (1978): On the value of the decay constant for spontaneous fission of uranium-238, Geol. Surv. Open-file Report 78-101, pp. 306–307.

Nishimura, S. (1981): On the fission-track dating of tuffs and volcanic ashes, *Nucl. Tracks* **5**, 157–167.

Nishimura, S. and Tosi, M. (1976): Fission track ages of the remains excavated at Shar-i Sokhta and Kangavar, Iran, *Mem. Vol. 6th Intern. Congr. Iran. Art and Archaeol.*, Teheran, pp. 281–285.

Omar, G., Johnson, K. R., Hickey, L. J., Robertson, P. B., Dawson, M. R. and Barnosky, C. W. (1987): Fission-track dating of Haughton Astrobleme and included biota, Devon Island, Canada, *Science* **237**, Reports, pp. 1603–1605.

Omar, G. I., Steckler, M. S., Buck, W. R., and Kohn, B. P. (1989): Fission-track analysis of basement apatites at the western margin of the Gulf of Suez rift, Egypt: evidence for synchroneity of uplift and subsidence, *Earth Planet. Sci. Lett.* **4**, 316–328.

Paretzke, H. (1977): On primary damage and secondary electron damage in heavy ion tracks in plastics, *Radiation Effects* **34**, 3–8.

Parrish, R. R. (1983): Cenozoic thermal evolution and tectonics of the Coast Mountains of British Columbia 1, fission-track dating, apparent uplift rates, and patterns of uplift, *Tectonics* **2**, 601–632.

Parrish, R. R. (1985): Some cautions which should be exercised when interpreting fission-track and other dates with regard to uplift rate calculations, *Nucl. Tracks* **10**, 425.

Parshad, R., Saini, H. S., and Nagpaul, K. K. (1978): Fission track etching and annealing phenomenon in phlogopite and their applications, *Canad. J. Earth Sci.* **15**, 1924–1929.

Perron, C. and Maury, M. (1986): Very heavy ion track etching in olivine, *Nucl. Tracks Radiat. Meas.* **11**, 73–80.

Petrov, Yu. V. (1977): The Oklo natural nuclear reactor, *Soviet Phys. Uspekhi.* **20**(11), 937–944.

Popeko A. G. and Ter-Akopian G. M. (1980): Measurement of the ^{238}U spontaneous-fission-half-life by detecting prompt neutrons, *Nucl. Instr. Meth.* **178**, 163–165.

Poupeau, G., Carpena, J., Chambaudet, A., and Romary, Ph. (1980): Fission track plateau-age dating, in H. François *et al.* (eds), *Solid State Nuclear Track Detectors*, Pergamon Press, Oxford, pp. 966–977.

Price, P. B. and Fleischer, R. L. (1964): Glass dating by fission fragment tracks, *J. Geophys. Res.* **69**, 331.

Price, P. B. and Fleischer, R. L. (1971): Identification of energetic heavy nuclei with solid dielectric track detectors: applications to astrophysical and planetary studies, *Ann. Rev. Nucl. Sci.* **21**, 295–334.

Price, B. P. and Salamon M. H. (1986): Advances in solid state nuclear track detectors, *Nucl. Tracks Radiat. Meas.* **12**, 5–17.

Price, P. B. and Walker, R. M. (1962a): A new detector for heavy particle studies, *Phys. Lett* **3**, 113–115.

Price, P. B. and Walker, R. M. (1962b): Observations of charged-particle tracks in solids, *J. Appl. Phys.* **33**, 3400–3406.

Price, P. B. and Walker, R. M. (1962c): Chemical etching of charged-particle tracks in solids, *J. Appl. Phys.* **33**, 3407–3412.

Price, P. B. and Walker, R. M. (1962d): Observation of fossil particle tracks in natural micas, *Nature* **196**, 732–734.

Price, P. B. and Walker, R. M. (1963a): Fossil tracks of charged particles in mica and the age of minerals, *J. Geophys. Res.* **68**, 4847–4862.

Price, P. B. and Walker, R. M. (1963b): A simple method for measuring low uranium concentrations in natural crystals, *Appl. Phys. Lett.* **2**, 23–25

Price, P. B., Lal, D., Tamhane, A. S., and Perelygin, V. P. (1973): Characteristics of tracks of ions of $14 \leqslant Z \leqslant 36$ in common rock silicates, *Earth Planet. Sci. Lett.* **19**, 377–395.

Rao, M. N. and Kuroda, P. K. (1966): Decay constant and mass-yield curve for the spontaneous fission of ^{238}U, *Phys. Rev.* **147**, 884–886.

Reimer, G. M. (1972): Fission track geochronology: method for tectonic interpretation of apatite studies with examples from the central and southern Alps, Diss. Univ. of Pennsylvania, Philadelphia.

Reimer, G. M. and Wagner, G. A. (1971): Fission track studies of alpine epidotes and garnets, *Ann. Soc. Geol. Belg.* **94**, 127.

Reimer, G. M., Storzer, D., and Wagner, G. A. (1970): Geometry factor in fission track counting, *Earth Planet. Sci. Lett.* **9**, 401–404.

Reimer, G. M., Wagner, G. A., and Carpenter, B. S. (1972): The thermal stability of fission tracks in the standard reference material glass standard (National Bureau of Standards), *Radiation Effects* **15**, 273–274.

Ritchie, R. H. and Claussen, C. (1982): A core plasma model of charged particle track formation in insulators, *Nucl. Instr. Meth.* **198**, 133–138.

Roberts, J. H., Gold, R., and Armani, R. J. (1968): Spontaneous fission decay-constant of ^{238}U, *Phys. Rev.* **174**, 1482–1484.

Roberts, J. H., Ruddy, F. H., and Gold, R. (1984): Optical efficiency for fission fragment track counting in muscovite solid state track recorders, *Nucl. Tracks Radiat. Meas.* **8**, 365–369.

Ross, R. J., Jr., Naeser, C. W., Izett, G. A., Obradovich, J. D., Bassett, M. G., Hughes, C. P., Cocks, L. R. M., Dean, W. T., Ingham, J. L., Jenkins, C. J., Rickards, R. B., Sheldon, P. R., Toghill, P., Whittington, H. B., and Zalasiewicz, J. (1982): Fission-track dating of British Ordovician and Silurian stratotypes, *Geol. Mag.* **119**, 135–153.

Saini, H. S. (1978): Solid state nuclear track detectors and their applications to earth sciences, PhD thesis, Kurukshetra University.

Saini, H. S., Nand Lal, and Nagpaul, K. K. (1975): Annealing studies of fission tracks in allanite, *Contrib. Mineral. Petrol.* **52**, 143–145.

Saini, H. S., Sharma, O. P. and Nagpaul, K. K. (1977): Analysis of fission track etching and annealing phenomena in muscovite, *Pure Appl. Geophys.* **115**, 951–959.

Saini, H. S., Sharma, O. P., Parshad R., and Nagpaul, K. K. (1978): Fission track annealing characteristics of epidote: applications to geochronology and geology, *Nucl. Track Detection* **2**, 133–140.

Saini, H. S., Srivastava, A. P., and Rajagopalan G. (1983): Fission track dating and uranium concentration in kyanite from the Himalaya, *Geol. Mag.* **120**, 341–348.

Schaer, J. P., Reimer, G. M., and Wagner, G. A. (1975): Actual and ancient uplift rate in the Gotthard region, Swiss Alps: a comparison between precise levelling and fission-track apatite age, *Tectonophysics* **29**, 293–300.

Schreurs, J. W. H., Friedman, A. M., Rokop, D. J., Hair, M. W., and Walker R. M. (1971): Calibrated U-Th glasses for neutron dosimetry and determination of uranium and thorium concentration by the fission track method, *Radiation Effects* **7**, 231–233.

Scott, B. G. (1976): The possible application of fission-track counting to the dating of bloomery slag and iron, *Hist. Metall.* **10**, 87.

Segré, E. (1952): Spontaneous fission, *Phys. Rev.* **86**, 21–28.

Seiberling L. E., Griffith, J. E., and Tombrello, T. A. (1980): A thermalized ion explosion model from high energy sputtering and track registration, *Radiation Effects* **52**, 201–210.

Seitz, M., Wittels, M. C., Maurette, M., Walker, R. M., and Heckman, H. (1970): Accelerator irradiations of minerals: Implications for track formation mechanisms and for studies of lunar and meteoritic minerals, *Radiation Effects* **5**, 143–148.

Seitz, M. G., Walker, R. M. and Carpenter, B. S. (1973): Improved methods for measurement of thermal neutron dose by the fission track technique, *J. Appl. Phys.* **44**, 510–512.

Sélo, M. and Storzer, D. (1979): Chronologie des événements volcaniques de la zone Famous, *Compt. Rend. Acad. Sci. Paris* **D289**, 1125–1128.

Sélo, M. and Storzer, D. (1981): Uranium distribution and age pattern of some deep-sea basalts from the Entrecasteaux area, south-western Pacific: a fission-track analysis, *Nucl. Tracks* **5**, 137–145.

Seward, D. (1974): Age of New Zealand Pleistocene substages by fission track dating of glass shards from tephra horizons, *Earth Planet. Sci. Lett.* **24**, 242–248.

Seward, D. (1975): Fission track ages of some tephras from Cape Kidnappers, Hawke's Bay, New Zealand, *N.Z. J. Geol. Geophys.* **18**, 507–510.

Seward, D. (1976): Tephrostratigraphy on the marine sediments in the Wanganui Basin, New Zealand, *N.Z. J. Geol. Geophys.* **19**, 9–20.

Seward, D. and Rhoades, D.A. (1986): A clustering technique for fission track dating of fully to partially annealed minerals and other non-unique populations, *Nucl. Tracks Radiat. Meas.* **11**, 259–268.

Seward, D., Wagner, G. A., and Pichler, H. (1980): Fission track ages of Santorini volcanics (Greece), in *Thera and the Aeqean World II*, London, pp. 101–108.

Sharma, K. K. and Nagpaul, K. K. (1978): Fission-track, K-Ar and Pb-U mineral ages from Danta mica mine pegmatite (Bhunas), central Rajasthan, India – a comparison, U.S. Geol. Surv. Open-File Report 78-701, pp. 390–392.

Sharma, O. P., Bal, K. D. and Nagpaul, K. K. (1977): Fission track annealing and age determination of chlorite, *Nucl. Track Detection* 1, 207–211.

Sharma, K. K., Saini, H. S., and Nagpaul, K. K. (1978): Fission track annealing, ages of apatites from Mandi granite and their application to tectonic problems, *Himalayan Geology* 8, 296–312.

Sharma, O. P., Bal, K. D., Saini, H. S., and Nagpaul, K. K. (1979): Fission track annealing characteristics and dating of vermiculite, *Phys. Chem. Minerals* 5, 133–139.

Sharma, K. K., Bal, K. D., Parshad, R., Lal, N., and Nagpaul, K. K. (1980): Palaeo-uplift and cooling rates from various orogenic belts of India, as revealed by radiometric ages, *Tectonophysics* 70, 135–158.

Shima, M., Okada, A., and Yabuki, H. (1969): Cross checking of fission track and K-Ar methods for age dating, *J. Japan Assoc. Min. Petrol. Econ. Geol.* 61, 100–105.

Shukoljukov, Ju. A. and Komarov, A. N. (1966): Possible palaeothermometry on the grounds of fission tracks of uranium in minerals (Russian), *Iv. Akad. Nauk. SSSR, Ser. Geol.* 9, 137–141.

Shukoljukov, Ju. A. and Komarov, A. N. (1970): Tracks of uranium fission in monazite (Russian), *Bull. Commiss. Determ. Abs. Age Geol Form.* 9, Akad. Nauk. USSR, Moscow, pp. 20–26.

Shukoljukov, Ju. A., Krylov, I. N., Tolstikhin, I. N., and Ovchinnikova, G. V. (1965): Tracks of uranium fission fragments in muscovite, *Geochem. Inter.* 2, 202.

Shukoljukov, Ju. A., Ashkinadze, G. S., Levchenkov, O. A., and Ovchinnikova, G. V. (1968): Determination of the rate constant of the ^{238}U spontaneous fission according to Xenon accumulation, *Geokhimiya* 3, 265–274 (Russian).

Silk, E. C. H. and Barnes, R. S. (1959): Examination of fission fragment tracks with an electron microscope, *Philos. Mag.* 4, 970–972.

Singh, S., Singh, D., Sandhu, A. S., and Virk, H. S. (1986): A study of etched track anisotropy in apatite, *Mineral. J.* 13, 75–85.

Sippel, R. F. and Glover, E. D. (1964): Fission damage in calcite and the dating of carbonates, *Science* 144, 409–411.

Smith, M. J. and Leigh-Jones, P. (1985): An automated microscope scanning stage for fission-track dating, *Nucl. Tracks* 10, 395–400.

Somogyi, G. (1980): Development of etched nuclear tracks, *Nucl. Instr. Meth.* 109, 21–42.

Somogyi, G. and Nagy, M. (1972): Remarks on fission-track dating in dielectric solids, *Radiation Effects* 16, 223–231.

Somogyi, G. and Szalay, S. A. (1973): Track-diameter kinetics in dielectric track detectors, *Nucl. Instr. Meth.* 109, 211–232.

Somogyi, G., Tóth-Szilágyi, M., Monnin, M., and Gourcy, J. (1979): Non-etching nuclear track visualization in polymers, fluorescent and dyed tracks, *Nucl. Tracks* 3, 151–167.

Spadavecchia, A. and Hahn, B. (1967): Die Rotationskammer und einige Anwendungen, *Helv. Phys. Acta* 40, 1063–1079.

Spagiarri, E. R. V. (1980): Determinacao do alcance efectivo de fragmentos de fissao no UO_2 e da Constante de Desintegracao para a fissao espontanea do Uranio-238, *Ann. Acad. Brasil Cienc.* 52, 213–233.

Srivastava, A. P., Saini, H. S., and Rajagoplan, G. (1983): Suitability of glauconite for fission track dating, *Current Sci.* 52, 866–869.

Staudacher, Th., Jessberger, E. K., Dominik, B., Kirsten, T. and Schaeffer, O. A. (1982): ^{40}Ar-^{39}Ar ages of rocks and glasses from the Nördlinger Ries crater and the temperature history of impact breccias, *J. Geophys.* 51, 1–11.

Staufenberg, H. (1987): Apatite fission-track evidence for postmetamorphic uplift and cooling history of the Eastern Tauern window and the surrounding Austroalpine (Central Eastern Alps, Austria), *Jb. Geol. B.-A.* 130, 571–586.

Steiger, R. H. and Jäger, E. (1977): Subcommission on geochronology: convention on the use of decay constants in geo- and cosmochronology, *Earth Planet. Sci. Lett.* 36, 359–362.

Steven, T. A., Mehnert, H. H., and Obradovich, J. D. (1967): Age of volcanic activity in the San Juan Mountains, Colorado U.S. Geol. Surv. Prof. Papers 575-D, D47–D55.

Steven, T. A., Cunningham, C. G., Naeser, C. W., and Mehnert, H. H. (1979): Revised stratigraphy and radiometric ages of volcanic rocks and mineral deposits in the Marysvale area, West-Central Utah, *U.S. Geol. Surv. Bull.* **1469**, 1–40.

Stöffler, D. (1973): Cratering mechanics, impact metamorphism and distribution of ejected masses of the Ries structure – an introduction, *Fortschr. Min.* **52**, 103–122.

Storzer, D. (1970a): Spaltspuren des 238-Urans und ihre Bedeutung für die geologische Geschichte natürlicher Gläser, PhD Thesis, Univ. Heidelberg.

Storzer, D. (1970b): Fission track dating of volcanic glasses and the thermal history of rocks, *Earth Planet. Sci. Lett.* **8**, 55–60.

Storzer, D. (1971): Fission track dating of some impact craters in the age range between 6000 y and 300 my, *Meteoritics* **6**, 319.

Storzer, D. (1978): unpubl. data

Storzer, D. (1985): The fission track age of high sodium/potassium australites revisited, *Meteoritics* **20**, 765–766.

Storzer, D. and Gentner, W. (1970): Spaltspurenalter von Riesgläsern, Moldaviten und Bentoniten, *Jber. Mitteil. oberrhein. geol. Ver.*, NF **52**, 97–111.

Storzer, D. and Poupeau, G. (1973): Géochronologie.-Ages-plateaux de minéraux et verres par la méthode des traces de fission, *Compt. Rend. Acad. Sci.*, Paris **D276**, 137–139.

Storzer, D. and Sélo, M. (1976): Uranium contents and fission track ages of some basalts from the FAMOUS area, *Bull. Soc. Geol. Fr.* **18**, 807–810.

Storzer, D. and Sélo, M. (1978): Fission track age of magnetic anomaly M-Zero and some aspects of sea-water weathering, in T. Donnelly *et al.* (eds), *Initial Reports of the Deep Sea Drilling Project*, Vols. 51–53, Washington, pp. 1129–1133.

Storzer, D. and Wagner, G. A. (1969): Correction of thermally lowered fission track ages of tektites, *Earth Planet. Sci. Lett.* **5**, 463–468.

Storzer, D. and Wagner, G. A. (1971): Fission track ages of North American tektites, *Earth Planet. Sci. Lett.* **10**, 435–440.

Storzer, D. and Wagner, G. A. (1977): Fission track dating of meteorite impacts, *Meteoritics* **12**, 368–369.

Storzer, D. and Wagner, G. A. (1979): Fission track dating of Elgygytgyn, Popigai and Zhamanshin impact craters: no sources for Australasian or North-American tektites, *Meteoritics* **14**, 541–542.

Storzer, D. and Wagner, G. A. (1980): Australites older than Indochinites, *Naturwissen.* **67**, 90.

Storzer, D. and Wagner, G. A. (1982): The application of fission track dating in stratigraphy: a critical review, in G. S. Odin (ed.), *Numerical Dating in Stratigraphy*, Wiley, Chichester, pp. 199–221.

Storzer, D., Horn, P., and Kleinmann, B. (1971): The age and the origin of Köfels structure, Austria, *Earth Planet. Sci. Lett.* **12**, 238–244.

Storzer, D., Wagner, G. A., and King, E. A. (1973): Fission track ages and stratigraphic occurrence of Georgia tektites, *J. Geophys. Res.* **78**, 4915–4919.

Stuckless, J. S. and Naeser, C. W. (1972): Rb-Sr and fission-track age determinations in the precambrian plutonic basement around the Superstition volcanic field, Arizona, U.S. Geol. Surv. Prof. Paper 800, B191–B194.

Sumii T., Tagami, T., and Nishimura, S. (1987): Anisotropic etching character of spontaneous tracks in zircon, *Nucl. Tracks Radiat. Meas.* **13**, 275–277.

Sutton, S. R. and Zimmerman, D. W. (1978): Thermoluminescence dating: radioactivity in quartz, *Archaeometry* **20**, 67–69.

Suzuki, M. (1970): Fission track dating and uranium contents of obsidian, *J. Anthropol. Soc. Nippon* **78**, 50–58.

Suzuki, M. (1973): Chronology of prehistoric human activity in Kanton, Japan. Part I: framework for reconstructing prehistoric human activity in obsidian, *J. Fac. Sci. Tokyo. Univ.* **IV**, 241–318.

Suzuki, M. and Yamanoi, T. (1970): Fission track dating of the Uonuma group, *J. Geol. Soc. Japan* **76**, 317–318.

Tagami, T. (1987): Determination of zeta calibration constant for fission track dating, *Nucl. Tracks Radiat. Meas.* **13**, 127–130.

Tagami, T. and Nishimura, S. (1989): Intercalibration of thermal neutron dosimeter glasses NBS-SRM 612 and Corning 1 in some irradition facilities: a comparison, *Nucl. Tracks Radiat. Meas.* **16**, 11–14.

Tagami, T., Lal, N., Sorkhabi, R. B., Ito, H., and Nishimura, S. (1988a): Fission track dating using the external detector method: a laboratory procedure, *Mem. Fac. Sci. Kyoto Univ., Geology & Mineralogy* **53**, 1–30.

Tagami, T., Lal, N., Sorkhabi, R. B., and Nishimura, S. (1988b): Fission track thermochronologic analysis of the Ryoke Belt and the Median Tectonic Line, Southwest Japan, *J. Geophys. Res.* **93**, 705–713.

Tagami, T., Ito, H., and Nishimura, S. (1990): Thermal annealing characteristics of spontaneous fission tracks in zircon, *Chem. Geol (Isot. Geosci Sect.)* **80**, 159–169.

Tahirkheli, R. A. K. and Naeser, C. W. (1975): Zircon fission track age of post Ranch ash bed near Benson, Arizona, *Arizona Acad. Sci.* **10**, 111–113.

Tamanyu, S. (1975): Fission-track age determination of accessory zircon from the Neogene-Tertiary tuff samples, around Sendai City, Japan, *J. Geol. Soc. Japan* **81**, 233–246.

Thiel, K. and Herr, W. (1976): The ^{238}U spontaneous fission decay constant re-determined by fission tracks, *Earth. Planet. Sci. Lett.* **30**, 50–56.

Thiel K. and Külzer H. (1978): Anisotropy of track registration in natural feldspar crystals, *Radiation Effects* **35**, 87–93.

Thiel K., Bradley J. P., and Spohr R. (1988): On the nature of latent nuclear tracks in cosmic dust particles, *Nucl. Tracks Radiat. Meas.* **15**, 685–688.

Thury, W. M. (1971): Die Bestimmung der Spontanspaltrate von U-238 mit Hilfe der Messung von Korrelationsfunktionen dritter Ordnung, *Acta Phys. Austriaca* **33**, 375–378.

Tingate, P. (1989): A fission track dating study of the Gosses Bluff impact structure, unpublished abstract.

Todorovic, Z., Antanasijevic, R., Jakupi, B., Vukovic, J., and Miocinovic, D. (1980): Fission track annealing of orpiment, *Proc. 10th Internat. Conf. SSNTD*, Pergamon Press, Oxford, pp. 805–809.

Togliatti, V. (1965): Distribuzioni dei ranges e distribuzione angolari delle trace di fissioni fossili di U^{235} in mica, *Boll. Geofis. Teor. Appl.* **7**, 326–335.

Tombrello, T. A. (1984a): The dimensions of latent ion damage tracks, *Nucl. Instr. Meth.* **B1**, 23–25.

Tombrello, T. A. (1984b): Track damage and erosion of insulators by ion-induced electronic processes, *Nucl. Instr. Meth.* **B2**, 555–563.

Tommasino, L. (1980): Solid dielectric detectors with breakdown phenomena and their applications in radioprotection, *Nucl. Instr. Meth.* **173**, 73–83.

Tommasino, L., Zapparoli, G., and Griffith, R. V. (1980a): Electrochemical etching – I. Mechanisms, *Nucl. Tracks* **4**, 191–196.

Tommasino, L., Zapparoli, G., Griffith, R. V., and Mattei, A. (1980b): Electrochemical etching – II. Methods, apparatus and results, *Nucl. Tracks* **4**, 197–201.

Uzgiris, E. E. and Fleischer, R. L. (1971): Amber: charged particle track registration, track stability and uranium content, *Nature* **234**, 28–30.

Van den haute, P. (1977): Apatite fission track dating of Precambrian intrusive rocks from the southern Rogaland (south-western Norway), *Bull. Soc. Belge Géol.* **86**, 97–110.

Van den haute, P. (1983): Bijdrage tot de studie van fissiesporen in glas en toepassing van de fissiesporen-dateringsmethode op apatieten uit precambrische gesteenten van Rwanda en Burundil PhD thesis, University of Gent.

Van den haute, P. (1984): Fission-track ages of apatites from the Precambrian of Rwanda and Burundi: relationship to East African rift tectonics, *Earth Planet. Sci. Lett.* **71**, 129–140.

Van den haute, P. (1985): The density and the diameter of fission tracks in glass with respect to age interpretation, *Nucl. Tracks* **10**, 335–348.

Van den haute, P. (1986a): Sphene fission-track dating of a precambrian alkaline pluton in Burundi (Central Africa), *Terra Cognita*, **6**, 165.

Van den haute, P. (1986b): Apatite fission–track dating applied to precambrian terranes, *Chem. Geol.* **57**, 155–165.

Van den haute, P. and Chambaudet, A. (1990): Results of an interlaboratory experiment for the 1988 fission track workshop on a putative apatite standard for internal calibration, *Nucl. Tracks Radiat. Meas.* **17**, 247–252.

Van den haute, P. and Vercoutere C. (1989): Apatite fission-track evidence for a Mesozoic uplift of the Brabant massif: preliminary results, *Ann. Soc. Géol. Belg.* **112**, 443–452.

Van den haute, P., Jonckheere, R., and De Corte, F. (1988): Thermal neutron fluence determination for fission-track dating with metal activation monitors: a re-investigation, *Chem. Geol. (Isot. Geosci. Sect.)* **73**, 233–244.

Vandenbosch, R. and Huizenga, J. R. (1973): *Nuclear Fission*. Academic Press, New York.

Vartanian, R. (1984): Spontaneous fission decay constant of ^{238}U: measured by the fission track technique, *Helv. Phys. Acta* **57**, 416–420.

Vincent, D., Clocchiatti, R., and Langevin, Y. (1984): Fission-track dating of glass inclusions in volcanic quartz, *Earth Planet. Sci. Lett.* **71**, 340–348.

Von Gunten, H. R. (1969): Distribution of mass in spontaneous and neutron-induced fission, *Actinides Rev.* **1**, 275–298.

Wadatsumi, K. and Masumoto, S. (1989): Colored images of fission-tracks for distinguishing three-dimensional track-geometry, *J. Geosci. Osaka City Univ.* **32**, 23–37.

Wadatsumi, K., Masumoto, S., and Suzuki, K. (1988): Computerized image-processing system for fission-track dating; system configuration and functions, *J. Geosci. Osaka City Univ.* **31**, 19–46.

Wagemans, C., Schillebeeckx, P., Deruytter, A. J., and Barthélémy, R. (1988): Subthermal fission cross-section measurements for ^{233}U, ^{235}U and ^{239}Pu, *Proc. Internat. Conf. Nuclear Data for Science and Technology (Mito, Japan)*, Saikon Publ. Tokyo, pp. 91–95.

Wagner, G. A. (1966): Altersbestimmungen an Tektiten und anderen natürlichen Gläsern mittels Spuren der spontanen Kernspaltung des Uran^{238} ("fission-track"-Methode), *Z. Naturforschung* **21a**, 733–745.

Wagner, G. A. (1968): Fission track dating of apatites, *Earth Planet. Sci. Lett.* **4**, 411–415.

Wagner, G. A. (1969a): Spuren der spontanen Kernspaltung des ^{238}Urans als Mittel zur Datierung von Apatiten und ein Beitrag zur Geochronologie des Odenwaldes, *N. Jb. Min. Abh.* **110**, 252–286.

Wagner, G. A. (1969b): Altersbestimmung an Gläsern und Mineralien mit der Spaltspurenmethode (unter dem Mikroskop), *Z. Deutsch. Geol. Ges.* **118**, 209–216.

Wagner, G. A. (1972a): Spaltspurenalter von Mineralen und natürlichen Gläsern: eine Übersicht, *Fortschr. Min.* **49**, 114–145.

Wagner, G. A. (1972b): The geological interpretation of fission track ages, *Trans. Amer. Nucl. Soc.* **15**, 117.

Wagner, G. A. (1973): Die Anwendung anätzbarer Partikelspuren zur geochemischen Analyse, *Fortschr. Min.* **51**, 68–93.

Wagner, G. A. (1976a): Spaltspurendatierung an Apatit und Titanit aus den Subvulkaniten des Kaiserstuhls, *N. Jb. Min. Mh.* 1976/9, 389–393.

Wagner, G. A. (1976b): Altersbestimmung und Urananalyse alter Objekte mittels Spaltspuruntersuchungen, *Accad. Nazion. Linc., Atti. Conv. Linc.* **11**, 503–513.

Wagner, G. A. (1977): Spaltspurendatierung an Apatit und Titanit aus dem Ries: Ein Beitrag zum Alter und zur Wärmegeschichte, *Geol. Bavarica* **75**, 349–354.

Wagner, G. A. (1978): Archaeological applications of fission track dating, *Nucl. Track Detection* **2**, 51–63.

Wagner, G. A. (1981): Fission track ages and their geological interpretation, *Nucl. Tracks* **5**, 15–25.

Wagner, G. A. (1986): Comments on the paper 'Fission track annealing in apatite: track length measurements and the form of the Arrhenius plot', *Nucl. Tracks Radiat. Meas.* **11**, 269.

Wagner, G. A. (1988): Apatite fission-track geochrono-thermometer to 60°C: projected length studies, *Chem. Geol. (Isot. Geosci. Sect.)* **72**, 145–153.

Wagner, G. A. (1990): Apatite fission-track dating of the crystalline basement of Middle Europe: concepts and results, *Nucl. Tracks Radiat. Meas.* **17**, 277–282.

Wagner, G. A. and Hejl, E. (1991): Apatite fission-track age spectrum based on projected track-length analysis, *Chem. Geol. (Isot. Geosci. Sect.)* **87**, 1–9.

Wagner, G. A. and Reimer, M. (1972): Fission track tectonics: the tectonic interpretation of fission track apatite ages, *Earth Planet. Sci. Lett.* **14**, 263–268.

Wagner, G. A. and Storzer, D. (1970): Die Interpretation von Spaltspurenaltern (fission track ages) am Beispiel von natürlichen Gläsern, Apatiten und Zirkonen, *Ecologae Geol. Helv.* **63/1**, 335–344.

Wagner, G. A. and Storzer, D. (1972): Fission track length reductions in minerals and the thermal history of rocks, *Trans. Amer. Nucl. Soc.* **15**, 127–128.

Wagner, G. A. and Storzer, D. (1975): Spaltspuren und ihre Bedeutung für die thermische Geschichte des Odenwaldes, *Aufschluss* **27**, 79–85.

Wagner, G. A. and Weiner, K. L. (1987): Spaltspurenanalysen an Obsidianproben, in M. Korfmann (ed.), *Demircihüyük. Die Ergebnisse der Ausgrabungen 1975–1978*,Bd. 2, Verlag Ph.v. Zabern Mainz, pp. 24–29.

Wagner, G. A., Reimer, G. M., Carpenter, B. S., Faul, H. Van der Linden, R., and Gijbels, R. (1975): The spontaneous fission rate of U-238 and fission track dating, *Geochim. Cosmochim. Acta* **39**, 1279–1286.

Wagner, G. A., Reimer, G. M., and Jäger, E. (1977): Cooling ages derived by apatite fission-track, mica Rb-Sr and K-Ar dating: the uplift and cooling history of the Central Alps, *Mem. Istit. Geol. Miner. Univers. Padova* **30**, 1–27.

Wagner, G. A., Storzer, D., and Keller, J. (1976): Spaltspurendatierung quartärer Gesteinsgläser aus dem Mittelmeerraum, *N. Jb. Min. Mh.* 1976, 84–94.

Wagner, G. A., Miller, D. S., and Jäger, E. (1979): Fission track ages on apatite of Bergell rocks from Central Alps and Bergell boulders in Oligocene sediments, *Earth Planet. Sci. Lett.* **45**, 355–360.

Wagner, G. A., Van den haute, P., and Hejl, E. (1989a): Apatit-Spaltspuruntersuchungen an Gesteinen der KTB-Vorbohrung: Ein Beitrag zur spätkretazischen und tertiären Hebungsgeschichte, KTB-Report 89-3, pp. 207–215.

Wagner, G. A., Gleadow, A. J. W., and Fitzgerald, P. G. (1989b): The significance of the partial annealing zone in apatite fission-track analysis: projected track length measurements and uplift chronology of the Transantarctic Mountains, *Chem. Geol. (Isot. Geosci. Sect.)* **79**, 295–305.

Wagner, G. A., Michalski, I., and Zaun, P. (1989c): Apatite fission track dating of the Central European Basement: Post-Variscan thermo-tectonic evolution, *The German Continental Deep Drilling Program (KTB)*, Springer-Verlag, Heidelberg, pp. 481–500.

Wagner, M., Altherr, R., and Van den haute, P. (1992): Apatite fission-track analysis of Kenyan basement rocks: constraints on the thermotectonic evolution of the Kenya dome. A reconnaissance study, *Tectonophysics* (in press).

Wall, T. (1986): Use of an alternative neutron dosimetry standard for fission track dating, *Nucl. Tracks Radiat. Meas.* **12**, 887–890.

Watanabe, K. (1988): Etching characteristics of fission tracks in Plio-Pleistocene zircon, *Nucl. Tracks Radiat. Meas.* **15**, 171–174.

Watanabe, N. and Suzuki, M. (1969): Fission track dating of archaeological glass materials from Japan, *Nature* **222**, 1057–1058.

Watt, S. and Durrani, S. A. (1985): Thermal stability of fission tracks in apatite and sphene: using confined track length measurements, *Nucl. Tracks* **10**, 349–357.

Weiland, E. F., Ludwig, K. R., Naeser, C. W., and Simmons, E. C. (1980): Fission track dating applied to uranium mineralization, U.S. Geol. Surv. Open-File Report 80-380.

Welin, E., Lundström, I., and Aberg, G. (1972): Fission track studies on hornblende, biotite and phlogopite from Sweden, *Bull. Geol. Soc. Finland* **44**, 35–46.

Westgate, J. A. (1988): Isothermal plateau fission-track age of the late Pleistocene Old Crow Tephra, Alaska, *Geophys. Res. Lett.* **15**, 376–379.

Westgate, J. A. (1989): Isothermal plateau fission-track ages of hydrated glass shards from silicic tephra beds, *Earth Planet. Sci. Lett.* **95**, 226–234.

Westgate, J. A. and Briggs, N. D. (1980): Dating methods of pleistocene deposits and their problems: V. Tephrochronology and fission track dating, *Geoscience Canada* **7**, 3–10.

Westgate, J. A. and Gorton, M. P. (1981): Correlation techniques in tephra studies, in S. Self and R. S. J. Sparks (eds), *Tephra Studies*, D. Reidel, Dordrecht, pp. 73–94.

Westgate, J. A., Walter, R. C., Pearce, G. W., and Gorton, M. P. (1985): Distribution, stratigraphy, petrochemistry, and palaeomagnetism of the late Pleistocene Old Crow tephra in Alaska and the Yukon, *Canad. J. Earth Sci.* **22**, 893–906.

Westgate, J. A., Easterbrook, D. J., Naeser, N. D., and Carson, R. J. (1987): Lake Tapps Tephra: an early Pleistocene stratigraphic marker in the Puget Lowland, Washington, *Quaternary Res.* **28**, 340–355.

White, S. H. and Green, P. F. (1986): Tectonic development of the Alpine fault zone, New Zealand: a fission track study, *Geology* **14**, 124–127.

Williams, I. S., Tetley, N. W., Compston, W. and McDougall, I. (1982): A comparison of K-Ar and Rb-Sr ages of rapidly cooled igneous rocks: two points in the Paleozoic time scale re-evaluated, *J. Geol. Soc. London* **139**, 557–568.

Yabuki, H., Yabuki, S., and Shima, M. (1973): Fission track dating of man-made glasses from Ali Tar Cavern vestiges, *Sci. Papers Inst. Phys. Chem. Res. Japan* **67**, 41–42.

Yada, K., Tanji, T., and Sunagawa, I. (1981): Application of lattice imagery to radiation damage investigation in natural zircon, *Phys. Chem. Minerals* **7**, 47–52.

Yada, K., Tanji, T., and Sunagawa, I. (1987): Radiation induced lattice defects in natural zircon (ZrSiO$_4$) observed at atomic resolution, *Phys. Chem. Minerals* **14**, 197–204.

Yadav, J. S. and Sharma, A. P. (1979): The effects of some pre-irradiation and etching conditions on the etching of glasses, *Nucl. Tracks* **3**, 119–124.

Yelland, A. J. (1990): Fission track thermotectonics in the Pyrenean orogen, *Nucl. Tracks Radiat. Meas.* **17**, 293–299.

Yim, W. W. S., Gleadow, A. J. W., and van Moort, J. C. (1985): Fission track dating of alluvial zircons and heavy mineral provenance in northeast Tasmania, *J. Geol. Soc. London* **142**, 3512–356.

Young, D. A. (1958): Etching of radiation damage in lithium fluoride, *Nature* **182**, 375–377.

Zaun, P. E. and Wagner, G. A. (1985): Fission-track stability in zircons under geological conditions, *Nucl. Tracks* **10**, 303–307.

Zeck, H. P., Andriessen, P. A. M., Hansen, K., Jensen, P. K., and Rasmussen, B. L. (1988): Palaeozoic palaeo-cover of the southern part of the Fennoscandian Shield – fission track constraints, *Tectonophysics* **149**, 61–66.

Zeitler, P. K. (1985): Cooling history of the NW Himalaya, Pakistan, *Tectonics* **4**, 127–151.

Zeitler, P. K., Johnson, N. M., Naeser, C. W., and Tahirkheli, A. K. (1982a): Fission track evidence for Quaternary uplift of the Nanga Parbat region, Pakistan, *Nature* **298**, 255–257.

Zeitler, P. K., Tahirkheli, R. A. K., Naeser, C. W., and Johnson, N. M. (1982b): Unroofing history of a suture zone in the Himalaya of Pakistan by means of fission-track annealing ages, *Earth Planet. Sci. Lett.* **57**, 227–240.

Zielinski, R. A. and Naeser, C. W. (1977): Fission-track dates from the White River formation, Shirley Basin, Uranium district, Wyoming, *Isochron/West* **18**, 19–20.

Zijp, W. L. and Baard, J. H. (1979): Nuclear data guide for reactor neutron metrology. – Part 2: Fission Reactions, ECN-71.

Zimmermann, R. A. (1980): Patterns of post-Triassic uplift and inferred Fall Line zone faulting in the eastern United States, *GSA Abstracts* **12**, 554.

Index

Accumulation of fission tracks 120, **126**, 136, 152
accuracy of age determination 63, **90**
actinides
–, spontaneous fission 11
added tracks in glass **17**, 18
–, density 19
–, diameter distribution 28
–, etching efficiency factor 20, 32
added tracks in crystals **44**
–, effect on track density 44
age calibration **66**, 90, 94
–, –, absolute approach **66**, 73, 91
–, –, zeta (ζ) calibration **71**, 73, 91
age correction 106, **156**, 172, 225, 226
–, plateau method 157, **158**, 171, 173, 188, 189, 218, 232, 234, 237
–, track diameter method in glass **157**, 171, 173, 188, 218, 233
–, track length method in minerals **157**, 178
age equation
–, fundamental **60**
–, practical **61**
age groups **88**
age profile
–, horizontal 141, 145, 165, 195, 207
–, vertical **145**, 165, 192, 196, 197, 202, 209
–, –, bend-in-slope **152**, 198
–, –, break-in-slope 149, 152.**154**, 198
–, –, complex **148**, 198
–, –, stratigraphic **145**
–, –, uplift **146, 151**, 192, 202
age spectrum 131
age standard 60, 70, 71, **90**, 135, 160
age type **134**
–, complex **142**, 196, 216
–, cooling 116, **137**, 143, 146, 159, 167, 171, 176, 191, 200
–, denudation 140
–, early cooling 135
–, formation 95, 120, **135**, 143, 145, 151, 159, 171, 218, 226
–, mixed **143**, 198
–, overprint **143**
–, stratigraphic 145, 187
–, undisturbed "volcanic" 135
–, uplift **136, 191**
allanite 162
Ally mine, Tanzania 182
Alofa obsidian, Bolivia 233
alpha (α) damage
–, effect on track etching 52, 75
–, effect on track stability 98, 168
alpha decay 52
–, of ^{238}U 60
alpha dose 98
alpha-interaction tracks 60
alpha particle 10, 60, 98
alpha-recoil tracks 59
Alps 116, 132, 139, 146, 154, 174, **192**
–, Bergell 194, 195, 196
–, Gotthard 194, 195
–, Gran Paradiso 151
–, Lepontine 117
–, Maggia Valley 193, 194, 195
–, Monte Rosa 195
–, Silvretta nappe 196
–, Simplon 195, 196
–, Ticino 195
Alshar Mine, Yugoslavia 179
Alto Ligonha, Mozambique 182
amazonite **163**
Andes 192
angle of incidence 14, 18, 27
–, frequency distribution 44, 48
annealing of fission tracks 2, 35, 96, **98**, 121
–, activation energy 2, 109, 121
–, ambient temperature 118, 121, 129, 135, 137, 160, 165, 166, 173, 218, 225, 226, 231
–, anisotropy 104, 107
–, effect on etchable length 52, 102, 121,

129, 139
–, in-situ 112, 148
–, kinetics **108**
–, natural **111**, 156
–, role of etching conditions 108, 122
–, track size-track density relation **104**, 111, 121, 157
–, zones **123**
annealing method 74
Aouelloul crater 220
apatite 75, 76, 135, 161, **163**
–, age plateau 159
–, age profiles 146, 151, 165
–, age standards 70, **91**, 160
–, closure temperature 115, 118, 125, 127, 139, 194
–, 60°C cooling age (t_f) 137, 148, 155, 198
–, detrital 144, 163, 207, 208
–, discontinuous tracks 5
–, etching characteristics 39, 40, 45, 52, 76, 163
–, –, basal plane 39
–, –, prism plane 39, 76
–, –, pyramid plane 39
–, in teeth and bones 229
–, in tephra 135
–, regional studies
–, –, Alps 132, 139, 145, 146, 154, 192
–, –, Central European basement 139, 199
–, –, Damara Orogen, Namibia 183
–, –, Marysvale ore deposit, Utah 213
–, –, Mississippi Valley type ore deposits 214
–, –, Northern England 144
–, –, Otway basin, Australia 210
–, –, Ries crater, Germany 144, 151, 221
–, –, Ryoke belt, Japan 206
–, –, Tejon oil field, California 208
–, –, Transantarctic Mountains 149, 155, 197
–, track length 51, 101, 105, 106, 130, 133, 137, 140, 157, 158, 165, 197, 198, 200, 202, 208, 211, 215
–, track stability 95, 98, 101, 104, 105, 106, **110**, 112, 115, 125, 135, 137, 139, 143, 151, 165, 191, 218
–, –, influence of chemical composition 101, 107, 114
–, zeta factor 72
Appalachian Mountains 199
archaeological objects **228**
–, authenticity testing 229
^{40}Ar-^{39}Ar dating 189
Arrhenius diagram **100**, 109, 121, 125
artifacts 161, 171, 228, **231**

–, geological provenance 171, 228
astroblem 217
atomic bomb 10
Auger process 8
australite 120, 133, 157, 223, 225
automated observation **57**

Baked soil 235
Banks Island Tephra, Canada 189, 191
barysilite 165
bastnäsite 165
Bavaria, Germany 183
bediasite 223
beryl 166
beta (β) particle 98
- , emission 10
B-factor 72
Bihar mica belt, India 170
billitonite 223
Bishop Tuff, California 91, 92, 191
biotite 157, **176**, 178, 183
Bohemian glass 229
Bohemian Massif 115, **199**
Boltysh crater USSR 221
Boltzmann's constant 109
Bosumtwi crater, Ghana 220, 225
Borchers ash, Kansas 191
bore-hole studies 96, **112**, 146, 191
–, Canning basin, Australia 216
–, Eielson, Alaska 112
–, KTB, Oberpfalz 115, 199, 202
–, Los Alamos, New Mexico 112
–, Otway, Australia 112, **210**
–, Ries crater, Germany 221
–, Tejon oil field, California **208**
–, Urach, Germany 114
Brabant Massif, Belgium 199
Bragg curve 6
British Columbian plutonic complex, Canada 139
British Paleozoic basement 199
Brunhes-Matuyama paleomagnetic boundary 189
Broken Hill, Australia 163
broken isochron 205
bromoform 77
bubble chamber 63, 65
bubbles in glass 18, 171, 172, 229
bulged isotherms 195
bulk etching **17**
–, in glass **17**, 35
–, –, effect on track density **19**
–, in crystals 36, 37, **42**, 76
–, –, effect on track density **42**

Buluk member zircon, Kenya 92, 238
bunte breccia 222
burnt stone 229

Cadmium
–, neutron absorption 78
–, ratio (CR) **77**
calcite **166**, 229
calibration (*see* age calibration)
californium fission fragments 18
–, energy spectrum 28
–, track etching rate in glass 33
Canning Basin, Australia 216
carbonatite 165, 182
Carpathian obsidian 234
Central European basement 139, **199**
ceramics 228, 229, 235
Cerro la Tefa obsidian, Columbia 232, 233
charged particle 1
–, energy loss **6**
–, interactions with solid matter **5**
charged particle track (*see* nuclear track)
Chassenon, France
–, –, Gallo-Roman bath 229
chemical etching (*see* etching)
Chihuahua, Mexico 166
Chinese napkin ring 230
chi (χ)-square test
–, grain-population methods 83
–, grain-by-grain methods 85, 94
–, age groups 88
chlorite 133, 157, **167**
Choukoutien cave, China 236
chronostratigraphy 137, 160, 182, 186, **187**
Clear Water Lake craters, Canada 221, 225
closure temperature 116, **117**, **125**, 127, 139, 146, 175, 178, 194
Coastal Plutonic Complex, Canada 116, 117, 167
co-existing minerals (dating of) 135, 139, 143, 136, 182, 189
common *k*-test 83
confined tracks **47**, **130**, 194, 198, 211
–, Arrhenius diagram 80, 110
–, measurement 80, 129
–, length distribution **49**, 103, 140, 143, 211
-, use in thermal history analysis **130**, 140, 143, 198, 201, 207, 211, 213, 216
contact heating 145
cooling
–, monotonous 118
–, path 126
–, rate 125, 126, 140, 191, 194, 195
cosmic ray induced tracks 59

counting of fission tracks 13, **79**, **161**, 170, 226, 228
–, automated 57, 170, 228, 229
–, electron microscope 53, 161, 180
–, optical microscope 54, 161
Cretaceous-Tertiary boundary 217
critical angle of incidence
–, in glass **18**, 32
–, in crystals 42
cross-section,
–, (n, γ) activation of metal monitors 69
–, ^{232}Th neutron induced fission 68
–, ^{235}U neutron induced fission 61, 62, **65**
–, ^{238}U neutron induced fission 68
crustal accretion 227
crystal surface
–, energy 39, 44
–, texture 44, 45
crystalline basement 125, 139, 145, **197**
crystallites 171

Damara orogen, SW Africa 167, 170, 183, 199
Darwin crater, Tasmania 225
Darwin glass, Tasmania 220, 225
dating procedure **73**, 91
–, procedure factor (Q) 62, 73, 76
dating techniques **73**
decay-constant 60
–, ^{238}U alpha-decay 60, 62
–, ^{238}U spontaneous fission 34, 36, **37**, 67, 73, 230
–, ^{238}U total decay 60
decoration techniques 16
deep-sea glass 135, 170, **172**, 226
density of etched tracks 17, 123, 161
–, in glass **18**
–, –, effect of varying V_t **32**
–, –, effect of annealing **99**
–, in crystals **42**
–, –, effect of annealing 105, 108
density distribution of etched tracks **81**
–, coefficient of variation 81, 83, 86
–, standard deviation 81, 86, 94
–, heterogeneity 56, 76, 83
–, variance 83
density of latent tracks
–, areal or planar **13**, 123
–, spatial 13, 61, 123
denudation 125, 140, 146, 151
–, uplift-induced 140, 191, 199
Deseadan Salla beds, Bolivia 147
detector **1**
–, mica 71, 72, 178

–, plastic 1, 2, 16, 56, 57, 76
devitrification 35, 174, 226
diameter of etched tracks in glass **25**
–, effect of annealing **103**, 105, 121, **133**,
    189
–, measurement 103
–, variation with prolonged etching **27**
diameter distribution of etched tracks in
    glass **28**
–, –, bimodal 143, 231, 232
–, –, effect of annealing 28, 103
–, –, effect of varying V_t 34
–, –, variation with prolonged etching **28**
diameter of latent tracks 1
diaplectic deformation 97
diffusion,
–, atomic 2, 109
–, of etchant and etch products 36
digitizing tablet 58, 80, 130
diiodomethane 77
dip angle (*see* angle of incidence)
discontinuous tracks 4, 8
dislocations 39, 166, 228
disthene (*see* kyanite)
drawing-tube 58
Durango apatite, Mexico 135
–, track annealing 101, 103, 110, 128, 160

East African Rift 199
effective retention temperature (*see* closure
    temperature)
Egergraben, Central Europe 202
Eielson bore-hole, Alaska 112
elastic relaxation 8
electrochemical etching 16
electrolytic etching 16
electronic collisions 6
electronic stopping power 6
electron microscope 8, 161, 180
–, scanning 53
–, transmission 2, 4, 16
–, –, high resolution 2, 5
Elgygytgyn crater, Siberia 159, 220, 225
El Inga obsidian, Ecuador 232, 233
Entrecasteaux area, Pacific Ocean 173
epeirogenic uplift **197**
epidote 163, **167**, 183, 185, 229
–, closure temperature 116, 118
–, fault dating 168, 206
–, regional studies
–, –, Alps 139
–, –, Coastal Plutonic Complex, Canada
    117, 167
–, –, Damara orogen 167

–, –, Sinai peninsula 168, 206
–, track annealing 96, 116
–, track length 133, 157
–, track stability 98, 117, 121
epithermal activation 69
epithermal neutrons 68
epoxy resin 77
equivalent time concept 110
error
–, on neutron fluence **86**
–, on track density ratio **86**
–, on age determination **86**, 170, 226, 229
etchable track length 17, 95, 102, 129, 139
etch-anneal-etch method 36
etched tracks in crystals 17, **36**
–, anisotropy 52, 75, 76, 98, 104, 180
–, density **42**
–, track etching rate 17, **36**, **42**
–, role of etching conditions 52, 108
–, shape 37, **39**, 54, 176
–, size **47**
etched tracks in glass **18**
–, cone angle 25
–, density **18**, **32**
–, role of etching conditions 35
–, shape 18, **25**, 33, 133
–, size **27**
etching 1, 16
etching conditions for track analysis 35, 47,
    54, 77, 108
etching efficiency factor 17, 19, 20, 32, 42,
    62
etching equipment 35, 77
etch product layer 35
etch time factor (*see* prolonged etching
    factor)
extended defect 4, 8
external detector method **76**, 79, 85, 163,
    102, 110, 113
external surface 15, 74, 76
–, track density 19, **20**, 46
–, track size 28
extra-terrestrial samples 54, 59

Fading of fission tracks 2, 65, 91, **95**, **121**,
    170, 175, 178, 189, 204, 207, 211, 218,
    224, 226, 231, 233 (*see also under*
    stability)
–, causes 96, 121
–, in-situ 96, 112, 148
Fall line, eastern USA 205
Faeroe Islands 184
fast neutrons 67
faults **203**

–, vertical displacement 205
Fe-ion tracks 4
feldspar 42, 163
Fennoscandian basement 199
Fichtelgebirge, Germany 202
fired bricks 235
firing ash 236
first-order kinetics 60, 109, 111
Fish Canyon Tuff, Colorado
–, –, apatite 91, 132, 135, 160
–, –, zircon 91
fission **9**
–, asymmetry 11, 13
–, energy release 10
–, induced 10, 69, 61
–, natural 59
–, spontaneous **11**
fission fragments **10**, 36
–, energy 10, 12
–, mass distribution 10
–, mass ratio 11
–, nuclear stability 10
fission tracks (latent) **10, 12**
–, areal density **13**
–, etchable length 17, 62, 102, 156
–, induced 59, 61
–, formation 10, 12
–, observation 2, 16
–, spatial density 13, 61
–, spontaneous 12, 59
–, structure 12, 95, 109, 111
fission-track tectonics **140**
–, active mountain belts **191**
–, basement massifs **197**
flat-bottomed tracks
–, in glass 18, 22, 27
–, in crystals 41
fluid inclusions 214, 228
Fort Selkirk tephra, Yukon 191
fossil tracks (*see* natural tracks)
fragmentation of fission tracks 97, 104
Fränkische Linie, Germany 202
Franklin deposit, New Yersey 165
Frantchi cave obsidian, Greece 234
frictional heating 204
frictionite 174
funnel-shaped tracks 39

Gambles cave, Elmenteita 232
"Gap" model of track formation **4**
gamma (γ) spectrometry 67, 79, 81
garnet 146, **169**, 192
–, closure temperature 116, 119, 170, 192
geometry (of track registration) 15, 62, 176

geometry factor (g) **15**, 19, 62
geometry ratio (G) **21**, 62
–, in glass 21
–, in crystals 46
Georgia tektites 223
geothermal gradient 141, 151, 192, 195,
    204, 208, 209, 211, 215
–, horizontal 141
–, reverse 151, 223
glass 76, 80, 146, 157, 162, **170**
–, artificial 28, 57, 228, **229**
–, impact 135, 157, 218, 222
–, inclusions 180
–, lunar 157
–, volcanic 18, 133, 135, 144, 157, 158, 159,
    171, 188, 237
–, –, deep-sea 135, 170, **173**, 226
–, –, hydrated 75, 188, 189
–, –, shards 75, 83, 135, 171, **173**, 187, 237
–, track stability 100, 103, 121, 124, 172,
    226
–, uranium distribution 83, 173
glass monitors **70**, 79, 87
–, use in absolute approach 70
–, use in zeta calibration 72
glauconite 175
Gosses Bluff crater, Australia 221
grain-by-grain method 73, **75**, 79
–, error calculation **86**
–, statistical analysis **85, 88**
grain-population method 70, **73**, 76, 79, 163,
    172, 183, 185
–, error calculation **86**
–, statistical analysis **81**
Greenland 139

Hadar formation, Ethiopia 238
half-life
–, of Th and U isotopes 12
–, ^{238}U spontaneous fission 11, **65**, 161, 228
Haughton dome, Canada 220
heated stones 228, 235
heavy-liquid separation 77
Heimefrontfjella Mountains, Antarctica 199
Henbury crater, Australia 220
heulandite 184
hillocks 44, 45
Himalaya 176, 192
hominid-bearing beds 166, 228, 237
hornblende **175**
hübnerite 175
hydrocarbon exploration 144, 208
hydrostatic pressure
–, effect on track stability 96

hydrothermal activity 183, 214
–, effect on track stability 107

Iceland spar 166
identification of fission tracks 41, 44, **54**
–, personal factor 47, 56, 108
idocrase (*see* vesuvianite)
image analysis systems 57, 103, 170
impact (*see* meteorite impact)
impactite 135, 217
indochinite 120, 133, 223
induced fission 10, 59, 61
–, by non-thermal neutrons 68, 77, 163
intergranular solutions 96, 107
internal surface 14, 73, 76
–, track density **19, 22, 46**
–, track size **28,** 29, 49, 63
ion-explosion spike theory **8**
–, thermalized 8
ionizing radiation
–, effect on track stability 96, 98
ion-sputtering 8
Iranian glass 229
irradiation
–, with, Cf. fission fragments 182
–, with thermal neutrons 61, 65, 77
isochronal annealing 99
isothermal annealing 99, 110
isotropic detector 17
isotropic track distribution 13, 46, 83
Ivory coast tektites 223, 225

Japanese obsidian 233
–, in pottery 232
Javanite 223

Kapton 76
K-Ar dating 66, 91, 120
–, deep-sea basalt 226
–, glass shards 187
–, meteorite impact 221
–, ore deposits 213
–, regional studies
–, –, Alps 116, 139, 193, 196
–, –, British Columbia 117
–, –, Damara Orogen 116, 170, 183
–, –, Ries crater 221
–, tektites 157, 224
–, tephra 187, 189, 237

Kasipatnam, India 183
Kazachstan, USSR 178
KBS-tuff, Kenya 238
knife-blade shaped tracks 41, 44

Köfels structure, Austria 174, 220
K-shell electrons 8
KTB bore-hole, Germany 115, 199, 202
kyanite **176,** 192

La Esperanza obsidian, Ecuador 233
Lake Tapps tephra, Washington 191
latent track (*see under* nuclear track *or*
    fission track)
lattice
–, defects 2, 39, 104, 111
–, image 2
–, restoration 109
–, strain 2, 4
–, stress 8
–, vacancies 8
length of etched tracks in crystals 41, **47,**
    102, 215, 216, 236
–, measurement 58, 80
–, effect of annealing 95, **102,** 104, 110, 121,
    126, 139, 156
length distribution of etched tracks in
    crystals 4, **47, 102,** 125, 128, **129,** 140,
    143, 165, 197, 207, 211, 218
Lennard Shelf ore deposit, Australia 215
lepidolite 176
Lexan polycarbonate 71, 76
Lipez obsidian, Bolivia 233
Los Alamos bore-hole, New Mexico 112
Lybian desert glass 218, 220

Macusani obsidian, Peru 233
Makapansgat, South Africa 166
Makrofol 76
Manicouagan crater, Canada 221, 225
Manson impact structure, USA 220
marker horizons 172, 182, 186, 187
–, paleolithic 237
Marysvale ore deposit, Utah 213
Maxwell-Boltzmann distribution 65, 67
metalliferous fluids 214, 217
metal monitors **67,** 78, 87
–, epithermal activation 69
–, neutron fluence determination **67**
–, neutron self-shielding 69, 79
metamictization 75, 179, 181
metamorphism 143
meteorite impact 97, 125, 150, 170, 174, **217**
–, crater 135, 174, 225
–, glass 137, 157, 218
–, heating 143
–, structures 174, 218
–, –, thermal evolution 218, **221**
mica 59, 76, **176**

–, alpha-interaction tracks 60
–, alpha-recoil tracks 60
–, Alps 196
–, external detector 16, 63, 71, 72, 76, 178
–, latent tracks 2, 4
–, track counting 56, 57, 98
–, track etching 39, 42
–, –, efficiency 42, , 63, 65
–, track length reduction 98, 133, 157
–, track stability 97, 98, 102, 106, 107, 157
microcline 163
microlite (pyrochlore) 178
microlites 187, 228
microscope (*see* electron or optical
    microscope)
microscopic observation 39, **53**, **79**
–, automated 57
–, electron microscope 2, 4, 8, 16, 53, 161,
    180
–, low magnification 16
–, observation factor (q) 45, **56**, 62
microtektites 173, 225
Mid-Atlantic ridge 173, 226
Middle Awash tuff, Ethiopia 238
Midnite deposit, Washington 180
Mien crater, Sweden 220
Milos obsidian, Greece 234
mimetite 180
mineral separation 77
mineralization veins 204, 206, 214
Mississippi Valley Type (MVT) ore deposits
    213
-, thermal history **214**
Mistastin crater, Canada 220, 225
Moiré interference fringes 4
moldavite 92, 223
monazite **178**, 229
Morocco 180
Mountain Pass carbonatite, California 165
Mount Doorly, S. Victorialand 149, 198
Mount Dromedary, Australia 92
Mount Jason, S. Victorialand 149, 198
Mullumica obsidian, Ecuador 233
Muong Nong glass 220
muscovite **176** (*see also under* mica)
–, closure temperature 178
–, latent track size 2
–, track etching efficiency 42, 65
–, external detector 76
–, track etching rate 36
–, track stability 97, 107
mylonite 206

Natural tracks in minerals

–, origin **59**
Nellore mica belt, India 170
neutron absorption 71, 179
neutron activation 65, 67
–, reaction rate, **65**
neutron capture standard 67
neutron dosimetry (*see* neutron fluence
    determination)
neutron energy spectrum 65, **67**
neutron flux **67**
–, gradient 79
–, thermal/fast ratio 67, 77
–, thermal/epithermal ratio 68, 78
neutron fluence (thermal) 61, 65, 79
–, determination **67**
neutron irradiation 61, 65, **77**
neutron scattering 2, 5
neutron temperature 66
New Zealand Alps 192
nuclear chain reaction 10
nuclear collisions 6
nuclear explosion 97
nuclear fission (*see* fission)
nuclear reactor 10, 61, 70
nuclear spike **7**
nuclear stopping power 6
nuclear track (latent) **1**
–, atomic structure **2**, 8
–, core zone 3
–, formation **5**
–, width 1
Nullabor plains, Australia 225

Oberpfalz, Germany 199, 201, 205
observation of fission tracks (*see* micro-
    scopic observation)
obsidian 158, 161, 170, 171, 223, 228, 230,
    237
–, artifacts 228, **231**
–, critical angle 19
–, in pottery 232
–, plateau age 158, 234
–, provenance 231, 234
–, track diameter 231, 232
Odenwald, Germany 199, 204
oil exploration (*see* hydrocarbon explora-
    tion)
oil immersion objectives 54, 73
oil-window 207
Oklo uranium deposit, Gabon 59, 61, 180
Old Crow tephra, Alaska 191
Olduvai Gorge, Tanzania 238
olivine
–, structure of latent tracks 4, 5

–, track etching 52
–, track stability 96, 98
optical microscope 1, 15, **54**, 161
–, micrometer ocular 80
–, objectives 54, 57
–, reflected light 55, 79
–, resolution 37
–, stage 54
–, –, motorized 58, 79
optical track revelation 1, 13, **16**, 36, 62, 83, 94
Ordovician-Silurian stratotypes 146
ore deposits **213**
–, thermal history 214
ore slags 229
orpiment **179**
Otway basin, Victoria **210**
–, bore-hole 112, 211

Palmar obsidian, Ecuador 233, 234
partial annealing zone (PAZ) 118, **123**, 126, 137, 146, 151, 192, 197, 208, 211
–, uplifted 149, 152, 154, 155, 198
Pearlette tephra, USA 189
Peking man 236
philippinite 223
phlogopite 157, 176, 183
phosphate glass 28
pillow lava 172, 226
Pine Point ore deposit, Canada 215
pitchblende 214
pitchstone 171
plagioclase 4
plastic deformation 97
plastic detectors 1, 2, 56, 57, 76
–, track revelation 16
–, track etching rate 33
plateau age 157, 158, 173, 188, 189, 218, 225, 232, 234, 237
–, isochronal 158
–, isothermal 158, 189
plutonium
–, natural abundance 11
–, spontaneous fission 11, 59
polishing scratches
–, observation 54
–, relation to bulk etching **43**, 76
Popigai crater, Siberia 220, 225
population method (*see* grain-population method)
point defects 2, 3
pointed tracks in glass (*see* sharp-bottomed tracks)
Poisson

–, distribution **81**
–, –, standard deviation 81, 94
–, error 86
–, probability paper 83
pre-annealing 171
Precambrian basement 199
pre-irradiation annealing 74
presentation of fission-track data **94**
primary ionization 7
–, rate 32
projected tracks
–, length 47, 103, **130**, 140, 157
–, –, distribution **49**, 103, 211
–, –, measurement 80
–, use in thermal history analysis 131, 137, 198, 213
prolonged etching factor **17, 21**, 36, 62, 76
prompt neutrons 10
Protactinium
–, spontaneous fission 11
pseudoplanes 39
pseudotachylite 174
pumice 135, 170, 172, 237
pyroclastics 88
pyromorphite 180
pyroxene 36
Pyrenees 192

Quartz 42, **179**, 214
Quito obsidian, Ecuador 233

Radiation damage
–, effect on track etching 52, 75, 181, 185
–, effect on track stability 98, 168, 185
–, healing 158
radial plot 88
radiocarbon dating 228.237
Rajastan range, India 170
range
–, of fission fragment track
–, –, etchable **17**, 62, 129
–, of nuclear track **5**
Rare-earth elements
–, in monazite 179
–, in glass monitors 71
Rb-Sr dating 66, 91, 120
–, Alps 116, 139, 167, 193, 196
-, Coastal Plutonic Complex, Canada 167
-, Damara orogen 116, 167
recoil nuclei 4, 60
Red Sea rift 199
re-etch method 75, 176
replicas of tracks 53
repolish method 76

reports of fission-track data **94**
retention-isotherm 140, 146
revelation of tracks (*see* optical track revelation)
Rheingraben 199, 201, 205
Ries crater, Germany 144, 151, 220, **221**, 225
rift
–, East African 199
–, Egergraben 202
–, Mid-Atlantic 173, 226
–, Red Sea 199
–, Rhein 200, 201, 205
Rochechouart impact, France 221
rock fluids 213
Rocky mountains 192
rounded tracks (*see* flat-bottomed tracks)
Ryoke belt, Japan 206

Sample preparation 77, 163, 170, 176, 181
San Joaquin valley, California 209
scheelite 180
Schwarzwald, Germany 199, 205
sea-floor spreading **225**
–, rate 173, 226
secondary lead minerals 180
sedimentary rocks 139, 144, 175
sedimentary basins 112
–, thermal evolution **206**, 214
semi-automatic track analysis 58
semi-cone angle 25
semi-tracks (*see* surface tracks)
shape of etched tracks
–, in glass 18, **25**, **34**, 54, 133
–, in crystals **39**, 44, 54
Shar-i Sokhta settlement, Iran 236
sharp-bottomed tracks in glass 18
–, density **22**, 33
–, size **27**
shear heating 205
shock wave
–, effect on track stability 97, 151, 218
–, metamorphism 151, 217
Sierra Nevada batholiths 139, 210
Sinai peninsula 168, 206
single grain method (*see* grain-by-grain method)
size of etched tracks
–, in glass **25** (*see also under* diameter)
–, in crystals **47** (*see also under* length)
size of latent tracks 1
skarn 183
sodiumpolytungstate 77
South American obsidian artifacts **232**, 234

sphene 76, 157, 161, **180**, 185, 192, 197, 218, 229
–, age profile 150
–, age standards 94
–, closure temperature 119, 125, 182, 192
–, detrital 207, 211
–, etching characteristics 52, 75
–, in tephra 237
–, radiation damage 52, 181
–, regional studies
–, –, Coastal Plutonic complex, Canada 117, 167
–, –, Greenland 139
–, –, Otway basin 211
–, –, Peking man, China 236
–, –, Ries crater, Germany 222
–, track length 133
–, track stability 96, 97, 98, 106, 113, 116, 121
spontaneous fission 10, **11**
–, in natural samples 59
spurious defects 54, 57, 76
spurious etch pits 162, 228
stability of fission tracks 95
–, natural 96, **111**
–, thermal **98**, 226
–, –, stability zones **123**
statistical analysis of fission-track data 58, **81**
–, age groups 88
–, grain-by-grain methods 85
–, grain population methods 81
Steinwald, Germany 202
stibiotantalite 182
stilbite 184
stopping power **6**, 12
–, electronic 6
–, nuclear 6
subsidence 126, 143, 148, 197, 200, 207, 209, 215, 216
subtraction method 74, 86
suevite 145, 222, 223
surface energy of crystal planes 39, 44
surface opening of etched tracks in crystals 37, **39**, 44, 46
surface tracks 13, 47, 61, 130
–, projected length **49**, 103, 105, 121, 140 (*see also* projected tracks)
–, true length **49**, 103, 133
Swartkrans, South Africa 166

Tanzanite 182
Tardree rhyolite, Northern Ireland 92
Tatra Mountains, Poland 192

tectonic interpretation model 141
Teflon 77, 185
Tejon oil field, California **208**
tektite 120, 133, 157, 158, 159, 161, **173**, **223**
–, cosmic ray exposure age 224
–, critical angle 19
–, origin 218, 223
–, strewn fields 219, 223
–, track stability, 96, 98, 121, 173
Temperature-time (*T-t*) path 102, 112, 116, 120, 126, 129, 133 (*see also under* thermal history)
tephra 18, 135, 163, 172, **187**, 237
–, mixed 187
tephrochronology **187**, 228
–, Japanese 88, 189
–, New Zealand 189
–, North American **189**
Tethyan geosyncline 192
textured surface 44, 45
Thera, Greece 237
thermal event 112, 115, 139, 144, 213
thermal fading (*see* annealing)
thermal history 95, 112, 126, 130, 133, 139, 148, 158, 182, 186
thermal neutrons 61, 65
thermal overprint 142
thermic age interpretation model 141
thermochronometry 120, 165
thermoluminescence dating 228
thorium
–, alpha decay 75
–, natural abundance 12
–, spontaneous fission 11
–, induced fission 68, 77, 163, 179
Th/U-ratio 77
Tibetean plateau 192
time-span of fission-track dating **161**
titanite (*see* sphene)
total annealing zone 124
total energy loss 7
total stability zone 124, 131
tourmaline 157
track channel 37, 39, 41
track core 2, 4
track etching rate (V_t)
–, in crystals 17, **36**, 42, 53
–, in glass 17, 18, **31**, 35
–, effect of annealing 95, 104, 109, 122
track-in-cleavage (TINCLE) 47
track-in-track (TINT) 47, 79
Transantarctic mountains **197**, 205

Unetchable gaps 4, 104
U-Pb dating 60
updoming 194
uplift 140
–, amount 151
–, epeirogenic **197**
–, model profiles **151**
–, post-metamorphic 145
–, post-orogenic **191**
–, rate 141, 149, 150, **151**, **191**, 197, 198, 200, 204
Urach bore-hole, Germany 114
uraninite 178
uranium
–, distribution 83
–, heterogeneity 75, 83, 85, 167, 173, 178, 181, 183, 185
–, migration 178
–, monitor 67
–, ^{234}U
–, –, natural abundance 12
–, ^{235}U
–, –, induced fission 10, 61, 65, 68
–, –, natural abundance 12
–, –, spontaneous fission 11
–, ^{238}U
–, –, alpha decay 60
–, –, induced fission 77
–, –, natural abundance 12
–, spontaneous fission 11, 59, 63
–, $^{235}U/^{238}U$ isotopic ratio 61, 62
–, –, in glass monitors 73
uranium glass 228, 229
uranium-mica sandwich 63
uranium tetrafluoride 8

Vanadinite 180
Variscan basement, Central Europe 139
vermiculite 133, 157, **183**
vesuvianite 116, **183**, 192
Viburnum Trend ore deposit, Missouri 215
Victoria, Australia
–, basement 199
–, sedimentary basin 112, 210
Victorialand, Antactica 197
vitrinite reflectance 207
Vosges, France 199

Wabar crater, Saudi Arabia 220
Wanganui basin, New Zealand 188
Westcott *g(T)*-factor 66
WN etchant 52

X-ray scattering 2, 4, 5

Zamanshin crater, Siberia 220, 225
zeolite 184
zeta (ζ)
–, calibration method **71**, 73, 87, 91, 94
–, –, factor 72
zircon 76, 146, 161, **184**, 196, 214
–, age profile 150
–, age standards 90, 92
–, chronostratigraphy 135
–, closure temperature 115, 118, 125, 185, 192
–, detrital 88, 186, 207, 208, 210, 237
–, etching characteristics 52, 75, 98
–, –, prism plane 76
–, in pottery 229, 236
–, in tephra 135, 146, 172, 186, 188, 189, 228, 237

–, latent tracks 2
–, radiation damage 52, 185
–, regional studies
–, –, Alps 151, 193, 196
–, –, Greenland 139
–, –, Marysvale ore deposit 214
–, –, Otway basin 210
–, –, Ries crater 222
–, –, Ryoke belt 206
–, –, Tejon oil field 208
–, –, Thera 237
–, –, Weald 88
–, track length 130, 185
–, track stability 96, 98, 105, 115, 121, 143, 185, 186
zoisite 182